THE SEARCH FOR A THEORY OF COGNITION

VALUE INQUIRY BOOK SERIES

VIBS

Volume 238

a volume in
Cognitive Science
CS
Francesc Forn i Argimon, Editor

THE SEARCH FOR A THEORY OF COGNITION

Early Mechanisms and New Ideas

Edited by
Stefano Franchi and Francesco Bianchini

Rodopi

Amsterdam - New York, NY 2011

Cover Image: www.morguefile.com

Cover Design: Studio Pollmann

The paper on which this book is printed meets the requirements of "ISO 9706:1994, Information and documentation - Paper for documents - Requirements for permanence".

ISBN: 978-90-420-3427-3
E-Book ISBN: 978-94-012-0715-7

Printed in the Netherlands

CONTENTS

ACKNOWLEDGMENTS

The early roots of the collection go back to a volume we edited for the *Discipline filosofiche* series in philosophy published by Quodlibet in 2007. Stefano Besoli, editor of *Discipline filosofiche*, supported our initial research and provided useful advice. The present volume extends the interpretation of cognitive science and Artificial Intelligence we started to articulate in the earlier work. Roberto Cordeschi provided helpful advice in the early phase of the project. Raffaela Serrani was the first translator of the chapter by Claudio Pogliano. Douglas Hofstadter provided precious help in the revision of the manuscript. His critical comments on the introduction were particularly helpful. Elena Pallottini provided significant help with proof-reading. Manuela Marchesini's editing skills were instrumental in turning the manuscript into a publishable work. Many thanks go also to the publishers of the *Journal of Philosophy* and to The University of Chicago Press, publisher of *Critical Inquiry*, for the permission to reprint the essays by Solomon Feferman and Peter Galison. Stefano Franchi would also like to thank the authors of the free software package LyX and the TeX collection of software, which made the typesetting of the volume possible and almost enjoyable.

FOREWORD

Douglas Hofstadter

The origins of humanity's ongoing attempt to realize intelligence, understanding, creativity, and consciousness in a technological artifact are many, including the disciplines of mathematical logic, cognitive and perceptual psychology, computer science and the theory of computation, the study of the details of brain cells and brain architecture, the philosophy of mind, cybernetics, information theory, linguistics, robotics, physics, evolutionary biology, and numerous other disciplines and schools of thought. To further complicate matters, over the decades, many different countries have contributed crucial ideas that reflect, in various ways, their particular cultures. Among the most important countries in this brew have been the United States, Great Britain, and Russia, but France, Italy, and Japan have also played significant roles.

In this book, two insightful and original young philosophers from Italy have brought together scholars from several countries and who collectively represent all of the disciplines mentioned above, in an attempt to shed light on the fascinatingly complex, international, multicultural, pluridisciplinary, and labyrinthine story of artificial intelligence, focusing on, but not limited to, the decades in the middle of the twentieth century. Thus there are articles about the attempts to use biological ideas to model the brain, about the role played by the science of self-organizing systems, about the raging philosophical debates caused by such results as Gödel's incompleteness theorems and Turing's results on the unsolvability of the halting problem, about the very nature of intelligence and how it might be detected in an agent, about the nature of simulations and models, and about what limits there are to computation posed by the nature of mathematics and by the laws of physics. Moreover, there are case studies describing how Italy, France, and the Soviet Union incorporated, reacted to, and idiosyncratically added to these swiftly moving streams of thought coming largely from abroad.

As we look back today at the origins of the movement now called "artificial intelligence," we are inevitably influenced by events that have happened in the more recent past, such as the rise of such paradigms as connectionism (sometimes referred to as "parallel distributed processing"), nonrepresentationalism, dynamic systems theory, robotics, cognitive linguistics, emotive computing, and others. And if anything, the debates over many classic philosophical questions have only heated up as artificial intelligence has made various kinds of progress. Such questions include the existence or nonexistence of free will, the potential role of randomness in the functioning of mind and brain, the controversial nature of supposed "qualia" (that is, the idea that certain physical systems are intrinsically capable of being a locus of experience, whereas others are intrinsically incapable of it), the difference between "primary intentionality" and "secondary intentionality" (that is, the question of the "aboutness," or lack thereof, of symbols that are manipulated by machines, which has given rise to a putatively strict dichotomy between "syntax" and "semantics"), the question of other minds (another way of looking at solipsism), the mind–body problem (the question of whether a mind is a direct result of physical processes or whether something about a mind transcends physics), the seeming paradox of rule-bound creativity, the realization of emotions in inanimate (or at least inorganic) objects, and of course the eternal riddle of consciousness (which, it must be added, is not a riddle at all, let alone an eternal riddle, for some thinkers, who argue that consciousness simply does not exist, or else that it is at best a kind of necessary illusion of any sufficiently complex perceptual system, which is caused by the drastic simplification that perceptual processes inevitably involve).

All of these burning questions are front and center in this stimulating international volume concerning a crucial period in the history of one of the most philosophically ambitious quests ever undertaken by our species—namely, the quest by *homo sapiens* to build a mechanical rival to itself. That is to say, artificial intelligence is our collective quest to build a rival to our very essence, which is our ability to think—and indeed to build a rival that may possibly overtake us and render us obsolete. This book therefore constitutes an important contribution to the literature seeking to understand the nature of this quixotic voyage that we humans have collectively embarked on.

— July 2010, Paris.

ON THE HISTORICAL DYNAMICS OF COGNITIVE SCIENCE: A VIEW FROM THE PERIPHERY

Stefano Franchi and Francesco Bianchini

In 1998, speaking from the privileged perspective of the East Coast researcher, Daniel Dennett presented a "logical geography of computational approaches," a conceptual map of cognitive science that demonstrated how, by simply locating theoretical proposals with respect to their spatial and geographical distance from MIT orthodoxy, we could shed a great conceptual clarity on the methods and goals of cognitive science. Concerns that are not too dissimilar lie also behind the present introduction. We investigate the same broad set of disciplines and approaches, with the difference that our focus has shifted from space to time: instead of a "logical geography," we aim at portraying a "historical dynamics." Like Dennett, we think useful to look at the discipline in terms of the opposition between a theoretical "center" and its "periphery." We set out to concentrate our attention on the periodical shifts that from time to time have reassigned the functional roles of "center" and "periphery" to different research programs, often thrusting research efforts that only yesterday were close to its core, out on the limb of the theoretical landscape.

From the very beginning of AI, a constant fluctuation among different focal points—the methodological, the ontological, and the epistemological—has characterized its historical development. Over time, the core interests of many researchers shifted towards the edge; what was once in the spotlight, slipped into a dark zone; the center became periphery. Conversely, returns to the center from the the periphery or even from the outer regions became the norm. Indeed, Artificial Intelligence has displayed a peculiarly non-straightforward history.

Is there a reason, we may ask, behind these rapid and frequent shifts? We believe that even a cursory examination of the temporal evolutions of the "science of the mind" may teach us some lessons about the forces at work behind its historical vicissitudes, and provide indications about the dynamics of the discipline that we may find useful for present and future reference. Cognitive science appears to be endowed with peculiar epistemological features that not only caused those oscillations, but also led to an unusual relationship between the current and the previous status of scientific research. Let us begin with a brief narrative.

Between the 1940s and the early 1960s, a large number of scientific disciplines began working on the mechanization of a range of cognitive processes that often included not just adaptive behavioral processes but also emotions and volitions. Their approaches varied. Some researchers in the field (including Alan Turing himself) tried to extend to mental operations the conception of abstract machine that Turing had just proposed. Others came up with computational architectures alternative to Turing's. Efforts to translate the concepts of the emerging school of cognitive psychology and well established behaviorist psychology into computer science went hand in hand with the software and hardware simulation of physiological processes that scientists often, but not always, targeted at the neural level. In spite of their theoretical incompatibilities, researchers in disciplines as different as psychology, anthropology, mathematics, formal logic, economics, physics, physiology, and others—all with their distinctive languages, methodologies, and histories—were all eager to discuss the strengths and weaknesses of their respective approaches. Researchers often met in person in the numerous conferences, seminars, and summer schools that dotted those years. The best known example of these public gatherings, though far from being the only instance of its kind, were of course the Macy's conferences that took place between 1946 and 1953.

Practical and theoretical reasons made these meetings possible. On the one hand, the great technical and political successes of the Manhattan project and, later, the panicked reaction to the Soviet Union's Sputnik launch, fueled a surge in government funding for scientific research—in the United States in particular, but also in the UK and in other European countries; in spite of the broad sample of disciplines present at the meetings, the limited number of participants facilitated personal discussions; last but not least, the close political and economic ties existing among the scientists' nations of origin worked in the same direction.

The theoretical reasons, on the other hand, were equally diverse. First, all participants to the multifarious enterprise shared one fundamental goal: namely, to provide a scientific theory of the mind, the entity "more ghostly than a ghost," —in the words of Charles Sherrington—whose understanding remained beyond the reach of modern science. Warren McCulloch used Sherrington's clause in a passage that is worth quoting in its entirety. It shows how the seekers of a theory of cognition saw themselves: as heirs to a long-standing philosophical tradition and, at the same time, as radical reformers engaged in a sweeping enterprise that a set of common tools had just made possible:

> Sir Charles Sherrington [the 1932 Nobel laureate physiologist who studied the role of synapses and neural integration in the nervous system] ... near the end of a life spent on studying the ways of the brain, was forced to conclude that "in this world, Mind goes more ghostly than a ghost." The reason for his failure was simply that *his physics was not adequate to the problem* that he had undertaken. That has so regularly been the shortcoming of scientists who would have approached this problem, that even Clark Maxwell[*sic*], who wanted nothing more than to know the relation between

> thoughts and the molecular motions of the brain, cut short his query with the memorable phrase, "but does not the way to it lie through the very den of the metaphysician, strewn with the bones of former explorers and abhorred by every man of science?" *Let us peacefully answer the first half of his question "Yes," the second half "No," and then proceed serenely.* (McCulloch, 1989[1948], p. 143, emphasis added)

With the serene confidence that is proper to the man of science, researchers felt in the position to take care of the problem that had afflicted the metaphysicians of the past. Not only does McCulloch, as many other mid-twentieth century scientists, identify the problem of his nineteenth century predecessors (metaphysics) but he also indicates that only the guidance of the natural sciences will unmistakably take us to the goal that had proved so elusive. Many researchers working in the natural and the social sciences shared, if perhaps not McCulloch's confidence, certainly his approach. Notwithstanding the frequent unease that the social scientists felt towards the methodological self-confidence of their colleagues in the natural sciences, the common allegiance to the methods of natural science provided a shared vocabulary that facilitated intellectual exchange among disciplines.The situation changed dramatically toward the end of the 1960s, when one particular approach gained the theoretical and institutional upper hand and consigned to the periphery methods once at the core of the enterprise. The winning approach has been variously labeled AI (*Artificial Intelligence* (McCarthy et al., 1956)), *cognitivism* (Neisser, 1967), GOFAI (*Good Old Fashioned* AI (Haugeland, 1985)), and even the *Boolean Dream* (Hofstadter, 1985).

For our use here, we will adopt *cognitivism,* a term that perhaps embrace a broader concept, and that we take to refer to a variety of approaches that more often than not overlap with one another. Haugeland defines *cognitivism* as the "position that intelligent behavior can (only) be explained by appeal to internal 'cognitive processes,' that is, rational thought in general" (1981[1978], p. 243). In turn, we interpret these "cognitive processes," Haugeland says, as information processes that connect inputs to outputs by virtue of their ability to manipulate internal representations. "Cognitivism can be summed in a slogan," Haugeland concludes: "the mind is to be understood as an I[nformation] P[rocessing] S[ystem]." The performances of actual information processing systems must, he adds, grant empirical confirmation to the thesis: "[cognitive] simulations can function as concrete checks on whether particular proposed IPSs in fact have the abilities they are supposed to explain" (1981[1978], pp. 261, 278). Because of the tight link between a view of the mind as an information processing system and a reliance on cognitive simulations of those processes, cognitivism, formerly and formally a psychological theory, turns *ipso facto* into a hybrid theory that spans the divide between psychology and theory of computation. Daniel Dennett, who later dubbed it "High Church Computationalism," reduced its theoretical content to three dogmas (1998, p. 217):

1. *Thinking is information processing*
2. *Information processing is computation (which is symbol manipulation)*
3. *The semantics of these symbols connects thinking to the external world*

Dennett's quasi-theological analysis has the merit of showing how, depending on which of the three dogmas we emphasize, the identification of *cognitivism* shifts from one domain to another. The traditional cognitivism that psychology (Neisser, 1967) develops, for instance, tends to put the emphasis on the first tenet; Fodor's Representational Theory of Mind (1987; sometimes called "representationism") emphasizes the third one; Herbert Simon and Allen Newell (defenders of the so-called "physical-symbol hypothesis," or "symbolicism" (1981[1976]) insist on the second one. In the USA, the philosophy of psychology has appropriated the term cognitivism; the philosophy of mind orbits toward "representationism" and the "RTM" (Representational Theory of Mind); the philosophy of Artificial Intelligence and of cognitive science prefers "computationalism" or "symbolicism." In Continental Europe (Varela, Thompson, and Rosch, 1991; Descombes, 2001; Dupuy, 2000), instead, scholars use *cognitivism* in a sense that is broader than their American counterparts', and that covers all of the approaches we just listed. Precisely because of this more extensive range of the term, of which we will say more shortly, we have opted—like the Europeans—for "cognitivism" and, alternatively, "the cognitivist paradigm."

Roughly at the same time, in the 1960s, scientists introduced another label that, like "cognitivism," was a convenient shorthand for the collection of disciplines working toward a scientifically rigorous and empirically testable theory of the mind: *cognitive science*. For the new research institute they founded at Harvard in 1960, George Miller and Jerome Bruner resorted to "cognitive studies." "Studies" slowly morphed into "science." By 1975, when Daniel Bobrow and Allan Collins used it in the title of their successful anthology (*Representation and Understanding: Studies in Cognitive Science*), the designation was popular enough that the editors did not feel compelled to explain it. Since George Miller was one of the main defenders of cognitivism, we may be tempted to simply define "cognitive science" as the kind of psychology that researchers of cognitivist persuasions practice. That would be a mistake. Even though for much of their history the two labels—*cognitive science* and *cognitivism*—have referred to the same research programs, to the same people, and to the same publishing ventures, we think that distinguishing them is crucial. *Cognitive science* (or, sometimes, "the cognitive science*s*") refers to the discipline or disciplines that study cognition. Cognition comes from the Latin noun *cognitio* (from the verb *cognosco*, "to become acquainted with"= "to know"). As the etymology indicates, we should understand "knowledge" and "thought" as the concrete, permanent results that lie at the end of a process—hence, an item of knowledge, a *cogitatum*, or a thought. We therefore could say that "cognitive science" is the discipline that seeks a theory of our thinking as knowledge-producing processes, and that keeps clear of other psychological processes like emotions, volitions, moods, drives, and so forth. But history, at least in this case, contradicts etymology. George Miller and Jerome Bruner introduced "cognitive science" to designate the discipline(s) that studied not just the smaller subset of rational processes, *but all* mental phenomena. Miller remarked that they had chosen "cognition" over the redundant "mental psychology" precisely in order to distinguish their approach from *behavioral* (non-mental) psychology:

> In reaching back for the word "cognition," I don't think anyone was intentionally excluding "volition" or "conation" or "emotion." ... [I]n using the word cognition, we were setting off from behaviorism. We wanted something that was *mental*—but "mental psychology" seemed terribly redundant. "Commonsense psychology" would have suggested some sort of anthropological investigation, and "folk psychology" would have suggested Wundt's social psychology. What word do you use to label this set of views? We chose "cognition."[1]

In short, researchers introduced "cognitive science" for the study of *all* mental processes, to signal a broad perspective that included classic psychological investigations and methods and procedures coming from other fields such as computer science, philosophy, linguistics, and the neurosciences. The aim was to provide not just a theory of the input/output correlations proper to mental processes, as behaviorist psychology was wont to do, but a theory of the mental processes themselves. In the beginnings, *cognitive science* nurtured an explicit *negative* theoretical commitment against behaviorism but, unlike *cognitivism* vis-à-vis the natural sciences, no equally explicit *positive* commitment to any specific theoretical framework. As the disciplines matured, the whole situation changed. The *cognitivism* we have defined above, with the theoretical position it encompassed, became such a dominant force within *cognitive science*, that in the short span of ten years the two formally distinct terms fused together. Against the grain of its historical unfolding, *cognitive science* had turned into the investigation of the mind seen through *cognitivist* glasses. This view is still popular. A recent textbook, for instance, offers the following definition:

> Here is the central hypothesis of cognitive science: Thinking can best be understood in terms of representational structures in the mind and computational procedures that operate on those structures. Although there is much disagreement about the nature of the representations and computations that constitute thinking, the central hypothesis is general enough to encompass

> the current range of thinking in cognitive science, including connectionist theories. (Thagard, 2005, p. 10)

We beg to disagree. Paul Thagard's definition may be historically correct for the period between the early 1960s and the late 1980s, but not earlier on. At any rate, even in the 1960s and 1970s, cognitivism's true golden age, notable differences were already present. For instance, Douglas Hofstadter, a well-known and respected researcher whose professional qualification—Distinguished Professor of Cognitive Science at Indiana University—designate him as a cognitive scientist, strongly disagrees with Thagard's identification of cognition with computation:

> In some domains, even in some relatively complex and technical ones, people have come up with programs that can [imitate chains of serial actions that come from verbal protocols of various experimental subjects]. But what about the simpler, noncognitive acts that in reality are the substrate for those cognitive acts? Whose program carries those out? At present, no one's. Why is this? It is because AI people have in general tended to cling to a notion that, in some sense, thoughts obey formal rules at the thought level, just as George Boole believed that "the laws of thought" amounted to formal rules for manipulating propositions. I believe that this Boolean dream is at the root of the slogan "*Cognition as computation*"—and I believe it will turn to be revealed for what it is: *an elegant chimera*. (Hofstadter, 1985, p. 654, emphasis added)

We are convinced that conflating *cognitive science* with *cognitivism*, or rather flattening the first term onto the second, is too narrow a move. It fails to provide an interpretive framework broad enough to include a remarkable wealth of theoretical positions that predated the onset of *cognitivism*, and that perdured during its hegemony—albeit in a more peripheral role—until, in more recent years, they staged an impressive comeback. The central hypothesis behind the work that went into the preparation of this volume is that Thagard's identification of *cognitive science* with *cognitivism* is a historically contingent fact. It happened, but it did not need to. Convinced as we are that a historical and conceptual analysis of these issues may prove theoretically relevant, we commit this book to an examination of how it happened and at what theoretical costs.

Let us go back to the historical narrative. By the mid 1960s, the ascendancy of cognitivism within cognitive science led to a disciplinary and institutional consolidation around that unique approach and to the coincidental marginalization of alternative strategies. Since 1967, the increasingly dominant cognitivist paradigm served as the main inspiration behind new journals, research institutes, Ph.D. programs, and publishers' series, and the skills of the researchers tended to be more narrowly focused on the engineering disciplines.[2] The relatively few exceptions to this pattern were mathematicians, logicians, and those amongst the psychologists who, unlike engineers, were eager to translate their cognitivist theories into computer simulations. Whereas only thirty years earlier an anthropologist (Gregory Bateson), a psychologist (Karl Lashley), a mathematician (Norbert Wiener), a neurophysiologist (Warren McCulloch), and a psychiatrist (W. Ross Ashby) could and had interacted face to face, the Artificial Intelligence of the 1970s strengthened its identity in international congresses (first and foremost the AAAI and the IJCAI series) where the sheer number of contributions was in inverse ratio to the breadth of their coverage.[3]

We could interpret this evolution as an unfolding that is typical of normal scientific progress.[4] When a new field of inquiry emerges, philosophers and, more generally, humanists belonging to different walks of intellectual life develop a multitude of varied and often mutually incompatible theoretical approaches. Then they test those approaches empirically. Little by little, they discard concepts and methods until the discipline—now past its initial "heroic" phase—enters a stage of normalization at the theoretical and institutional levels. An almost inevitable consequence of this second stage is the narrowing of interests and the standardization of the methods and techniques. Several examples corroborate this model. We might recall, for instance, the development of the science of linguistics: after its beginnings in the age of the Enlightenment, the Romantic philosophers and humanists laid the basis for the study of language as an autonomous field, even though their theories were still at the crossroads of many different disciplines, from philosophy and psychology, to philology and literary criticism. Only with the Neogrammarian revolution of the late 19th century, though, was linguistics to emerge as an autonomous science that supposedly obeyed laws as inflexible as those of the natural sciences. Later on, the Structuralist and Chomskian frameworks would refine and extend such an approach. A similar path we could also detect in the development of natural sciences such as biology and chemistry, two disciplines that underwent a long period of theoretical and methodological instability before reaching a normalization on the basis, respectively, of a biochemical reinterpretation of evolutionary theory and of a physical understanding of chemical properties such as valence.

This model construes scientific development as a linear, self-correcting, and quantitatively increasing process. The development of a science begins with the invention of novel yet necessarily imprecise concepts and methods, and it reaches maturity when a widely accepted methodological and conceptual framework—which the leading textbooks of the discipline typically canonize—comes to constitute its hard-core. While philosophical work can help scientific inquiry by clarifying difficult concepts and by removing theoretical blocks, historical analysis can never contribute to actual scientific progress. History studies past theories and concepts that were definitively refuted or, at best, were included as limit cases into newer and more powerful theories (the canonical example being the subsumption of Newtonian physics into Einsteinian relativity). This model of scientific progress downplays the scientific interest of any research approach that is historically and conceptually peripheral to the mainstream. Even though an accurate historical reconstruction might revive the past, its results would be irrelevant to the actual path of present-day research.[5]

On the contrary, we find that the scientific study of the mind did not follow this pattern. We began assembling this collection with the assumption that a historical study of the research pathways that fell off the mainstream is indeed a necessary component of the scientific study of cognitive and sub-cognitive behavior. The motive behind our assumption is not our lack of confidence in the general validity of the standard model of scientific progress we described above. Instead, we are convinced that some precise and distinctive features proper to the scientific study of the mind force us to reject an assumption of historical and theoretical linear progress. As Roberto Cordeschi has shown (2002), the scientific field of "artificial studies" is non-traditional and non-cumulative. Why is that the case? Which peculiar features of a theory of cognition make the study of its theoretical and historical periphery relevant to contemporary research? Briefly stated, as far as we are concerned, two: the intrinsic self-reflexivity that pertains the study of cognition and its essential dependence on traditional philosophy. More precisely, these two aspects, which in fact summarize the dominant double feature of any philosophical reflection, boil down, like in the proverbial two-side coin, to a single one.

Cognitive science was the discipline that would provide a rigorous and complete theory of the most distinctive aspects of human nature grounded in the natural sciences. However, since its main targets were language and thought (as well as emotional and adaptive behaviors), the discipline would never be able to leave behind its "heroic" phase and acquire full autonomy from traditional philosophical inquiries, because its central concerns were exactly the same as philosophy's. The new science of the mind could only aim to replace the old speculations with new and scientifically more adequate theories. Content-wise, however, the new science and the old philosophy would be equivalent, thereby preventing scientific autonomy according to the traditional model.

Let us expand this argument, by taking care of a possible objection first. We could say that a similar attitude has often been present in most scientific disciplines. After all, the history of science has witnessed many forms of reductionism like biologism, psychologism, sociologism, physicalism, just to name a few. In each case, researchers made the effort to explain a variety of human phenomena—at the individual and the social levels—as the surface manifestation of an underlying reality whose complete theory only the discipline operating the scientific reduction, be it biology, psychology, sociology, and so on, could provide.[6] Some of the research projects working toward the mechanization of the mind had reductionist leanings. We could mention, for instance, some interpretations of the early cybernetics in the style of Wiener (Richter, 1976), as well as later versions of Ashby's theory of the brain (J. G. Taylor, 1962). In the first case, we were confronted with a rather strict form of physicalist reduction, while in the second case biology came to the forefront. Nonetheless, the most common attitude in the early years—an attitude that AI would adopt—was deeply anti-reductionist in the following sense: researchers did not reduce the multiplicity of phenomena that they could directly or indirectly classify as distinctly human to any single aspect—be it psychological, physical, or biological. Instead, they interpreted them as different manifestations of an essentially human nature whose complete theory the future discipline would eventually provide. How broad is this theoretical goal? Perhaps more than we usually acknowledge. We will try to characterize the scope of cognitive science with the help of a terminology that Kant made popular first, when discussing the architectonics and method of philosophical research. Our reference to Kant is not gratuitous, we think, since at least one of the key figures in the history of cognitive science, Herbert Simon, has recognized the close affinities between his project and Kant's philosophical system.[7]

Kant divides the scope of philosophical reflection into three main categories. In order of increasing generality, they are: *special metaphysics*, *general metaphysics*, and *propaedeutic metaphysics* or *critique*. Special metaphysics is the analysis and clarification of the specific objects and concepts that a particular discipline uses. In physics, for instance, it would clarify the concepts of force, motion, mass, and so on. General metaphysics is the analysis and clarification of objects in general (of what there *is*). We nowadays call it "ontology": an analysis of the general conditions under which human beings can think of objects and their properties and make successful references to them in their communicative acts. Finally, the *critique* is the preparatory and most general philosophical inquiry into the capabilities of human reason, independently of any particular object to which humans may apply them (Kant, 1998, pp. 696ff.=A841/B869ff.). Nowadays, when biologists discuss the ontological priority of the gene over the organism (Dawkins, 1976; Eldredge, 1995), or when physicists argues over strings and quarks as the ultimate constituents of physical reality (Greene, 2000; Penrose, 2005), we witness instances of scientists that engage in special metaphysics. General metaphysics and critique, on the other hand, tend to be the specialized province of philosophers. Labels might have changed since Kant's time, but the theoretical tasks are not so different. Even though general metaphysics has become ontology, it struggles with some of the same problems, including the realism/anti-realism controversy, the debate over the nature of causality, the definition of objects and parts, and so on. Today hardly anyone, except hard-core Kantians, still calls the general inquiry into the capacity of human reason a *critique*. Nonetheless, it has become one of the prevalent modes of philosophical discussion, especially—but not only—in the so-called Analytic tradition that flourishes in the USA and other English-speaking countries. Much contemporary work in the philosophy of mind, in epistemology, in the philosophy of language and in the philosophy of logic is a reflection—very similar to Kant's own—upon the structural features and overall limits of rationality and its application to the real world.

Here is also where the peculiar character of cognitive science lies, however: the scope of its investigation was (and still is) sufficiently broad to cover, in addition to empirical research, not only its special metaphysics, but also general metaphysics and critique. Examples abound. We cited above McCulloch's declaration about a science of the mind that would finally solve the puzzles of metaphysics. Likewise, in the works of Herbert Simon and Allen Newell (1990) we may find similarly all-comprehensive attitudes toward the methods and goal of cognitive science.[8] A similar perspective endures in more recent works in sub-fields as diverse as cognitive neuroscience (Bennett and Hacker, 2003), evolutionary robotics (Di Paolo, Rohde, and Jaegher, 2011), and the biology of cognition (Weber and Varela, 2002). Patricia Kitcher's critique of cognitive science (1992) springs from the same claim: like Freud's, Kitcher argues, we may interpret cognitive science as a broad and interdisciplinary research program whose efforts aspire to reach an overall theory of how the mind works and how it produces theories. Phrased in traditional philosophical terms, we claim that cognitive science is intrinsically self-reflective because it produces a theory of its conditions of possibility (a theory of itself, in other words). A geological theory is not a rock, nor is a physical theory a billiard ball: their results and predictions do not apply to themselves. A physical theory may provide an account of how a billiard ball came into being, or at least an account of how it achieved its current position and momentum. But that theory does not have to provide an account of how the physical theory *itself* came into being. On the contrary, a theory of cognition *is* cognition, and as such it must include an account of how it was produced. This is why we cannot draw a clean-cut separation between the theory's objects on the one hand, and the theory's methods and concepts on the other: they are cut from the same cloth. And this is why we cannot apply the simple—and arguably simplistic—model of inexorable, linear, and cumulative scientific progress to cognitive science. In principle, every new theory of cognition necessarily rethinks, to a greater or lesser degree, what a theory is, what its proper objects and its proper methods are, an so on. With the exception of a few momentous paradigm shifts, such as the transformation of the teleologically-oriented closed world of Aristotelian physics into the infinite universe of Galilean dynamics that Alexandre Koyré (1957) documented, a new scientific theory does not *usually* change the definition of its objects. The new theoretical apparatus is external to its domain, for it provides new explanations for the same phenomena.[9] Cognitive science does not follow the pattern of the other scientific disciplines.

Kant's terminology lets us grasp a peculiarity in the historical evolution of cognitive science that has important repercussions on any historical analysis of the discipline. Because AI must produce—alongside its specific, testable theories concerning particular aspects of human activities—the theory of those theories (namely, its special metaphysics), and the theory about the general possibility of a theory (the general metaphysics as well as the critique of "Reason"), AI is intrinsically unable to become an autonomous discipline by following the same path of physics or biology. In Kuhnian terms, we might even say that cognitive science is almost always operating in a paradigm-shifting mode: its historical development is less linear, less cumulative, and not at all monotonic. This is why history becomes an important tool for actual research, and not only for an antiquarian reconstruction of the past. If the development is not linear and not cumulative, then we may find actual scientific progress in the reactualization and rediscovery of past lines of inquiry.[10]

The simultaneous presence of intertwined epistemological levels reveals several ontological questions about the objects that the discipline studies. In order to sanction the explicative power of the scientific theories that the discipline produces, we must properly address those ontological questions. Two different and opposite needs have always inhabited research in AI—and rightly so, for the structural reasons we just sketched. Narrowly focused, hyper-specialized testable projects always, unsurprisingly, go hand in hand with vast, often too vast, generalizations. In the golden years of classic AI—roughly, between the mid-1960s and the late 1970s—these two components entered into a virtuous circle. The positive, although limited results that researchers obtained at the particular level (early semantic systems, expert systems, heuristic programs, and so forth) provided an empirical verification of the general symbolic approach, which, in turn, refined and renewed the general concepts ("heuristic navigation," "semantic representation," etcetera) to be tested in the field. On the one hand we could interpret the circle as vicious, the symptomatic behavior of a discipline that fails to become a science (Matteuzzi, 2005), or on the other as the virtuous mixture of a series of related disciplines that would produce a meta-science of the mind (Gardner, 1985).

Everything changed, however, when the cognitivist paradigm entered a prolonged period of crisis. The advent, in the 1980s, of the so-called "winter of AI" (Winston, 1987; Crevier, 1993) broke the virtuous circle between narrow empirical projects and wide theoretical inquiries. The only element left from the original unity was interdisciplinarity, although often the naturalistic reductionism of neuroscience was responsible for it. The history of AI's crisis has already been told, sometimes in considerable detail (see, for instance, Boden, 2006). We would like to emphasize some of its epistemological consequences. Clearly, breaking the tight link between special inquiries and general investigations causes a problem that we can solve by developing the discipline in two different and incompatible directions. First, we can sever the link once and for all by focusing attention on narrow pieces of research that researchers will no longer see as contributions to a general science of the mind. They will reinterpret their work as solutions to specific technical problems that arise within computer science or within software engineering. Second, replacing the discarded cognitivism with a different approach could reestablish the link (in Kant's language) between general and special metaphysics that used to guarantee the existence of a virtuous circle in the first place. In the two decades that separate us from the "winter of AI," both options have been repeatedly tried. Stuart Russell and Peter Norvig's successful textbook (1995) canonized the first alternative. It reinterprets Artificial Intelligence as a technical discipline that is firmly located within the field of computer science and that is oriented solely toward the solution of specific technical problems bound to produce usable artifacts—as does every other branch of engineering. The second option, on the contrary, has generated a continuous search for an adequate theoretical framework that could fill the epistemological role that cognitivism once played. We might mention, roughly in chronological order, some candidates for the position: the connectionism that Rumelhart and McClelland relaunched in the early 1980s; the focus on sub-cognitive abilities that Hofstadter championed; the *nouvelle AI* that Brooks's behavior-based roboticsinspired; the dynamicist approach to cognition, evolutionary robotics, all the way up to artificial life and research paradigms that theoretical neuroscience brought forth.[11] As Margaret Boden shows in her historical monograph, we have no reason to believe that the search for cognitivism's heir will be over any time soon. Her comprehensive review ends with a chapter that lists all the candidates that, by possibly producing a theoretically unified field comparable with the one cognitive science had once enjoyed, might revitalize the study of cognitive and sub-cognitive phenomena.

The present work aims to contribute to the historical and theoretical reconfiguration that cognitive science is currently going through. We have chosen the concept of "periphery" as a privileged standpoint from which to look at the current situation. Now that the theoretical center of gravity of Artificial Intelligence has for all practical purposes, dissolved, historians and philosophers of cognitive science must direct their interest toward the formerly marginal approaches that were at an (often considerable) theoretical, geographical, and cultural distance from the center. We say "must" because we are convinced that a careful study of the periphery is crucial for several different purposes. Anyone wishing, like Margaret Boden, to reestablish a new unifying core for the science of the mind should designate the periphery as a privileged object of study. A historical and theoretical study of it is a prerequisite for all discussions about the possibility and necessity of a new center that might replace the current fragmentation. We view such an analysis as preliminary to a fruitful practical and theoretical approach. Finally, a dispassionate look at the periphery is crucial for all philosophical discussions about the possibility of a project that, like AI, turned out to be an updated and naturalized version of Kant's system.

The present volume does not exhaust all possible peripheral approaches, but it offers a collection of studies that examine four different ways to gain a perspective on the theoretical center of AI research. Part I deals with the approach that we might consider as the most important periphery of classic AI research. So much so, in fact, that it constituted a second center or even an altogether autonomous suburb. We refer to cybernetics. Firmly established as the driving force behind the new science of the mind in the 1940s and 1950s, cybernetics became the main rival of classic AI research until after a well-known institutional fight it was ultimately marginalized.[12] Stefano Franchi, Peter Galison, and Peter Asaro are the authors of the three essays in part I.

Franchi focuses his attention on a cybernetic artifact, Ashby's Homeostat. He closely follows its development from the biological theories of the early 20th century to Ashby's full-blown discussion in *Design for a Brain* (1960). Then he examines its disappearance in the following decades, until he bears witness to the unexpected forays of the Homeostat into some of the most recent research projects in cognitive science (evolutionary robotics in particular). The Homeostat is reborn. This rebirth promises a radical reorientation of the field that would reintroduce the philosophical subject—endowed with needs, freedom, and self-generated goals—within the scope of cognitive science. Franchi concludes by suggesting that such a perspective, albeit thrilling, may be also, at least for now, perhaps too hasty—unless psychoanalytic conceptions of the self will come to supplement homeostasis-rich theories. In his seminal essay on the historical evolution of cybernetics' theoretical framework, *Galison* traces the development of Norbert Wiener's thought from its formation during the war years to the end of Wiener's life. Galison shows the crucial role that the metaphor of the "enemy," which allowed cybernetics to develop ideas and artifacts focused on reaching a target that might be (non-linearly) moving across time and space, played. *Asaro* discusses the concept of "scientific model" in cybernetics, in cognitive science, and in the neurosciences, and shows how the different referents—the mind, the brain, the physical world—force the concept to play radically different theoretical roles.

Part II focuses on the geographical periphery of AI research. It examines the translations into countries that had different cultures and traditions—Italy, France, and the Soviet Union—of the theories and techniques developed in the United States, AI's home ground. Claudio Pogliano, Patrice Maniglier, and Slava Gerovitch are the authors of the essays.

Pogliano discusses the diffusion of cybernetics and proto-AI ideas in Italy in the 1950s and 1960s. Thanks mostly to the efforts of US-educated intellectuals and businesspeople, Italian culture was quick to absorb the new ideas and the computing devices that were coming from the United States. Yet, while welcoming the new theories, Italy's solid humanistic and metaphysical tradition brought into question some of the most radical epistemological and philosophical claims that those ideas accompanied. The result was an innovative integration between classic humanistic values and new technological artifacts that aimed at providing the foundations for a new "mechanical civilization." *Maniglier* examines the development of the main tenets of cognitive science from the point of view of French Structuralism, a theoretical framework that was being developed at the same time as Artificial Intelligence. He claims that the theories that were part of Artificial Intelligence's background, such as von Neumann and Morgenstern's game theory and Shannon's theory of information, partly inspired also Structuralism. Maniglier shows that AI and Structuralism offer complementary solutions that we could integrate with a positive outcome, if we were to combine them: the static view of cognitive processes that classic AI developed may fruitfully complement the dynamic view of evolution of cultures and semiotic systems that classic Structuralism articulated. *Gerovitch* examines the diffusion of cybernetic and Artificial Intelligence ideas in the Soviet Union. In spite of the cultural and political rivalry that separated the two countries, researchers in the USSR were eager to adopt some of the new insights that were coming from the USA. Their different cultural and political background, however, led scientists of the Soviet era to interpretations that were often new and different from their sources'. In particular, Gerovitch shows how Soviet researchers tended to sideline the category of "rational choice" that was crucial to North American cybernetics and to AI (especially in heuristic programming), in favor of the category of (not necessarily rational) "creativity."

Part III focuses on AI's periphery understood in the cultural and historical meaning of the term. It includes essays that locate some of the central concepts of AI, such as representation and computability, within a broader background, be it philosophical (Descartes, Aristotle, Leibniz) or technical (programming theory and practice). Francesco Bianchini, Maurizio Matteuzzi, and Christopher Kelty are the contributors to this part.

Bianchini examines the debate around the experimental validation of computer simulations of intelligent behavior. Under which conditions can we consider a machine intelligent? Since Turing's original test, researchers have developed several criteria that have often generated lively debates about their adequacy. Bianchini offers a new perspective. He recommends that we locate the evolution of Turing's ideas in the historical development of the epistemology of AI. By showing that we might interpret contemporary criteria such as "behavior," "formalism," and "realism" as technical and pragmatical reformulations of philosophical conceptions of the mind that philosophers such as especially Descartes and Leibniz developed during the modern era, Bianchini offers some reflections toward a hopefully fruitful new awareness and reemployment of those concepts. *Matteuzzi* discusses the concept of Turing-computability from the point of view of mathematical constructivism and with the help of Leibniz's conception of computation. He suggests that even within the domain of constructivist approaches to mathematics (as Bernard Bolzano and Karl Weierstrass defined them), the use of random choice produces computations that bypass the limitations of Turing-computability. Ever since its inception, AI has identified computation with Turing's formalization of it, while the notion that mainstream computer science uses, Matteuzzi suggests, has been more flexible and more aware of its limitations. The close relationship between Turing-computability and the simulations of intelligent behavior that Artificial Intelligence attempts raises the possibility that a broader conception of computability may substantially renew the theoretical framework we use to model cognitive behavior. *Kelty* reconstructs the history of regular expressions and shows that this powerful programming technique has played a notable practical role in the development of AI software, in particular in the sub-fields of knowledge representation and problem solving. Kelty shows that we can consider the transition from formal theory to algorithm construction as a magnifying glass revealing the process of identification of thought and logic that was characteristic of the years immediately preceding the birth of AI. Such an identification still exerts a strong influence even on the most recent theories of cognition that Artificial Intelligence has developed.

Part IV focuses on the limitation of classic forms of computations and possible alternatives to them. Solomon Feferman, Rossella Lupacchini, and Giorgio Sandri are the authors of the essays.

Feferman discusses the issue of the intrinsic limitations of the mind on the basis of the debate around the publication of Ernst Nagel and James Newman's *Gödel's Proof*—the first popular book about Gödel's work on incompleteness. A close examination of the correspondence between Gödel and Nagel and Newman allows Feferman to show how the debate on the limitation of human thought originated and developed over the decades and how it influenced the philosophy of mind, AI, and cognitive science. Feferman suggests that a renewed appreciation of that debate can lead us to a new understanding of the formal/informal dichotomy of mathematical thought and of the closely associated notion of computability. *Lupacchini* examines a more peripheral notion of computation. She discusses the possibility that we might use concepts derived from quantum theory to broaden the notion of effective computability. As John von Neumann himself realized in the 1920s, the intrinsic non-determinism of quantum mechanics makes it quite difficult for researchers to provide a satisfactory mathematical axiomatization. Since it appears to require non-deterministic choices, which are quite difficult to formalize within classic computability theory, the simulation of intelligent behavior faces a similar difficulty. We could resolve the difficulty, Lupacchini suggests, through the development of quantum computers that would rely upon the lack of determinism that is characteristic of quantum theory. *Sandri* examines the relationship between, on the one hand, the classic theory of computability that Alan Turing formulated in the years immediately preceding AI's birth and, on the other, the objections that Gödel raised a few years later. Sandri argues that we can trace the difference between Turing's and Gödel's notion of computability back to a different conception of mental states. He stresses how cybernetics' emphasis on information and real-time processing had a profound effect on the notion of computability.

NOTES

1. Baars, 1986, p. 210. Later in the interview, Miller gives an even more salient example of the breadth of cognitive psychology as he intends it, by declaring: "I read Freud's *The Interpretation of Dreams* and chapter 7 of that book [the theoretical chapter about the unconscious, repression, etc.] is to me one of *the* great essays in cognitive psychology" (*ibid.*, p. 214, Miller's emphasis).

2. For a snapshot of the publishing industry that sprouted around the newly minted "Cognitive Science," from the early 1970s (when the label entered common use) onward, see Boden, 2006, pp. 354–365. Boden does not mention that several among the researchers themselves pioneered some publishing ventures. For instance, Nils Nilsson had a leading role in the establishment of Morgan Kauffmann, one of the main publishers of technical literature in the field (including the proceedings of its major conferences, AAAI and IJCAI). 1967 is also the year that saw the publication of Marvin Minsky and Seymour Papert's *Perceptrons*, the final and (provisionally) winning shot in the debate with cybernetics.

3. This claim is hard to prove in strict numeric terms, but a quick look at the table of contents of the early *vs.* the later conferences may offer a useful indication. The 1952 Macy's conference had ten contributions coming from eight different disciplines spanning a spectrum that goes from logic to anthropology, and from computer science to anatomy (von Foerster, Mead, and Teuber, 1953). Thirty years later, the *8th International Joint Conference on Artificial Intelligence* (Bundy, 1983) had over 240 contributions divided among technical sub-fields such as "Natural Language [Processing]" (30), "Expert Systems" (24), "Knowledge Representation" (20), "Logic Programming" (10), "Robotics" (10), "Vision" (15), and so on. "Cognitive Modeling" (with 10 contributions, less than 5% of the total) is the only subdivision to reach out of the strict confines of computer science.

4. This is how Patrick Winston (1987) reads it.

5. For a detailed analysis of the reconstructive vs. interpretive views of history our discussion relies upon, see Gadamer, 1996, pp. 164–169.

6. A small sample of such attempts would include at least Comte's and Durkheim's sociological theories of science, along with their more recent incarnations in the so-called "strong program" of the sociology of science (Bloor, 1976; Latour and Woolgar, 1986; Latour, 1987); the biological reductions that evolutionary psychology attempted (Tooby and Cosmides, 1992); the psychologism in logic and mathematics that the German school at the turn of the 19th century carried out (see Kusch, 1995); Quine and his followers, who

continued and revived psychologism's program by naturalizing epistemology (Kornblith, 1985; Kornblith, 2002); Jean Piaget's genetic epistemology (1950; 1971); and recent variations on the same theme provided by the emergence of forms of reduction that are rooted in the neurosciences—such as, for instance, the "Neurophilosophy" that Patricia Churchland, among others, practices (1986; 2002).

7. Even though Carnap's interpretation (Carnap was Simon's teacher at The University of Chicago) of Kantism was "epistemologized"; see Proust, 1987; Franchi, 2006. Kant adopted his terminology from Christian Wolff and from his follower, Baumgarten, whose textbook on metaphysics Kant routinely used when lecturing.

8. In a personal communication to one of the Editors, Simon once remarked: "Franchi is right in considering Artificial Intelligence as an effort to produce a metaphysics [even though I prefer the term 'epistemology'] with non-philosophical means. The reason why AI can do so, and has already started to do so, is that we must rephrase epistemology's questions as empirical questions (how mind works and interacts with its environment)." See Franchi, 2007, for a detailed discussion.

9. See Matteuzzi, 1981, for a sophisticated discussion of the complex relationship between theories and their objects.

10. Gould, 2002 offers a vivid example of a practicing scientist who integrates historical research with scientific inquiry.

11. See Rumelhart and McClelland, 1986a; Hofstadter, 1985; Hofstadter and Fluid Analogies Research Group, 1995; Brooks, 1999; Beer, 1990; van Gelder and Port, 1995; Harvey et al., 2005; Nolfi and Floreano, 2000.

12. See Dupuy, 1994; Dupuy, 2000; Heims, 1991; Franchi and Güzeldere, 2005a; Minsky and Papert, 1967.

I

THE CYBERNETIC SUBURB

One

LIFE, DEATH, AND RESURRECTION OF THE HOMEOSTAT

Stefano Franchi

The Homeostat, a device that W. Ross Ashby put together of military surplus equipment just after WWII in order to demonstrate his theory of cognitive behavior, enjoys a peculiar reputation within the history of AI and cognitive science. Scholars have regularly cited it as one of the most famous machines that early cybernetic tradition built. Yet, they often remark that, because the Homeostat's main behavioral goal is to keep its interactions with the surrounding environment to a minimum, it embodies a quite strange—if not outright weird—model of cognitive functions. Since Ashby explicitly presented the Homeostat as a prototypical example of *all* behavior (indeed, of all living functions), the resulting model has been at once fascinating and disturbing. Fascinating, because Ashby was able to set up his odd device to perform tasks appearing to indicate that the machine had the ability to learn and to adapt to its environment. Disturbing, because the theoretical model, and its physical implementation, were so at odds with customary models of cognition that observers of the Homeostat's feats found them to be beyond comprehension.

The distinguished audience on attendance at one of the 1952 Macy's conferences where Ashby presented his machine granted his creature a less than enthusiastic reception. Cybernetic concepts like negative feedback were crucial to the design of the Homeostat, but its performance genuinely puzzled several among the cyberneticians and cybernetic-inspired scientists assembled in the room—a group that included Warren McCulloch, Gregory Bateson, Margaret Mead, Karl Lashley, Julian Bigelow, among others. Cybernetics sought to understand behavior, both in animals and in artifacts, as the complex set of mechanisms that allow the organism to achieve its goals by taking effective steps that incrementally reduce the distance between the current state of the organism and the goals it pursues. Complex, "intelligent" behavior, they claimed, was goal-seeking. Cybernetics was an effort to explain in rigorous scientific terms the mechanisms that make goal-seeking possible. From the cybernetic standpoint, the Homeostat represented a puzzling contraption: while it implemented the physical mechanisms of goal-*seeking* in its electronic circuitry, it lacked any externally-assigned or internally-generated *goal*—in essence, the Homeostat strove to do nothing. Indeed, it was a *goal-less goal-seeking machine*, a characterization that partially captures one of Ashby's perhaps most famous achievements, namely the so called "Law of Requisite Variety" (1956, p. 207). The law states that, in order to "destroy" the variety of disturbances that comes from the environment, a system needs a wide variety of possible responses, of goal-seeking strategies. The ultimate goal is no goal at all. We can quantitatively express the law as follows: for the system to become completely impervious to the environment—which, for a biological organism, would mean achieving immortality—the system needs at least as many possible responses to the many possible disturbances that come from the environment. In a remark that neatly encapsulated the uneasy fascination the Homeostat stirred in its observers, Julian Bigelow, one of the authors of the seminal paper on goal-seeking that founded cybernetics, expressed his frustration: "It may be a beautiful replica of something—Bigelow concluded after repeatedly rejecting Ashby's identification of the Homeostat with a living creature—but Heaven knows what" (von Foerster, Mead, and Teuber, 1953, p. 95).

In spite of practicing scientists' oblivion that followed the late 1960s demise of cybernetics, the Homeostat has recently made a remarkable comeback. The British school of evolutionary robotics, for instance, has built some effective devices that rely on the mechanism of ultrastability that Ashby introduced, of which I will soon say more. Meanwhile, the biological work that Francisco Varela and his colleagues have carried out recovers some of the fundamental intuition of Ashby's (and von Foerster's) cybernetics (Di Paolo, 2000b; Di Paolo, 2005; Iizuka and Di Paolo, 2007; Di Paolo, 2000a; Harvey et al., 1997; Weber and Varela, 2002; Varela, 1979). In her recent comprehensive history of the cognitive sciences, Margaret Boden goes as far as listing "Homeostat-based cognitive science" as one of the new developments that may be able to replace cognitivism as the main research paradigm of the future.[1]

What prompted such an abrupt change of fortune that transformed an admired but strange device into a source of inspiration for a renewed foundation of cognitive science? Homeostat's new admirers propose a standard interpretation: the cognitivitist paradigm that prevailed at the time, the insufficiency of available hardware and software technology, and the univocal conception of the brain as an "active" machine prevented a proper understanding of Ashby's revolutionary ideas for modeling an *autonomous, need-based* creature. But times have changed. The cognitivist theoretical paradigm has been showing its weaknesses for some time; technology has greatly progressed; and neurophysiology has been providing us with a more complete picture of the neuronal substrate. Finally, Ashby's time may have arguably come.

I find this narrative of the death and resurrection of the Homeostat too hastily drawn. Although recent shifts in the prevailing paradigms about cognition make some of Ashby's ideas easier to accept, a careful look at his intellectual background and to the contemporary theories of homeostasis and of cognitive functions, however, reveals a more complex picture. Today the Homeostat may even be more at odds with Varela's and Di Paolo's non-cognitivist view of embodied cognition than with the earlier dominant functionalist one. Moreover, I suggest that the Homeostat offers a glimpse of an interpretation of natural and artificial life that is genuinely alternative to both approaches. In the next section I will discuss the main principle behind the Homeostat, and I will provide some details about its reception. Then I will examine the physiological theories of homeostasis that Claude Bernard and Walter Cannon advanced and that served as inspiration to Ashby. In the following section I will analyze the rebirth of the Homeostat, which started in the year 2000, within the theoretical framework that really supports it: Humberto Maturana and Francisco Varela's theory of autopoiesis, and Hans Jonas's phenomenology of the living. In the final section I will offer some suggestions on alternative frameworks that may be capable of accommodating the disturbing and yet fascinating intuition lying behind Ashby's creature.

1. Life and Death of the Homeostat

Ashby's main work—*Design for a Brain*—contains not only a rigorous theoretical discussion of the key concepts of his theory and a detailed description of the Homeostat and of the experiments that Ashby carried out on it, but also the most thorough discussion of the device.[2] In spite of the remarkable novelty of some of the concepts that this book introduced to the contemporary scientific audience, we can reduce Ashby's theory to a small number of theses that I will sum up as follows:

1. A complex system running to equilibrium generates a complex behavior.
2. A system is in equilibrium when it keeps its essential variables at stable levels.
3. A system runs to equilibrium through ultrastability: when its essential variables are unstable, it rearranges the connections between its subparts.
4. A system achieves ultrastability through a stochastic step function that operates on the internal connections of the system by randomly stepping through the parameters connecting together its various parts.

The first thesis is a compact formulation of chapter 4 of *Design for a Brain* ("Stability"), where Ashby discusses the main properties of systems with feedback. He concludes: "it is true that the stable system usually has a state of equilibrium at which it shows no change; but the lack of change is deceptive if it suggests rigidity: if displaced from the state of equilibrium it will show active, perhaps extensive and complex movements." Toward the end, the work introduces the real scope of that assertion, hence the thesis itself:

> what this book attempts to display is that when a *complex* system runs to equilibrium, the equilibrium necessarily implies a complex relationship

> between the states of the various parts. When the relationship between the states of the part is examined, they will show unusual and striking features, features that are of peculiar interest to the student of behavior. (Ashby, 1960, 4/17, 55–56; 14/1, p. 184, Ashby's emphasis)

The second thesis encapsulates Ashby's interpretation of biological adaptation. First, he states that "a form of behavior is *adaptive* if it maintains the essential variables within physiological limits," and then, after a discussion of homeostasis, he concludes that "adaptive behavior is equivalent to the behavior of a stable system" (*ibid.*, 5/3, p. 58; 5/7, p. 64); hence, my compact formulation. The third thesis presents what Ashby calls, after using a more abstract terminology, the *principle of double feedback*. Notice that "rearranging connections," by changing the sensory-motor pattern that a reaction to a stimulus produces, (*ibid.*, 7/2–7/8, pp. 80–84) may have direct effects on the environment. The fourth one contains the principle of random search, which I will discuss in more detail later (*ibid.*, 7/13–7/25, pp. 87–98).

The physical device that Ashby built and refined since the middle 1940s implemented these four theses. Relying on the experiences he had carried out on it, he eventually derived the following further set of contentions:

5. A system with many parts (subsystems) that are fully connected to each other will need an exceedingly long time to reach equilibrium (*ibid.*, chp. 11, "The Fully-Joined System").
6. A real system will then need weak connections among its parts—the system will be *dispersed* (*ibid.*, chp. 16/11, p. 214).
7. Typically, the subsystems will not be mutually connected on the organism side: (its interactions with) the environment will mediate the communication among subsystems (*ibid.*, sec. 16/5–16/6, pp. 208–209).

Ashby could provide empirical confirmation for the fifth thesis when he conducted a series of experiments to estimate the connectance threshold value that would separate a stable system from a hopelessly unstable one. His data indicated that even small systems become immediately unstable when their connectance goes beyond a relatively low value of 13% (Ashby, 1970). The sixth thesis is the principle of *Dispersion* that Ashby discussed in the first edition of *Design for a Brain*, where he claimed that it was applicable to sense-organs and to the brain (Ashby, 1952b, pp. 103ff). Unfortunately, the later edition keeps the concept but drops the effective label, "dispersed": "evidence has been given ... that the *actual* richness [of connections within a system] is by no means high. [We] have shown that when it is not high, adaptation by ultrastability can occur in a time that is no longer impossibly long." Ashby uses this insight to elaborate his notion of a "multi-stable system" that several loosely connected and ultrastable systems compose. He claims that it represents the best model of real brain functions. A footnote in the first edition references a concrete multistable device (DAMS,

Dispersive and Multistable System) that he was building at Barnwood House (*ibid.*, p. 171), but Ashby does not provide any details about it. Philip Husbands and Owen Holland (2008, pp. 125–126) mention a presentation of the DAMS system at one of the last meetings of the *Ratio Club*, a British cybernetics dining club that met regularly between 1949 and 1958, which took place at Barnwood House in 1955. Although Husbands and Holland neither provide any details on the device nor they reference any descriptions of it, they notice that "Ashby was not able to develop it sufficiently to fully demonstrate his theories."[3] We may label the seventh thesis the *principle of deep situatedness of living beings*. The diagram that illustrates the descriptions of the "poorly joined environment" and the "multistable system" in *Design for a Brain* shows "the organism's reacting part [...] divided into subsystems *between which there is no direct connexion*" (Ashby, 1960, p. 208, emphasis added). Since Ashby argues that brains are multistable systems, this general thesis applies to the brain too.

The epistemological status of theses number 1, 2, 3, and 4 differs remarkably. The first one—let us call it the *generalized homeostasis thesis*—is Ashby's most original and, at the same time, least warranted idea.[4] It embodies Ashby's fundamental claim about the behavior of *any* complex system—be it a physiological, psychological, or even a social one. Demonstrating the breadth of his position, in *An Introduction to Cybernetics* Ashby proposes the Homeostat as a model, in principle, of a complex economic system. Besides *Design for a Brain*, we may find examples of cognitive applications of the generalized homeostasis thesis also in his discussion of a "homeostatic" chess-player (Ashby, 1952a). Ashby presented this idea in a separate contribution paper that he read at the 1952 Macy's conference. The transcript shows that Ashby met a perplexed resistance from his audience (*ibid.*; Ashby, 1953). The breadth of the first thesis turns it from a scientific statement *strictu sensu* into a general philosophical and metaphysical position that underlies all of his work. More importantly, the thesis as such is usually absent from contemporary discussions of his work. For instance, Norbert Wiener praised the Homeostat as "one of the great philosophical contributions of the present day" and often made reference to it in his writings (Wiener, 1954, p. 38), but I doubt that he appreciated, let alone shared, Ashby's philosophical assumption. Wiener's discussion of the Homeostat never addressed the foundational role that homeostasis plays within Ashby's conception of living and of cognitive functions, thus reducing Ashby's contribution to an elegant and especially powerful implementation of the idea of feedback. Conversely, Wiener's approach to homeostasis, like Claude Bernard's and Walter Cannon's, reduces its scope to the physiological domain.

Briefly touching on homeostasis at the end of *Cybernetics'* chapter on feedback, Wiener presents it as "another important physiological application of the principle of feed-back," whose main difference with our "voluntary and postural feed-backs" is that "it tends to be slower."[5] Wiener is interested in the mechanism

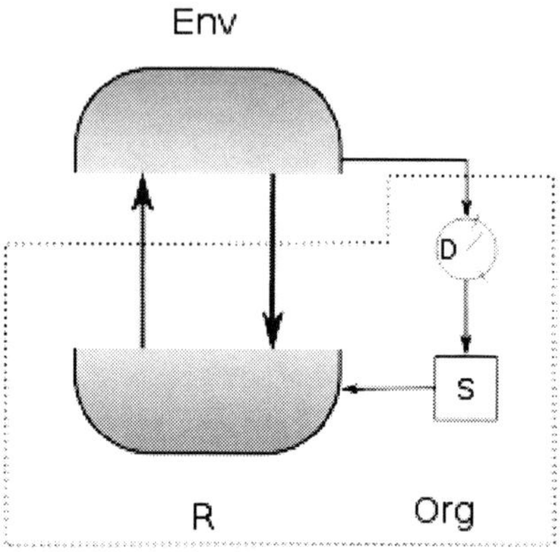

FIGURE 1.1: A diagram of the Ultrastability mechanism as depicted in *Design for a Brain* (p. 83), and corresponding to one of the units in the Homeostat. *Env* is the Environment (in the Homeostat, the other three units). *R* is the part of the organism that interacts with it; the box *S* represents the set of parameters affecting the organism's behavior. The dial *D* stands for the "essential variable" monitor whose out-of-bound (unfavorable) value triggers the switch to a new set of randomly chosen parameters in *S*, ensuring "ultrastability." The thick arrows stand for the sensory and motor channels. The area within the dotted line is on the organism side, the area outside is on the environment side. The shaded areas in *Env* and *R* represent the "fine networks of linkages" existing within each of them. Notice that Ashby identifies *R* in purely functional terms and it does not need to correspond to any anatomical or histological structure.

that makes the process possible, not in what the process enables (homeostasis), or in the condition that it seeks (stability or ultrastability). He exactly reversed Ashby's priorities. Later in the book, while discussing psychological processes, Wiener mentions one among Ashby's favorite references, Ivan Pavlov's work on conditioned reflexes, and introduces the idea of pleasure and pain, or "affective tones," that we associate with a pattern of action favorable or unfavorable to the organism (1948, p. 151–153). Even in this instance, though, Wiener, unlike Ashby at the time and Michel Cabanac now, fails to associate pleasure with the achievement of stability.

For Wiener, homeostasis remains a particular instantiation of the mechanism of feedback at the physiological level. Homeostasis's goal—stability—plays no significant theoretical or explanatory role. Nor did Wiener's opinion on the matter change once Ashby's results became better known and appreciated. In his last published lecture, Wiener first defines biocybernetics as that branch of the discipline that emphasizes living systems, and then splits it into neurocybernetics and medical cybernetics. He declares the second to be the branch in which "homeostasis is the main consideration," whereas the first is mostly concerned with "the pathways of actions via sense-organs, neurons and effectors." This classification confirms that while Ashby put homeostasis at the center of his *Design for a Brain*, Wiener effectively took it out of it.[6] Ashby's remaining theses and the Homeostat itself, by showing the insights that we can gain through their broad application and by virtue of the conclusions stemming from them, make the generalized homeostasis plausible.

The second thesis—the equilibrium of an organism equals the stability of its essential variables—is Ashby's least original but most solid one. It amounts to a restatement of the principle of homeostasis that by the late 1940s was to become widely accepted in the physiological community. The third and fourth theses, instead—the principle of ultrastability is a double feedback that we achieve through random step functions—represent the core of Ashby's technical contribution. They are directly involved in the concrete make up of his machine, whose performance in turn immediately influenced them. The fourth thesis may appear, as it did to the participants of the Macy's 1952 conference, especially counter-intuitive. It claims that all successful adaptations are the result of a random process. All *successful* cognitive or non-cognitive achievements proceed from the random selection of a possible behavioral choice among a pool of options that are randomly assembled. Yet Ashby was quite insistent on the crucial link between randomness and creativity—or, in biological terms, between random choices and survival chances. When the environment presents a new, previously unseen challenge, only randomness can ensure that the organism will try a genuinely new behavioral option that may be crucial to its survival. Ashby's discussion of a chess-playing algorithm, which he more carefully argued in *An Introduction to Cybernetics*, extended the argument to cognitive functions.[7]

Ashby also strongly believed that a complex system had to produce parameter changes through discontinuous stepping functions: an organism, he claimed, could never adapt to the environment by way of small, incremental changes to its behavioral responses. In "The Nervous System as a Physical Machine; with Special Reference to the Origin of Adaptive Behavior" (a precursor to the formal theory of biological systems that he will later present in the mathematical appendix to *Design for a Brain*), Ashby defends his choice on the basis of a mathematical analysis of dynamical systems. As supporting evidence, he points to contemporary neurobiological work on synaptic growth that occurs in response to developmental and environmental stimulation. Notice that the new parameters that the stepping function induces may, and most often do, affect the environment as well as the organism that dwells in it. Any given change directly affects only the organism's internal parameters, but the new parameters may result in actions by the organism that cause changes in the external environment. For instance, the ultra-stable mechanism can trigger a change that may result in the organism changing the value of the parameter that determines how to deal with an element of the environment from "avoid" to "eat." The ultrastable mechanism triggers *directly*

FIGURE 1.2: Two of the Homeostat's four units, from *Design for a Brain* (p. 101). The switches visible on the front panel allow manual control of the unit's connections and polarity. The needle unit and its fluid-filled trough (which together model the *essential variable* whose constant value the unit tries to defend) are visible on top.

a change internal to the organism and *indirectly* a change in the environment, in this case switching the organism from a prey into predator role.

I will now take a brief look at the machine itself.

A. The Homeostat

The Homeostat comprises four units, each one of which consists of a needle connected to a magnet, some electro-mechanical circuitry, and other physical components. The position of each of the needles can deviate either side of a central point. The positions of the magnets represent the essential variables that the system must keep within bounds. All four units are connected to each other. The torque that operates on each magnet is approximately proportional to the sum of its own output current plus the currents that the other three units produce. The current that a unit sends out and then feeds back to itself is proportional to the deviation of the magnet from its central position: the needle of each units dips into a trough of water below it and it picks up a potential that is dependent on the position of the magnet relative to the water. The viscosity of the fluid affects the stability of the Homeostat: a higher viscosity dampens the system and makes it more stable; lower viscosity leads to a more fluctuating and dynamic behavior. We can picture each unit as portraying the environment or the sensorimotor system of an ultrastable mechanism; the interactions between the units simulate the primary feedback. When a magnet deviates from its central position by more than the limit we have specified, it triggers a secondary feedback mechanism that randomly changes the parameters that affects the movement of the magnet. By virtue of this simple and random mechanism, the Homeostat behaves as though it were seeking to keep the magnets in their central positions in a goal-oriented fashion. As I mentioned above, the text of *Design for a Brain* is not completely explicit on how Ashby's Homeostat implemented this random search. Most likely, the sub-mechanism that is responsible for the parameter search stepped sequentially through the same hardwired sequence of random connection values. The limitations of technology available at the time most likely dictated this solution. Elsewhere (for instance, in "Design for an Intelligence Amplifier"), Ashby is adamant that we should *not* limit the randomness that the machine requires to the set of values that the designer of the machine has provided, lest the device be confined to the designer's preconceptions and limitations.

Ultrastability (see figure 1.1) implies a double feedback loop that is time-delayed: the primary feedback loop takes care of the system's reactions to the environment in ordinary circumstances; the secondary time-delayed feedback loop changes the parameters of the primary loop in order to accommodate the organism to extraordinary changes in the environment. Ashby's favorite example of a single feedback loop system is the automatic pilot that keeps a plane flying level by forcing the ailerons to roll in the one direction when reacting to a roll in the opposite direction. Although such a system works fine in ordinary circumstances, in extraordinary circumstances it can run into troubles. If, for instance, the connections between the autopilot and the ailerons were reversed, the autopilot would still react the same way, hence worsening the roll until the plane crashed. Instead, a double feedback or ultrastable system, would intermittently check if the primary loop is working effectively. If not, it would reverse the connection between the ailerons and the autopilot, preventing the system from crashing. Ashby found evidence of a double feedback system in living beings in the experiments that researchers had conducted on monkeys. Although scientists had surgically reversed the neural connections that joined their forearms' flexor to their extensor muscles, the monkeys could nonetheless regain their motor skills, thus proving that an internal restructuring of the neural connections had taken place.

First, Ashby tested the Homeostat letting it stabilize. Then he took control of one of the units and reversed the commutator by hand, thereby causing an instability. He then observed the system as it adapted its configuration until it found again a stable state. Most importantly, the restoration of stability required only relatively few restructuring changes. He also carried out several more sophisticated experiments involving a varying number of units in different configurations. In all cases, Ashby found out that to reach homeostatic stability the Homeostat needed only a few changes. These results authorized him to conclude that the Homeostat displayed an elementary power of self-organization that was equivalent to the behavior of the nervous system: "first the established reaction, then an alteration made in the environment by the experimenter, and finally a reorganization within the nervous system, compensating for the experimental alteration" (1960, p. 107).

FIGURE 1.3: A detail of the needle-magnet part resting on top of the Homeostat's units, from *Design for a Brain* (p. 101). When the needle hits the walls of the arc-shaped trough, the secondary feedback process is triggered and the unit will rearrange its internal connections.

This brief description of the Homeostat elucidates the relation between Ashby's device and the four general principles I mentioned above. The performance of the device attests to the validity of the three concepts of ultrastability, random search through stepwise functions, and double feedback loops between the organism and the environment. The implementation of such mechanisms lets the Homeostat learn by modifying its behavior in response to external circumstances that have changed, but the device itself cannot validate the first premise, the so-called *generalized homeostasis thesis*. Instead, Ashby built his Homeostat around it. Whereas the other principles descend from the device's behavior, the axiom of generalized homeostasis lies at the core of its very construction. To put it differently, the first thesis is the philosophical—and philosophically quite unorthodox—foundation of everything else. By building a simplified system—the Homeostat—that demonstrates the physical plausibility of his first premise, Ashby aimed at proving a claim that essentially concerns *life* at large.

This first thesis is distinctly unorthodox, and as such it met, at the Macy foundation's meetings where Ashby first made his Homeostat known, the resistance of the cybernetic community. The reason is clear: the Homeostat entails that life is fundamentally a process (or a series of processes) that aims at minimizing the impact of external disturbances on the organism. Grey Walter devised a tag that aptly suits Ashby's Homeostat: because its ideal function is to go back to sleep as quickly as possible, he called it a *machina sopora*.[8] Walter's characterization, failing to take into account Ashby's insistence on the complex behavior that the systems he discussed could cause, may well be superficial, but it does point to the source of the main resistance that scientists opposed to Ashby's thesis about life: his effort to derive action from non-action. Ashby's homeostatic organism accords philosophical and theoretical preeminence to non-action—to sleep, in Walter's mocking formulation. This means not that the Homeostat will always be sleeping but, even more disturbingly, that the Homeostat will act as much as needed solely in order to return to sleep. It will go to extraordinary lengths, it will take every action it can, just to revert to rest (or to equilibrium):

> The constancy of some variables may involve the vigorous activity of others .... The point of view taken here is that the *constancy* of the essential variables is fundamentally important, and that the activity of the other

variables is important *only in so far as it contributes to the end.* (Ashby, 1960, p. 67, emphasis added)

The ever possible failure of such compensatory system, the device's inability to go back to "sleep", will unfailingly lead to its demise or death. For this reason, the Ashbian organism will always try to accommodate itself to the environment by whatever means necessary. First and foremost, the Homeostat is a *passive machine whose activity is a by-product of its search for inactivity.*

B. Physiology's vs. Ashby's Homeostasis

Ashby's approach raises the basic question whether his *generalized homeostasis* is simply an extension of the concept that physiologists had so far mainly applied to lower level automatic functions or whether it is instead a significant departure from it. Ashby lifted the term mostly from Walter Cannon, whose successful 1932 book, *The Wisdom of the Body*, had systematized the theory—which Claude Bernard advanced first and then later Ernest Henry Starling and John Scott Haldane extended—by supplementing it with additional experimental evidence. .[9] After studies first in Berlin and later in France with Charles Richet—who, in turn, had been a student in Claude Bernard's laboratory under Paul Bert—Arturo Rosenblueth, who had became the collaborator of Walter Cannon at Harvard and Norbert Wiener's close colleague, supplied the direct connection between the physiological and the cybernetic conceptions of homeostasis.

Since its inception, a peculiar tension, almost a blind spot, marks the physiological theory of homeostasis: it lies in the relationship between the mechanisms that preside over the regulation of the lower level and the higher level conscious functions. Claude Bernard introduced the basic concept of homeostasis by determining that the *"constancy of the internal environment is the condition for free and independent life"* (1974, p. 84, emphasis added). Because life cannot thrive in air or any other gaseous medium, Bernard stressed that its environment must be internal to the organism. The fundamental chemical exchanges that make life possible need a liquid medium; over time, the obvious variability of atmospheric conditions forbid a biological process from thriving. Life requires a constant and *liquid* medium that we must also keep within strict physico-chemical parameters in order to guarantee the possibility of chemical reactions. This liquid, in higher animals, is the blood, the genuine internal environment that is proper to every single cell of the organism. The organism must keep blood's properties constant through a "continuous and delicate compensation established as if by the most sensitive of balances," that is, by the processes researchers will later call homeostatic.[10] The primary goal of an independent and free organism is to ensure that the internal environment (*le milieu intérieur*) that makes it possible remains constant. Higher organisms, Bernard concludes, thrive because they have placed themselves in a hothouse that needs a constant and delicate maintenance.

But what does Bernard mean by saying that forms of life capable of building their unique constant internal environment are *free?* The concept is not only laden with philosophical overtones that exceed the strict boundaries of Bernard's physiological theory, but also under-determined among several different meanings. Are the higher organisms free *from* something—the vagaries of external conditions, for instance—or are they free *for* something—the pursuit of goals forbidden to lower life forms*?* Or perhaps both statements are true, and higher organisms are free to pursue higher goals precisely because they are free from the stricter constraints that nature imposes on lower forms? By calling "free" all life forms equipped with a constant internal environment, Bernard inscribed the issue at the very core of the theory of homeostasis, but he gave it only a tentative answer. In his classic and often quoted *Leçons sur les phénomènes de la vie communs aux animaux et aux vegetaux*, he contrasted a *free, independent* life form with two others: *the latent*, and *the oscillatory* or dependent life. Bernard's three categories stem from his conception of life as an intrinsic conflict between, on the one hand, "the external physico-chemical conditions," and on the other, "the organic or pre-established laws that regulate the sequence, the concert and the harmony of the biological phenomena" (1974, p. 251). The three kinds of life forms spring from the more or less close relationship that these two sets of conditions entertain: when the external conditions dominate, life become latent—as we see in seeds and in simple animals such as infusoria, which are in a state of chemical indifference and display a pure potentiality for life. When the variability of the external elements alters life, it becomes oscillatory or dependent—as we see in all plants, invertebrates, and cold-blooded vertebrates. Finally, when the relationship is constant and in a state of (fragile and delicate) equilibrium, life becomes free and independent—as in the warm-blooded vertebrates that, for Bernard, are the highest life forms. Leaving aside the flaws of a classification that all too easily maps morphological complexity onto evolutionary advance, Bernard's configuration manifestly turns the freedom of the highest forms of life into a paradoxical concept. Ironically, free life is the most strongly tied to its environment; it can earn independence only by way of continuous and subtle adjustments. Bernard is then correct when he declares that the independence of free life from its internal conditions is merely "an illusion, and it is especially in the mechanism of constant or free life that the close relationship of the two kinds of conditions are exhibited in the most characteristic manner" (*ibid.*, p. 251). Free life is the least independent; only latent potential life, by virtue of its chemical indifference, achieves its freedom from the environment. Why, then, we may wonder, did Bernard choose to label as "free" the highest, most interconnected, and homeostasis-based kind of life? Processes that we will later call homeostatic yield the exact opposite of freedom: as homeostasis increasingly ties the organism to its environment, it also ensures that the life it regulates is as physico-chemically determined and un-free as possibly imaginable. The option is twofold: either Bernard commits to a latent form of biological determinism—and his terminology appears to counter the idea—or the regulatory mechanisms that he places at the center of all physiological functions do not exhaust the complex biological make up of higher animals, including animals like us. His most explicit statement on this point claims that even though the physico-chemical relationships that we find at the physiological level provide the conditions of possibility of cognitive functions, they do not explain them:

> We will not understand better, from the physiological point of view, the essence of intelligence, than we could understand, from the chemical point of view, the essence of fire or of the flame. The physiologist must not linger, therefore, on these interpretations; it will be enough for him to know that the manifestation of the phenomena of intelligence and of consciousness, no matter how extraordinary they may appear to us, need organic or anatomic conditions which are accessible to his inquiries, and it is precisely within these limits that he circumscribes his domain. (1869, p. 20)

Here are Bernard's final remarks on the relationship between philosophy, psychology, and physiology:

> there is no contradiction between physiology and metaphysics: they just approach the problem of cognition from opposite sides. Physiology links the study of cognitive functions to the organic and physical conditions that express them, whereas metaphysics neglects these relationships to focus on the manifestations of the mind that appears in the progressive march of humanity or in the eternal aspirations of our sentiment. (*ibid.*, p. 23)

The two passages articulate a solution to the philosophical implications of homeostasis that is all too precarious and point to the basic problem that the generalization of homeostasis, which will be later attempted by Ashby and thematized by Haldane and Cannon, raises. Haldane, perhaps the most influential physiologist to spread Bernard's conception of the *milieu intérieur* in Britain and in the United States, explicitly addresses the issue. He resolutely denies that we can possibly generalize upward Bernard's conception of the stability of the internal environment to include conscious activities. In the 1916 Silliman lecture, Haldane is keen on stressing that the regulation between the organism and the environment is a necessity—to the point of dismissing any meaningful distinction between the two. But he is as keen to point out that organic regulation differs from knowledge and cognition:

> It is unmeaning to treat consciousness as a mere accomplishment to life, or to ignore the difference between blind organic activity and rational behavior. Conscious personality is far more than mere organism, and the conception of life is just as inadequate in connection to personality as the conception of matter and energy is in connection to life. (1917, p. 115)

A few years later, in his major work on respiration, Haldane extends his argument and places it within a broader Kantian philosophical framework.[11] He finds that Bernard's statement on the internal environment is the most "pregnant sentence ever framed by a physiologist" and that, with respect to the blood reaction and oxygen supply, his own book follows the principle Bernard had devised. Moreover, Haldane indicates that the fundamental distinction between conscious and unconscious activity lies in the role that memory plays in perception; even in a simple action such as handwriting, the "movements of my muscles are determined by the actual presences to me of innumerable past, present and anticipated future events in both my own individual history and that of mankind" (1922, p. 397). The facts that we face in our everyday experience of conscious activity, Haldane remarks, are beyond the reach of both physical and biological hypotheses. Hence he concludes his book by sharply reducing the domain of Bernard-like inquiries: "we have reached a limit to the sphere of physiology when we come to deal with conscious activity" (1922, p. 399). In a more philosophically oriented work of a few years earlier, Haldane had provided an extended account of the issue that he articulated on four possible epistemological levels. We could, he argued, look at a human being at the pure *phenomenal* level, as a "coloured extended patch or form"; at the *physico-chemical* level, as "about seventy kilogrammes of material with a certain configuration, internal structure, and movement"; at the *biological* level, as a "living organism, blindly maintaining its structure and activity and fulfilling its function in relation to the species"; and finally, at the *person* level, as an individual who is "in conscious relation through perception and volition to his environment." Each level presupposes and extends the previous one, so that the person level "includes all that was left out in the previous modes" (1913, p. 124). He concludes: "it was shown that the relation of a person to his surrounding world which he is in contact through perception and volition is not a mere external relation, since his surrounding world is teleologically determined in relation to his organic life" (1913, pp. 138–139).

The tension between consciousness and physiological regulation reaches a resolution (albeit differently than in Ashby) in the work of Walter Cannon, and the general theoretical framework (which Bernard had identified and Haldane expanded) reaches a larger audience beyond the borders of physiology. The great success of Cannon's book—*The Wisdom of the Body*, published in 1932—and his personal connection to Arturo Rosenblueth, his last collaborator at Harvard and one of the main collaborators of Norbert Wiener in the post-WWII years of classic cybernetic theory, were the main factors behind the wide dissemination of homeostatic ideas. Cannon introduced the concept of homeostasis within an almost engineering framework: it is "a steady state produced by the action of forces." Four are the conditions that make such framing necessary. First, our bodies are open, unstable open systems that require agencies to achieve the constancy that we observe. Second, the resistance to change—through negative feedback, as Rosenblueth, Bigelow and Wiener will say—produces this constancy. Third, by acting simultaneously or in succession, several cooperating agencies reach this regulation. Lastly, agencies that are capable of shifting the homeostatic state in one direction will often, if not always, co-exist with agencies that are capable of effecting the reverse shift (1929, p. 401 and 1939[1932], pp. 299–300, respectively). Only recently physiologists have started to question some of the assumptions that are implicit in Cannon's formulation, whose four conditions constituted the backbone of all interpretations of homeostasis for the decades to came. Researchers point out, for instance, that the base levels of the essential variables that homeostatic processes defend are often dependent on cyclic external factors.[12]

Eventually, the contradiction between homeostasis and freedom comes to the fore in the final chapter of *Wisdom of the Body*. What do we mean, Cannon asks following up on the work of Joseph Barcroft, by saying that the constancy of the internal environment frees the organism from the limitations that both internal and external agencies impose on it or from the conditions that could disturb it? What is this freedom *for?* Cannon's answer is unequivocal:

> It is chiefly freedom for the activity of the higher levels of the nervous system and the muscles that govern it .... We find the organism liberated for its more complicated and socially important tasks because it lives in a fluid matrix, which is kept in a constant condition.[13]

Anticipating cybernetics' strategy, Cannon even extends his argument from the physiological domain to the body politic. As the supreme value of the "body physiologic" is the integrity of the individual organism, so the supreme value of the body politic is social stability. And similarly to the physiological body, in order to achieve the stable society that frees us to the pursuit of higher purposes, careful planning, in addition to regulation of production and distribution, is of the essence:

> with homeostatic devices ... we are free to enter into agreeable relations with our fellows, free to enjoy beautiful things, to explore and understand the wonders of the world about us, to develop new ideas and interests, and to work and play, untrammeled by anxieties concerning our bodily affairs. (1939[1932], p. 323)

As the following passage demonstrates, the connection between the two domains is tight:

> The main service of social homeostasis would be to support bodily homeostasis. It would therefore help to release the highest activities of the nervous system for adventure and achievement. With essential needs assured, the priceless unessentials could be freely sought. (1939[1932], p. 323)

Cannon negotiates the conflicting insights of Bernard and Haldane with an almost Hegelian move that potentially resolves the dialectical tension that the constancy of the *milieu intérieur* as a condition of free life implies. Cannon's homeostasis becomes the defining feature of free, independent life, as Bernard had claimed, not because it constitutes freedom, but because it provides the preconditions for a free life that is to all ends and purposes homeostasis-free, as Haldane had stressed. Moreover, homeostasis fulfills this difficult act by virtue of its complex interplay in the two different domains that I have called body politic and "body physiologic." The body politic provides the means for the realization of physiological homeostasis which, in turn, allows society to pursue its loftier goals. This said, I could hardly conclude that Bernard's original word choice—a symptom perhaps of his deeply seated uneasiness about the relationship between *les sciences physiologiques* and *les sciences métaphysiques*—strictly necessitates Cannon's solution. On the contrary, I believe that the conceptual tension that Bernard's work and his terminology encapsulate could have been resolved in different directions. Although Cannon's solution represents one possible solution, alternatives strategies are possible, as a discussion of Ashby's confrontation with Cannon's theory will make clear.

If homeostatic processes are what the organism needs in order to gain stability, could we not assume that stability is its own reward? Homeostatic theory has no theoretical need to transform the stability of the internal environment into a precondition for a higher form of life. We could just as well reach the opposite conclusion: namely, that life *just is* homeostasis, and that the different kinds of life we encounter, as Bernard had originally posited, are explainable as different realizations of homeostatic processes.. As the following remarks from his *Introduction to Cybernetics* show, Ashby was aware of the difference between his position and Cannon's:

> The subject of regulation in biology is so vast that no single chapter can do it justice. Cannon's *Wisdom of the Body* treated it adequately so far as internal, vegetative activities are concerned, but there has yet to be written the book, much larger in size, that shall show *how all the organism's exteriorly-directed activities—its "higher" activities—are all similarly regulatory, i.e. homeostatic*. In this chapter I have had to leave much of this to the reader's imagination, trusting that, as a biologist, he will probably already be sufficiently familiar with the thesis. The thesis in any case has been discussed to some extent in *Design for a Brain*. (1956, pp. 195–196, emphasis added)

It is worth noting that Ashby's phrase—"the thesis has been discussed to some extent,"—is downright misleading. Although *Design for a Brain* certainly implied the sharp divergence between his position and Cannon's—as I have been attempting to show— Ashby had never explicitly addressed it.

I believe that now we are in the position to appreciate how Ashby's generalization of a physiological theory, its apparent innocence notwithstanding, drove him onto a path that is the exact opposite of the one that physiologists had taken and later had bequeathed to cyberneticians. I may even say that Ashby, wanting to push homeostasis all the way up, was closer to physiology than the physiologists. His move is exactly the reverse of the one Haldane, who insisted on the incompleteness of the biological description, was recommending. While Ashby's view of homeostasis contradicts Haldane's—and, consequently, Cannon and Rosenblueth's—it is still as logically consistent as theirs, and as consistent with the physiological data. It genuinely represents an independent conceptualization of a phenomenon—homeostasis—which I see driven by nothing short of an alternative view of the status of the regulatory mechanism itself—an alternative view of what constitutes the "mechanical" altogether.

When Haldane, as I indicated above, pitted "blind organic activity" against "rational behavior," he provided the best formulation of the traditional view of the subject. Organic activity, blind because it lacks the freely set goals that should guide it, is mechanical; rational activity is just the opposite, not blind and unmechanical. We could then view the homeostatic mechanisms that Bernard described for the blood, and Cannon likewise for other physiological functions, as ways in which the body behaves as a machine. But the reduction of the body, including the human body, to the status of a machine is undoubtedly not a new idea: European scientists were familiar with it at least since Descartes and Gassendi.[14] Things are remarkably different, however, with Ashby's proposed extension of homeostatic processes to cognitive functions, a move entailing that not just the body, but the mind, behaves "mechanically." The problem, though becomes: what do we mean by behaving mechanically? And even more saliently: what is a machine? The first possible answer is that we call a system mechanical when it predictably produces the same behavior in response to the same set of stimuli. The mechanical will be:

1. *passive-reflexive*: the machine will never initiate a behavior on its own, therefore we need stimuli to produce it.
2. *predictable*: the same set of stimuli produces the same kind of behavior.
3. *heteronomous:* as the environment, not the machine, sets the machine's goals, the regularities of the environment, not the machine's, will dictate its rules.

In the case of the Homeostat, Ashby claims that thesis 2 is false: the device's behavior is mechanical by definition (state-determined, in Ashby's terminology), but the complexity of its behavior makes it not predictable. Scholars have usually focused their attention on this well-known aspect of the question. But arguably a deeper problem—perhaps Ashby's further challenge—lies ahead: we can hardly say if the Homeostat is passive-reflexive at all. On the one hand, the Homeostat is passive—it never does anything unless disturbed. On the other hand, it perpetually stands always ready to do something, constantly (albeit minimally) readjusting itself to the surrounding environment. On occasion, it can even produce major internal readjustments. The behavior of the Homeostat may well resemble the behavior of a machine, but to reconcile its comportment with a machine's is difficult. By extending homeostasis upward, Ashby scores a double result: he equates cognitive functions with mechanical functions and at the same time redefines (or throws into doubts) what a mechanical function is. The following schema may clarify his position. We could summarize the traditional "Cartesian" framework as follows:

1. A machine is a predictable passive-reflexive heteronomous device.
2. A body → is a machine.
3. A mind → is not a machine → but an active goal-setting autonomous entity.

At first it might look as if Ashby's upward generalization of homeostasis gets rid of (3), reducing body and mind (more precisely, physiological and cognitive functions) to the status of machines that we take to be predictable, passive-reflexive, and heteronomous devices. But Ashby's move, while accepting the reduction of the cognitive to the mechanical, gets rid of (1) too—and this is an all too different step. Ashby redefines the status of the machine as an *unpredictable*, *not* passive-reflexive, and perhaps *not* even heteronomous device. This is why, in my opinion, Ashby's conceptualization of homeostasis is of interest to cognitive scientists and researchers in computability theory alike: because it questions both sides of the equivalence.

To assess Ashby's complex series of claims, I will now take a closer look at the status, in classic cybernetics and in the Homeostat, of goals and goal-seeking behavior. This will lead me to discuss the second of the four Ashbian theses that I listed at page 5 above.

C. Cybernetic Purposiveness

Peter Galison has remarked how crucial Wiener's wartime work on the development of an anti-aircraft automatic gun was to his conception of cybernetics.[15] A designer-created goal was essentially at the core not only of Wiener's anti-aircraft gun but also of often-cited precursors such as Watt's governor. Wiener's main interests was focused on the feedback-based mechanisms that would allow his device to accomplish a specifically set goal: the automatic tracking of an airplane. Later, Wiener extended his approach to natural organisms. He came to see those organisms as essentially feedback-based, goal-seeking "machines." And yet, the original distinction between designer-generated goals and the device-generated behavior that seeks those goals remains at the root of the classic cybernetic approach. We have difficulty understanding how a living organism could be functionally similar to a cybernetic device. A living organism generates its own goals, whereas Wienerian cybernetics offers few resources toward a satisfactory account of goal generation. Ashby's approach, however, takes a quite different path. Paradoxically enough, goal-seeking is not a problem at all, as Ashby explicitly remarked in a late paper:

> Looked at this way [as the final product of extremely protracted physical interactions] natural evolution and the emergence of forms such as the guinea pig, with its well-developed goals, are in no way unusual: they merely exemplify the fact that almost all state-determined systems tend to preferred regions. It is the exceptional system that does not show such preferred regions.
>
> "Making a system that seeks a goal" is thus trivially easy: one forms a state-determined dynamic system at random. One is then almost certain to have a system that, like a guinea pig, will show that it is actively goal-seeking for some preferred state.[16]

Whereas Wiener faced the problem of how to design a mechanism that, once the designer assigned it a goal, would achieve goal-seeking behavior, Ashby faced an opposite problem: with a goal-seeking machine like the Homeostat, the question is *how to set it up so that it will seek the goal that the designer wants*. Where Wiener takes goal-seeking as a serious engineering tasks, Ashby reduces it to a fairly trivial given. Since Ashby conceives all living beings as homeostasis-driven devices that naturally generate their own goals as a by-product of their overall behavior, his approach, which may turn out to complicate the design of actual machines, will instead allow a better account of goal-generation in organisms. Moreover, since such goals match the organism's ultimate needs (its essential variables being the indicators of its biological viability), it follows that they may well serve also as philosophical *ends*, as behavior-informing *values*. Hence the difference between Wiener's and Ashby's conception of goals—a difference that Ashby's generalization of homeostasis made possible—widens to a true philosophical chasm. The best way to appreciate this is to turn to Hans Jonas's criticism of cybernetics, a position that as of lately, in the context of recent reappraisal of Ashby's projects, has generated much interest..

Before doing so, let us briefly recall the features that distinguish a prototypical and archetypal Wienerian device such as the automatic anti-aircraft gun or, even better, one of its analogues such as the automatic target-seeking torpedo. In "Behavior, Purpose, and Teleology," an article that Arturo Rosenblueth, a student of Cannon, coauthored with Norbert Wiener and Julian Bigelow, and which we may consider the founding document of classic cybernetics, the authors define the torpedo a machine "with intrinsic purposeful behavior." They compare it to the clock, which is orderly but not purposeful, and to a man-operated weapon, which has a human provided purpose that is not intrinsic to the device (Rosenblueth, Wiener, and Bigelow, 1943, p. 19). In which sense, Jonas asks, can we meaningfully say that the observable, goal-seeking behavior of such devices is purpose-driven? What do we mean by *purpose*?

> Once we have abandoned the original meaning of "purpose" as the *propositum*, that which someone sets *before* himself as the whereto of his action, we are reduced to the necessity of granting purpose to *all* actions—thereby depriving the definition of all defining force. (1966, p. 115)

Consider, he says, a human-driven torpedo. We would not want to say that the purposiveness lies in the whole torpedo+sailor, because the sailor can step out of the contrivance and take the purpose with him, "complete and unabridged .... We may say, of course," Jonas continues, "that the whole contrivance, judged from without, behaves as if it was purposeful; only, we must immediately add, 'but we know better'" (*ibid.*, p. 119). Jonas goes on to claim:

> The feedback combination of receptor-effector system lends itself to purposive action precisely if and when it is *not* a mere feedback mechanism—that is, if the two elements are not coupled directly, but if interposed between them there is will or interest or concern. (1966, p. 120)

Jonas concludes that "everything turns on purpose itself" and asks what this purpose is about. With a typical Heideggerian move (an analysis he will later carry out in *The Imperative of Responsibility*), Jonas links the immediate purpose of the torpedo+sailor—keeping the torpedo on target—to larger and larger frameworks. Why does the sailor want to keep it on target? To sink the enemy ship. Why does he want to sink the enemy ship? For honor, for glory, to live by his duty, or even to win a bet. This line of questioning leads him directly to the sailor's ultimate concerns, literally to "the ends by which he lives." Jonas concludes that we may identify the pilot-less torpedo with the torpedo that a sailor *who is following orders* guides. The sailor's purpose truly belongs to someone else: namely, to the sailor's commander. The cybernetic device does have a purpose, but this purpose is completely extrinsic to it, being always the designer's. In the end, Jonas's analysis reveals that the distinction between the non-automatic manned weapon and the automatic goal-seeking torpedo, which Rosenblueth, Wiener, and Bigelow had found meaningful, is instead philosophically irrelevant. It turns out we have based the whole cybernetic doctrine of teleological behavior on a semantic mistake—on the confusion between "having a purpose" and "serving a purpose," and between "*carrying out* a purpose" and "*having* purpose." We must find the rationale for this confusion in the being of the cybernetician and more specifically in her attitude toward her own life and the "life" of her creatures. The disarray stems from the lack of self-reflexivity that is typical of the scientist: "The cybernetician considers behavior, except his own; purposiveness, except his own; thinking, except his own. He views from without, withholding from his objects the privileges of his own reflective position" (1966, p. 124). This attitude, admissible and perhaps even necessary in some special sciences, is fatal in philosophy. It forgoes a crucial experiential truth that is common to all human and non-human existence alike: "*living things are creatures of needs*." Jonas concludes:

> Only *living things* have *needs* and act *on needs*. Need is based both on the necessity for the continuous self-renewal of the organism by the metabolic process, and on the organism's elemental urge thus precariously to continue itself. This basic self-concern of all life, in which necessity and will are bound together, manifests itself on the level of animality as *appetite*, *fear*, and all the rest of *emotions*. The pang of hunger, the passion of the chase, the fury of combat, the anguish of flight, the lure of love—these, and not the data transmitted by the receptor, *imbue objects with the character of goals, positive or negative, and make behavior purposive*.
>
> The cybernetical model reduces animal nature to the two terms of sentience and motility, while in fact it is constituted by the triad of perception, motility and emotion. Emotion, more basic than the two it binds together, is the animal translation of the fundamental drive which, even on the undifferentiated animal level, operates in the ceaseless carrying-on of metabolism. There is no analogy in the machine to the instinct of self-preservation—only to the latter's antithesis, the final entropy of death. (*ibid.*, p. 126, emphasis added)

Since the philosophical notion of purpose, which is essential to all living beings, is extrinsic to the feedback-regulated device, the feedback between the sensory and the motility apparatuses can never recover or translate it: the machine is as blissful when reaching the target as when overshooting it. The notion of feedback is inadequate because it ignores the ever constant needs of the organism—that is to say, I will point out, exactly what Ashby had intimated.

Jonas's insistence on the fact that emotions—the "animal translation of the fundamental drive"—are missing from the classic cybernetic approach helps us appreciate a third fundamental difference between Ashby's approach and the prevailing (contemporary and future) research trends in cybernetics and Artificial Intelligence. In order to keep the organism's essential variables within acceptable limits, the Homeostat will generate goals as needed. These essential variables correspond to the fundamental biological needs of the organism (sugar content of the blood, internal temperature, an so on). It follows that the behavior of any Homeostat-like machine will exhibit a need-based, radically non-cognitivist behavior. In the context of Ashby's model, we consider all brain activity and functions, from the most modest and compulsory to the highest and cognitive-like such as thinking and willing, forms of regulation between the organism and the environment. This approach is sharply at odds with the cognitivist conception that views the mind as an information processing device able to variously transform internal representations and then translate the results of those transformations into actions to carry out in the external environment.

Jonas's critique does not stop at the classic cybernetic understanding of purposes and goals. In *The Imperative of Responsibility*, he directly tackles the Ashbian homeostatic thematization of purposes we are familiar with. He tries to show that it falls short of doing justice to the conception of purpose of a living being. Since that second Jonasian critique is part and parcel of the recent reevaluations of the Homeostat, I will postpone its discussion after a survey of such recent efforts. But before I get there, I need to conclude my historical narrative of the Homeostat.

D. Death of the Homeostat

Ashby's generalization of homeostasis as the key biological goal, both at the physiological and at the psychological and social levels, along with his explicit insistence on need-based self-generated goals, drove his creature, the Homeostat, to theoretical and practical isolation. Ashby's contraption was widely admired, but few took note that it spawned almost no progeny. The rare attempts to replicate and extend its capabilities were systematically aborted. Ashby did conduct digital simulations of some particular aspects of the Homeostat, such as the influence of connectance on stability. These setups, however, fell short of the range of experiences that a real and interactive Homeostat made possible, let alone the more powerful, multi-unit, and multi-stable (multi-Homeostat) devices that the final chapters of *Design for a Brain* envisioned. In the 50 years between the publication of *Design for a Brain* and the end of the millennium, the literature reports only four other attempts.

Zemanek (1956), the earliest on record, appears to have just built replicas of a variety of cybernetic devices of the 1950s, including Ashby's Homeostat and Grey Walter's turtles. He did not, at any rate, further the experimental or the theoretical work that lay behind Ashby's creature. Then, as Ashby reports in a footnote to the second edition of his *Design for a Brain*, Earl Kletsky at the Technische Hogeschool in Delft, Holland, replicated on different hardware a device that is functionally identical to the original Homeostat. Two more ambitious projects followed, but failed. George G. Haroules and P. F. Haire (1960) developed a larger, sixteen units Homeostat for the US Air Force, called "Jenny." At the practical level, one of Jenny's expressed goals was, via electrical inputs and output of different kinds, to connect the Homeostat to the real world so as to allowing its deployment in the field as a data processor. Jenny had also serious theoretical goals. Robert Williams (1961) presented empirical results documenting some experiments he carried out to test the general mathematical analysis of stability that scientists in his laboratory had proposed.[17] After 1961, though, any mention of Jenny disappears from scholarly literature. The project appears to have been terminated before it achieved its stated goals. The most promising successor that the original Homeostat spawned was the device that Michael Wilkins (1968) designed at the University of Illinois's Biological Computer Laboratory. Ashby himself, who worked at BCL between 1960 and 1970, presumably collaborated with him. The "New Homeostat" was to be a fully interactive electromechanical device like the original one, but it used a slightly different mechanism to operate the stepwise random functions that implemented ultrastability. Moreover, by allowing the four units to receive an additional input, it connected them to the outside world. The timing of this "New Homeostat", however, was unfortunate as it coincided exactly with the brutal funding war that cybernetics and Artificial Intelligence engaged—a war that cybernetics eventually lost. BCL closed its doors not long after the publication of the first report, and this project too was most likely never completed.

Finally, the only historical attempt toward a digital simulation of the Homeostat I am aware of is a FORTRAN program written in the mid-1960s. Barney Lee Capehart reports of a digital simulation of a Homeostat on an IBM digital computer. Although he extended Ashby's original design to 10 units instead of 4, as an experimental device his software simulation was decidedly inferior to the original one, because it did not permit any interactivity between the machine and the observer. Capehart's FORTRAN program modeled the Homeostat as a set of differential equations and then used numeric routines to solve them as a batch program. As a consequence, Capehart had to focus his attention on a few general features of the Homeostat—such as the relationship between number of units and convergence, etcetera—instead of conducting, as Ashby did in *Design for a Brain* and elsewhere, specific experiments on specific learning experiences, such as reorientation behaviors, etcetera.[18]

In short, two of the four physical implementations of the Homeostat were simple replicas that did not produce any experimental or theoretical results; two never achieved their stated goals; the only digital simulation that researchers attempted could not provide the needed experimental setup. The unavoidable conclusion is that the Homeostat was a *unicum* to its one and only scientist-designer is —a feature that hardly endears it to contemporary scientific communities. But the Homeostat was also, I should note, a unicum for its widespread popularity in those same scientific environments. The situation I am describing could apply, readers may say, mostly to the American cybernetic landscape. Andrew Pickering has documented that Ashby's influence, in spite of his becoming a full-time researcher in cybernetics only after he left Britain to accept Heinz von Foerster's offer at the University of Illinois, was stronger on his home soil. Similarly, Slava Gerovitch reports on the extent to which researchers in the Soviet Union too felt Ashby's influence. It remains true, nonetheless, that such influence (with the partial exception of Stafford Beer's attempted transplant of Ashby's results from the domain of cognitive activities to social organizations) descended from the rigorous analysis of stability, of hierarchical feedback loops, and of state-based abstract machines that *Design for a Brain* conducted, and far less from the actual performances of the Homeostat, the only artifact implementing those otherwise abstract analysis.[19]

The Homeostat's peculiar ontological status led it to an almost unnoticed, quiet death that gave way to a long period of oblivion. Its disappearance is consistent with the general demise of cybernetics. By the late 1970s, Ashby's approach along with his Homeostat had largely disappeared from the scientific and philosophical debates about cognitive science. The ontological singularity of the Homeostat made its fate, and the fate of the theoretical approach that had produced it, even worse. First, the machine itself disappeared. Ashby left it behind when he returned to England after his retirement from the Biological Computer Laboratory. BCL ceased to operate soon after, and the Homeostat ended up in storage in the basement of the Electrical Engineering building in Urbana-Champaign, until it finally disappeared in a flood a few years later. I argued that the performance of Ashby's Homeostat supplied the main evidence in support of a broad thesis about life that was at odds with (neuro-)physiology, cognitive science, and computer science. Does the Homeostat's final vanishing act guarantee also the death of the fascinating and yet disturbing theoretical framework it supposedly exemplified?

2. The *Resurrection*

As intriguing as the vicissitudes of the life and death of the Homeostat undoubtedly are, the most surprising twist in the history of the device remains by far its recent rebirth. The revival of Ashby's original model is most often associated with the work that Ezequiel Di Paolo and his associates at Sussex University carried out in evolutionary robotics.[20] The rediscovery of the Homeostat, however, is not confined to evolutionary robotics. The convergence of a large array of research efforts spread over multiple and varied disciplines such as physiology, neuroscience, and theoretical and applied computer science prepared and facilitated the Homeostat's return. Moreover, other ambitious theoretical efforts, whose overall goal is a general reorientation of research in the discipline, present the biological model of the Homeostat as the crowning piece of their projects: the cognitive sciences (through the adoption of the *enaction paradigm*, a strong version of situated cognition); biology (through the reintroduction of a naturalized version of teleological causes); and philosophy proper (through the promotion of a biologically-informed notion of phenomenological experience as constitutive of subjectivity). They say that Ashby's ideas are revolutionary (and partly responsible for the "weird" status of his contraption) because they explicitly fashioned an autonomous, need-based creature, capable of independently setting goals that would ultimately turn out to fulfill its own *bodily* needs. Hence Ashby's Homeostat models a creature that is, they say, in the veritable, Kantian sense *teleologically oriented* and *autonomous*. To express it differently, Ashby's model breaks free from the stranglehold of the body-mind dualism.

Hence the Homeostat, this weird contraption long since consigned to the dustbin of history and theoretically at odds with most of the dominant research paradigms in cognitive science and Artificial Intelligence, is now cast to play an important role in a series of impressively ambitious programs. We may ask if the revived Homeostat can bear the weight of this new theoretical role. In particular, is the original Homeostat—which I take here metonymically as a token for Ashby's general approach—compatible with the new philosophical, scientific, and technical frameworks we have grafted it upon? And conversely, is the new Homeostat compatible with Ashby's original vision? In the remainder of this chapter I will argue that the answers to both questions are negative. Although the reborn Homeostat does indeed recover some of the essential characteristics of the original one, Ashby's original device remains at odds with the Kantian-Jonasian philosophical framework in which scholars have currently relocated it. As a consequence, the new philosophical paradigm appears to eliminate from the Homeostat's new incarnations some essential aspects of the old one that would otherwise make it ill-matched to their purposes. I will then address the next logical question: which philosophical framework, in lieu of the apparent misfit of the enaction paradigm, is most compatible with the original Homeostat? The answers to this question rely on a comparison between the original and current historical conditions. So far I have investigated the Homeostat's beginnings; next I will survey the present mutated environment that made the return of the Homeostat possible. These recent mutations have allegedly removed from physiology (both general and neurophysiology), from cognitive science, and from computer science (especially in the theory of artificial neural networks) the conditions that had so far prevented a full understanding of Ashby's creature.

A. The Homeostatic Role of Pleasure

First, and somewhat unexpectedly, the concept of homeostasis left its secure position within physiology to make a foray into several disciplines connected to cognitive science. Michel Cabanac, for instance, broadens the scope of physiological homeostasis to provide the foundation of a redefinition of pleasure that is quite radical: the organism recognizes a sensation as pleasant only if the stimulus it generates is useful to restore the organism's equilibrium. Pleasure is consequently the index of homeostatic success. An organism may interpret the same stimulus as pleasant or unpleasant depending on its internal states and their relative distance to their optimal values. Cabanac sees the organism's need to achieve physiological homeostasis as the central biological mechanism. He identifies pleasure (or its arithmetic reciprocal) as the measure of the actual distance from full equilibrium. With a move strongly reminiscent of Ashby and Freud, Cabanac then extends his homeostasis-based theory of pleasure upwards to the origin of consciousness.

He claims that the affective dimension of a sensation—its potential homeostatic usefulness—gives rise to consciousness:

> Activity is what makes consciousness useful and different from robots. Before sensory pleasure, behavior may be understood as a combination of reflexes. Pleasure introduces cognition. The evolutionary advantage of sensory pleasure, for the animals that first had it, is that it saved the nervous system from storing astronomical numbers of potentially useful or noxious stimulus-response reflexes. Sensory pleasure added flexibility to the behavioral pattern of those which possessed it and passed it to their offspring.[21]

Cabanac's reevaluation of pleasure within a homeostatic framework fits nicely within Jonas's general thesis about the tripartite structure of life. The void that traditional distinctions between perception and action left, pleasure now fills. Cabanac introduces an emotional dimension, even though under an hedonistic guise. He defines emotions as the particular class of mental experiences that shows "high intensity and high hedonicity."[22] In spite of failing to make the emotional axis completely independent of the other two as Jonas had intimated, Cabanac's definition still works in the same direction as his. Michel Cabanac's work belongs squarely within the domain of physiology, but his hypothesis about the origin of consciousness brings the concept of homeostasis back into the purview of cognitive science.

B. Neuronal Homeostasis

The concept of homeostasis has also made an important comeback in the neurosciences. We now recognize it as one of the most important mechanisms that allow a neuron to keep its response to the environment within an optimal range. Starting with the work of Gina Turrigiano, several researchers have provided ample evidence for the phenomenon and have distinguished a variety of different mechanisms (Turrigiano, 1999; Turrigiano, 2007; Maffei and Turrigiano, 2008; Desai, 2003). For instance, researchers have observed that, in order to keep their firing rates within an optimal range and prevent complete saturation or complete inactivity, neurons can effectively scale the strengths of the synaptic inputs up or down. By multiplying their weights by a single factor, such a synaptic scaling acts on all the incoming (excitatory) connections at once. Another phenomenon that scientists have extensively studied is the change in the intrinsic excitability of a neuron through a modulation of its electrical properties that are, yet again, a response to the firing range of its target. Although the biological mechanisms that realize intrinsic excitability is very different from synaptic scaling—the first affects the cell as a whole, the second only the synapses—from the functional point of view the result is similar. The homeostatic mechanisms change the global

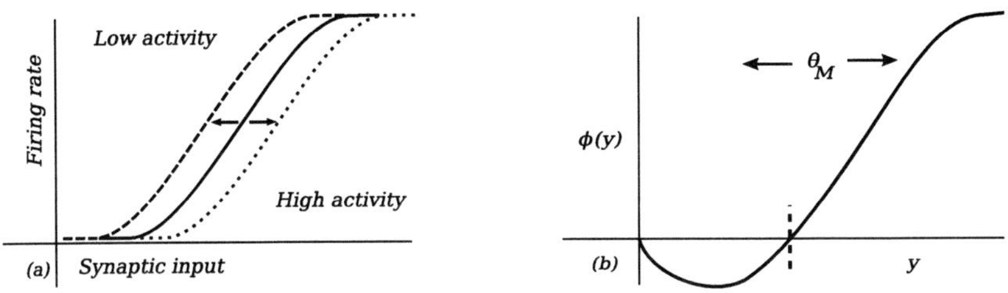

FIGURE 1.4: Homeostatic plasticity at the neural level: (a) *synaptic scaling* and *intrinsic plasticity*, (b) *BCM-plasticity*.
(a) The solid line represents the standard, instantaneous correlation between a neuron's synaptic input and its firing rate. When the average firing rate over a longer period of time tends to reside on the lower left end of the spectrum (neuron deactivation) or on the upper-right corner (neuron saturation), homeostatic plasticity may shift the function toward the left (in case of deactivation, dashed line), or toward the right (in case of saturation, dotted line). Synaptic scaling would actually multiply the inputs by a single constant (thus translating the x-axis itself), while intrinsic excitability would translate the neuron's function.
(b) In the BCM model, the synaptic strength is the product of the pre-synaptic activity and the non-linear function $\phi(y)$, represented here, and of the activity y of the post-synaptic neuron. θ_M is the threshold value that divides depression from potentiation. It depends on the previous activity of the cell, and it slowly slides leftward or rightward depending on the post-synaptic neuron long-term average firing rate (adapted from Desai, 2003).

function between the incoming inputs and the firing rate of a neuron, effectively translating the function along the x-axis (see figure 1.4).

Although this neuroscientific work is crucial to the new homeostats, the mechanisms of neuronal homeostasis differ from how Ashby implemented ultrastability in his original device. Neuronal homeostatic plasticity is always *global*, because it operates on all the synapses of a given neuron or on the electrical properties of the neuron itself, thus maintaining the relative differences among synaptic weights. Homeostatic regulation may change the absolute weight of a particular synaptic connection, but it will never change the relative strengths of synapses. Homeostatic regulation operates, mathematically, as a multiplicative constant that applies to a partially ordered set. Turrigiano and Nelson confirm that neural homeostasis *must* operate globally, otherwise it would conflict with the Hebb-like mechanisms of long term potentiation and depression of synaptic weight that we consider crucial to the functioning of neuronal networks. Differently put, homeostatic plasticity is complementary to Hebbian plasticity. As I discussed above, Ashby's homeostatic mechanisms are also global, but since they would (randomly) change the *relative* weight of all the connections at once, they are antagonist to and potentially incompatible with Hebb-like plasticity. The contemporary equivalent to Ashby's interpretation would be to let homeostatic regulation occur locally at the synaptic level.[23]

In turn, the BCM theory of synaptic plasticity (by Elie Bienenstock, Leon Cooper, and Paul Munro, 1982) acts at the local level of neuronal connections (Figure 1.4b). The researchers suggest that synaptic activity may weaken or strengthen depending on whether post-synaptic activity is below or above a designated threshold, which is itself a slowly evolving function of post-synaptic activity. The threshold then moves, making potentiation more likely when the average activity of the neuron is low, and, conversely, less likely when average activity is high. Although scientists had originally proposed the theory as a model of synaptic competition, since it will tend to regulate—locally and slowly—the value of the individual synapses on the basis of average neural activity, we can adapt it to model homeostatic stability. Even in this case, though, the model is different from Ashby's. The change, instead of being the immediate result of a step-wise function that a neuron's activity falling outside of the preferred range triggers, is slow and gradual.

The recent emphasis on homeostatic plasticity raises questions that are still unanswered. For instance, we do not know why so many different mechanisms are present when, theoretically speaking, one should suffice. Nor is it clear if any kind of interaction among them exists, so that some plasticity functions may invalidate or potentiate others. We do not know whether the multiplicity of mechanisms simply indicates that neural network stability is of such importance to the organism to require redundant strategies. Also, and perhaps most importantly from the point of view of a high-level theory such as Ashby's, all these mechanisms operate at the level of a single neuron, whereas we know that neural networks may exhibit stereotyped patterns of activity like synchronized bursting. Experimental evidence suggests that homeostatic regulation may occur at this higher level, for example preserving the bursting activity of the network as a whole against a decrease in excitatory activity (Turrigiano, 2007).

The abundance of physiological results on neural plasticity that scientists have brought forth over the last decade, together with our relative ignorance about higher level integration or cooperation among them, help explain the influence that neuroscientific results have exerted on the resurrected homeostat. Some researchers include a number of homeostatic rules in their simulations, resorting for example to evolutionary strategies to select which rule a certain neuron will apply at run time. For instance, one of the three plastic weight-change rules that Di Paolo (2000b) uses closely derives from the BCM theory I discussed earlier. Another one consists of a symmetric reversal of BCM: it depends on a pre-synaptic activity threshold, a phenomenon for which no biological evidence exists. Other researchers select rules that seem to be best grounded in physiology, firmly rejecting any attempt to deviate from homeostatic principles for which we have a well-understood biophysical explanation (Williams and Noble, 2007). These issues notwithstanding, Turrigiano and her associates' neurophysiological evidence still provides an important stimulus for the reintroduction of the long-forgotten homeostat-like feature in cognitive science studies.

C. The Evolution of Artificial Neural Networks

At a more formal level, interest in the Homeostat has recently emerged in computer science, where researchers see Ashby's machine as an alternative hybrid model of computing device that allows information and interferences from the outside to modify its purely mechanic (Turing-like) procedures. Giorgio Sandri, for instance, suggests that Kurt Gödel's critique may affect the concept of Turing-computability. Instead of a fundamental limitation of the human mind, Gödel's interpreted Turing's original results as indicating an intrinsic limit of the formalism that his machines adopted. In this context, Sandri recovers Ashby's homeostatic model as a possible alternative to the formalism of Turing machines, connecting it with Turing's later reflections along those same lines (Turing, 1948). Sandri suggests that we may capture the Gödelian insight in formal machines that are capable to modify their internal configuration through procedures that let information enter from the outside and are able to deal with external interferences.[24]

We find the best indicator of the wide gap that scholars had to bridge before Ashby's highly-esteemed theoretical work could regain a viable measure of historical efficacy in the history of neural networks. In the very title of his major work Ashby declared his intention: he was after the fundamental principles that we must use in the design of natural and artificial *brains*. If we equate the units of the Homeostat to neurons and their connections to synapses, we could interpret Ashby's device as one of the first attempts to realize a neural network. This is not the only interpretation we may draw: we could equally recast the units that are at higher levels as neuron complexes or even as sensory-motor functional units. Ashby, in any event, was careful to avoid any particular physiological interpretation of the details of his machine, preferring instead to cast his model at the highest level of generality. But a straightforward neuronal interpretation of the Homeostat remains possible and it would generate a blueprint for a neural network architecture and a "learning" algorithm (random choices of step-wise connection values within the ultrastability homeostatic paradigm). Yet, the neural network architecture that emerged in the 1950s and later became dominant could not have been more distant from the Ashbian Homeostats.[25]

The main evolutionary line stems from Donald Hebb's work, it intersects Warren McCulloch and Walter Pitts's first theoretical article, goes through Frank Rosenblatt's *Perceptron,* and reaches its climax—along with a completely unforeseen explosion of academic and commercial interest—with the (re)discovery of the back-propagation algorithm and the publication by David Rumelhart and James McClelland of their systematic explorations. Their architecture is different from the Homeostat's on several counts. Topologically, since no recurrent path leads from the output of a unit back to a previous stage, classic feed-forward neuronal networks are directed acyclic graphs. Since any unit of the Homeostat can have a feed-back loop to other units, its topology, which in current terminology is an example of a *recurrent* neural network, is cyclic. Within Ashby's conception of nervous activity, the idea that a directed acyclic graph(s) may constitute a network makes no sense. At the heart of his model of nervous activity lies the continuous two-way, double-feedback, and mediated exchanges that take place between organism and "environment" and between information and operational cycles. Even Hopfield's networks—the most widely studied recurrent network architecture that emerged in the early 1980s, after the scattered attempts by Oliver Selfridge (*Pandemonium*), Stephen Grossberg, and others—displayed a substantial difference: they did admit feedback loops, but were *homogeneous,* fully connected networks that also placed some crucial limitations on the kind of connections they allowed. On the contrary, the Homeostat that Ashby used in his experiments—and even more so the purely speculative Homeostats that *Design for a Brain* and later publications presented as the subject of possible experiments—were partially connected, highly *heterogeneous* networks.[26] Ashby insisted that real nervous systems must be highly "dispersed": in order to achieve stability, they must have, as I discussed above, a very low level of intrinsic connectance. If we were to translate it into the language of neural networks, Ashby's principle of dispersion would require the connection of a single node (neuron) to no more than about 10% of the other nodes (neurons). This requirement obviously implies that the sets of connections among the nodes will all be different, and that the network will be highly heterogeneous.

Because classic feed-forward neural networks lack the notion of an environment in which they would "live" (homeostatically or otherwise), these networks do not self-generate goals—instead, we need to train them from the outside via learning methods that we supervise. In contemporary terms: feed-forward networks of the classic type do not possess an intrinsic *dynamics*. They are static systems. Each neuron is timeless, in the sense that its past behavior does not influence its current state. Only recently have scientists started paying serious attention to the temporal aspect of the behavior of neural networks, and to the related possibility, which brings about the whole field of *neurodynamics*, that we may study networks in terms of classic dynamical systems (Jordan, 1997; Elman, 1990). The Homeostat is a classic example of a neurodynamical system. In the final chapters of *Design of a Brain*, Ashby couched the (incomplete) formal discussion of his model in the preferred language of dynamics—differential equations—while he presented the qualitative analyses he scattered throughout the book in terms of phase spaces. The formal structures that are closer to Ashby's original formulation are *Continuous Time Recurrent Neural Networks* (CTRNN), a model similar to Hopfield's, but without any restriction on topology or on synaptic weights. CTRNNs have become one, if not *the*, tool of choice for the investigation of the adaptive behavior from a dynamical perspective. Although CTRNNs are not fundamentally dissimilar from Grossberg's networks and are, arguably, a generalization of Hofpfield's networks, they have come to prominence only recently. They are strongly associated with the work of Randall Beer. Beer was perhaps the first to champion their use within a broad framework that aimed at the understanding of adaptive behavior from a neurocomputational perspective. In a recent survey article that does not explore the possibility of recovering, besides Ashby's dynamicist approach, also his detailed theory of generalized homeostasis, Beer none the less explicitly acknowledges that Ashby was the first advocate of a dynamical understanding of behavior (Beer, 1990; Beer, In Press).

Strictly related to the topological incompatibilities between classic neural networks and Ashby's architecture is the difference in learning algorithms typical of self-organizing networks that, like the Homeostat, are tuned into the environment they interact with and not to the behavior the experimenter wishes for. Donald Hebb provided the basic rule of self-organizing networks in *The Organization of Behavior* (1949): *neurons that fire together will wire together*. This rule, although widely adopted in self-organizing artificial networks and still providing the fundamental principle for our understanding of the interaction between real neurons, is entirely at odds with Ashby's understanding of neural interaction. First, Hebb's rule acts locally: the rule will increase the strength of single synapses on the basis of the activity levels of the pre- and post-synaptic neurons. Instead, Ashby's "rule" (random step-wise change) is global: when it reaches a critical state, a unit (a neuron) will modify the weights of *all* of its connections. Because the Homeostat's self-organizing strategy may undo any, and possibly all, synaptic weight changes that the associative rule produced, Hebbian learning is incompatible with it. Second, Hebbian associative learning is slow and cumulative, whereas the Homeostat proceeds by leaps and always restarts from zero. Finally, and perhaps most importantly from Ashby's perspective, no randomness occurs in Hebbian learning, whereas, as I pointed out, Ashby believed that any adaptive behavior that did not include a random element would not be flexible enough to respond to unpredictable environmental challenges and would be bound to fail. The authors of the recent rediscovery of neural homeostasis are careful to reformulate it in such a way to avoid outright conflicts with Hebbian learning, while the advocates of the new homeostats strive to rephrase Ashby's ultrastability rules around an outright conflict with Hebb's.[27]

The long evolution of neural network theory I just outlined—from McCulloch and Pitts's original models to the recent CTRNN—is an essential and necessary condition for a reappraisal of Ashby's original paradigm. Ashby explicitly relied upon and forcefully backed a dynamicist framework that, to reach its full theoretical potential, needed technical tools such as recurrent neural networks. Recent findings in neurophysiology and in the physiological role of pleasure, which

I mentioned above, motivate and empirically corroborate these computational advances.

I think that we should consider these developments as more than mere technical advances, though. They are at once the products and the main contributors to a general reorientation of research within cognitive science, whose focus has shifted from an emphasis on cognition understood as the symbolic manipulations that a detached agent enacts, to cognition as the performance that arises out of the interactions of an embodied participant with its environment.[28] Precisely because of their larger epistemological import, the developments I have outlined may be the necessary conditions for a Homeostat's renaissance, but they are not sufficient for a genuine return of it. Ashby's theory of generalized homeostasis (and its associated mechanisms of ultra- and multi-stability) is not co-extensive with the Situated Embedded Dynamic paradigm (or SED, to use Beer's terminology). Though consistent with SED, Ashby's approach builds a much more narrowly focused theory on top of it. His theory offers *both* a general view of how we should interpret the "situatedness" and the "embodiment" of the cognitive agent *and* a detailed view of the mechanisms that regulate agent-environment interactions. A profound agreement and at the same time a potential for friction exist between Ashby's approach and the situated paradigm. This state of affairs translates into several significant technical differences, for instance in the all-important subject of learning. Tensions arise because Ashby's approach, insofar as it translates into a specific interpretation of SED, inevitably clashes with other interpretations. More importantly, though, tensions arise because, if I may put it this way, Ashby's approach is *too* technical: if we want to bring out its implicit framework, we need to apply an act of conceptual interpretation. Such interpretation, in my view, should distinguish Ashby's overall approach and prevent us from identifying it as a collection of purely technical and largely outdated techniques or, alternatively, as a general and altogether vague appeal in favor of a dynamicist approach to the problem of (cognitive) behavior modeling. I will now discuss how the new reincarnations cope with this issue.

D. Twenty-First Century Homeostats

Beginning in 2000, several researchers in Artificial Intelligence and in the cognitive sciences started rediscovering some of Ashby's ideas about adaptivity. Initially, they grafted them upon the emerging CTRNN paradigm in neural network research and upon the rediscovery of homeostasis in physiology and neuro-physiology that I discussed above. Most of this work took place within the boundaries of evolutionary robotics, the discipline that attempts to gain insights on human and non-human cognitive behavior by mimicking the biological evolution of organisms through devices that are either abstract simulations or real-life robot (Nolfi and Floreano, 2000; Harvey et al., 1997; Harvey et al., 2005). Although many different ideas have emerged in the last few years, we should credit the work of Ezequiel Di Paolo for being the most sustained and far-ranging attempt to revitalize Ashby's dynamical approach. In his first paper on the subject, Di Paolo successfully replicated on simulated robots the experiments on the inversion of the visual field that James Taylor had originally carried out in the 1960s.[29] First, Di Paolo developed artificial organisms that could perform the simple phototactic task of navigating in their environment and reaching a visual source on the basis of the stimuli that two light sensors located on the opposite side of the robot's body provided. Then, after artificially switching the input coming from the two sensors, thus giving rise to the robotic equivalent of a left-right inverted visual field, he investigated the behavior of the robots. Following closely Ashby's theory of ultrastability as a second-order, time-delayed feedback loop, Di Paolo was able to show that artificial organisms that could rearrange their internal connections (the weights of the neural network's synapses) could adapt to the new situation and successfully perform phototaxis again. In Di Paolo's wake, the application of homeostatic principles to robotic devices has been replicated, extended, and modified in several different experiments.[30]

Di Paolo's approach is unique, however, because he greatly extended his at first quite technical approach. He linked his interpretation of the role of homeostasis not only to the biological theory of autopoiesis that Humberto Maturana and Francisco Varela had put forward, but also to the philosophy of biology of Hans Jonas. Di Paolo's (simulated) devices are more than a mere application of Ashby's homeostatic principles to contemporary technical concerns; they beg us to consider them contemporary but genuine reinventions of the original Homeostat. As in Ashby's case, Di Paolo's devices are concrete implementations of a broad, dynamicist, and homeostasis-based theory about human and non-human cognition that, wishing to provide an overall account of the fundamental features of life, transcends the limits of biology, psychology, and computer science. Yet, how close is Di Paolo's new homeostat to Ashby's?

The biological theory of autopoiesis by Maturana and Varela, born out of the authors' profound dissatisfaction with the (still largely prevalent) physico-chemical interpretation of life processes at the molecular-genetic level, is the first and most building block of Di Paolo's new homeostats. Maturana and Varela recast the metabolic interactions as the set of self-producing processes that the cell—the minimal living organism—uses in order to continuously build the elements that constitute the cell itself and those processes. Cellular life, life *tout court*, becomes a property of autopoietic systems that

1. continuously regenerate the network that produces them, and
2. constitute themselves as a distinguishable unity in the domain in which they exist.[31]

This characterization of autopoietic systems entails a strong emphasis on the situated and embedded character of any biological process, from the most elemental to the most advanced living function. Autopoietic systems build *themselves*, and therefore they are essentially autonomous individuals—in the Kantian, strict, though naturalized sense of autonomy as self-legislation. Necessarily dependent on the surroundings as they strive to maintain a necessary distance from it, living organisms must constantly *interpret* the world around them by casting a web of meaning onto it. Life perforce enjoys a "sense-creation purpose whence meaning comes to its surrounding, introducing a difference between environment (the physical impacts it receives), and world (how that environment is evaluated from the point of view established by maintaining an identity)" (Weber and Varela, 2002, p. 118). This new interpretation of fundamental metabolic processes at the cellular level leads the theorists of autopoiesis to plausibly claim that they, while avoiding the vitalist trap and keeping a solid empirical grounding, have been able to reintroduce the philosophical subject in biology. Since "self-production is always, and inevitably, a self-affirmation that shows the organism…involved in the fundamental purpose of maintaining its self-identity" (*ibid.*, p. 116), it follows that in biology we must abandon once and for all the urge to reintroduce values and purposes on the basis of a neat separation between the realm of pure natural science on the one side, and the realm of values on the other. "Life is thus strongly subjective in the strong sense of the word," they declare (*ibid.*, p. 118), and this subjectivity implies that life is always teleological. The purposes that human beings consider essential to their conscious life and the will that they deem instrumental to realize those purposes do not belong to a different non-biological level of being. We need not to explain those purposes away, teleonomically, as mere epiphenomena of deeper biological processes that, beyond the transmission of the genetic code, have supposedly no direction. On the contrary, the intrinsic purposiveness of human and vertebrate life is just one realization of a phenomenon that all biological creatures share.

Maturana and Varela's theory of autopoiesis is complementary to the phenomenology of life Hans Jonas proposed. Jonas's work has in turn exerted a profound influence on Varela, and remains a deep influence on Di Paolo and on other contemporary advocates of the Homeostat. Jonas argues that if we interpret biological processes from the purely physical standpoint, then physical laws reduce the organism to a mere fiction, hence life remains a mystery: no physical difference exists between a sea wave and a living organism. Both are complex physical phenomena that occur in specific time and space as a result of a multitude of physical and chemical processes that are all intertwined. The wave has no identity—it is a fiction that our epistemological position creates—whereas the living organism has a first-hand experience of its own identity. The toil and labor it incessantly goes through in order to preserve itself dramatically confirm this experience. But a strictly physicalist approach equates the ontological status of the organism to that of a sea wave and it denies the veracity and authenticity of the experience of self-identity that the living body brings forth, hence making the organism fictitious. Since the experiential source of the organism's autonomy is the biological body, the argument applies to all kinds of living beings. From bacteria to humans, we all share, in different forms and to different degrees, the same fundamental experience: the experience of self-identity as *needful freedom*, as the partial detachment of a *subject* from the environment that engenders a dialectics of need and self-affirmation. Because we can maintain the sameness of the subject arising from that relation only through material exchanges that would, left unattended, negate it, it follows that the organism experiences itself not as matter, but as a *form* that keeps and reproduces itself. The sameness of the organism—its identity—is lived as a *purpose*: the purpose of keeping the organism in existence for a definite amount of time inside a well-defined environment that simultaneously grants and threatens its survival. Each and every single life form exemplifies the ontological status of a teleological structure of behavior that we all experience daily: "The teleological structure and behavior of organisms is not just an alternative choice of description: it is, on the evidence of each one's own organic awareness, the external manifestation of the inwardness of substance. To add the implications: there is no organism without teleology; there is no teleology without inwardness; and: life can be known only by life" (Jonas, 1966, p. 90). Purposiveness becomes the fundamental feature of life. Jonas sees it as the fundamental structure of any subjectivity: "While I can distance myself from single ends, I cannot distance myself from the principle of having ends as such" (Hösle, 2001, p. 38).

Varela and Weber, in turn, claim that the theory of autopoiesis—which shows at the cellular level how we must conduct the scientific study of metabolic processes from the outside and from the inside—provides the empirical grounding for Jonas's theory:

> Thus autopoiesis is a singularity among self-organizing concepts in that it is on the one hand close to strictly empirical grounds, yet provides the decisive entry point into the origin of individuality and identity, connecting it, through multiple mediations with human lived body and experience, into the phenomenological realm. These are the mediations that Jonas addresses with so much force, and that make these two lineages of thought not only contemporaneous but fully *complementary*. Both seek a hermeneutics of the living, that is, to understand from the inside the purpose and sense of the living. (Weber and Varela, 2002, p. 116, emphasis in the orig.)

Di Paolo's revival of Ashby's conception of homeostasis comes into play in the transition from Jonas's phenomenology of the living to Maturana and Varela's theory of autopoiesis. While accepting wholeheartedly Jonas's approach, Di Paolo argues that autopoiesis is necessary but insufficient to ground it on a scientific basis. Because organisms constantly interpret the world from their internal point of view, Jonas endows them with a *sense-making* capacity. But autopoiesis, Di

Paolo argues, is not sufficient for sense-making. Maturana and Varela's theory may well explain how a cell conserves its identity by virtue of its structural coupling—via metabolic processes—to the trans-membrane environment that surrounds it. Structural coupling, however, is a necessary but not sufficient condition for self-identity: what the organism needs is "the capacity to regulate itself with respect to the boundaries of its own viability" (2005, p. 430).

Referencing the double feedback loop structure of Ashby's theory of ultrastability (see fig. 1.1 above), Di Paolo claims that autopoiesis targets the first loop but forgoes the crucial second-order interaction that causes ultrastability. Without active regulation, though, no *adaptivity* is possible: organism are either adapted to their environment—they have the right kind of structural coupling to it—or they are not—in which case they die. As a result, we can say that an "autopoietic organism" does not *act* in the strong sense of the word, but only that it *reacts*, via robust physico-chemical or sensory-motor loops, to the stimuli coming from the outside. Such a weak conception of action does not grant to the organism a sense-making capability or a real temporal structure, because only "adaptivity is a structured event, with defined stages of onset (the sensing of negative tendency), acceleration (the activation of the adaptive mechanism), consummation (the overturning of the negative tendency), and cadence (the de-activation of the adaptive response)" (*ibid.*, p. 444).

Takashi Ikegami and Keisuke Suzuki, authors of another re-evaluation of the Homeostat, perhaps best express the point (2008, p. 398). First, they cast Maturana and Varela's autopoiesis theory as a decisive step forward from the classic conception according to which extrinsic goals drive machines. Then, they contend that Ashby's ultrastability theory provides the essential and still missing link between, on the one hand, the self-reproducing capabilities of an organism upon which autopoiesis focuses and, on the other, the essential needs and motivation that drive the organism's adaptation to the environment and cause its complex range of behaviors. Along similar lines, Di Paolo has recently broadened his original approach to an even more comprehensive reformulation of the enactive paradigm. He claims that the whole paradigm in its five and mutually implicating concepts—autonomy, sense-making, emergence, embodiment, and experience—requires a process of active regulation of the conditions of the exchange occurring between the organism and the environment. It requires, differently put, an Ashbian model of regulation.[32] Di Paolo therefore sees the new homeostat as incorporating Ashby's ultrastability mechanisms (which Di Paolo usually calls "homeostatic regulation") inside the Jonas/Varela enactive framework.

From my point of view, Di Paolo's extension and retooling of Ashby's homeostat runs in direct contradiction with Jonas's theory. *The Imperative of Responsibility*, the work where Jonas stages his comprehensive phenomenological account of life as an ethical and political theory, contains also a trenchant argument against the homeostatic interpretation of life as purposiveness that goes well beyond Jonas's previous critique of cybernetics I discussed above. To exemplify the traditional account of animal purposiveness, consider—Jonas's argument goes— the sequence of actions a cat may be involved in when chasing a mouse. First, we say that the cat, which has no high-level planning abilities, has no final goal in its consciousness. A sequence of immediate goals confines the animal. Second, some fundamental bodily needs, or drives, provide the unifying force that keeps this sequence together, so that "feeling associate with need is the psychic proxy of purpose in the voluntary behavior of prerational life."[33] Third, semi-autonomous behavioral schemata (pouncing, killing, and eating), which are pre-programmed and come automatically into force when the appropriate cue triggers them, are the means to fulfill the eventual "goal" (*satiation*, or hunger's satisfaction). Jonas stresses how this explanation, which fits Ashby's adaptive schema quite well, is really a-teleological. It expels purposes from the explanatory account: the animal acts *from* hunger, not *for* satiation. The cybernetic explanation of prerational behavior thus becomes:

> physiological stress (homeostatic "gradient"); internal secretion and nerve stimulation; readiness for selective triggering of a behavioral pattern; external (sensory) nerve stimulus triggering the appropriate behavior pattern; sensorimotor feedback guiding the course of behavior; subject of action eliminating the initial stress (homeostatic equilibrium), this itself being the "goal." (1984, p. 62)

Jonas's point is that the aforementioned "homeostatic goal" *is no goal at all*: "the *bonum desideratum* would be the *subjective representation* of the state of indifference or quiet that waits at the end" (*ibid.*, p. 62, emphasis added). The crucial observation, Jonas argues, is that the first-person subjective experience of the goal-directed behavior is in direct contradiction with "the subjective representation of indifference." This alleged goal cannot work as such: we have negatively defined it as a lack, hence it cannot have the causal power that motivates a behavior. The subjective experience of purposes, which is undeniable, flatly contradicts the homeostatic explanation. We must trace the sole reason why such explanations have enjoyed some success back to incompatibility that allegedly exists between the a-teleological and necessary structure of the physico-chemical world on the one side, and the teleological structure of consciousness on the other. The cybernetic explanation became appealing, Jonas continues, because by reducing high-level purposiveness to low-level mechanistic explanation it promised to work around the psycho-physical problem. The price it exacted, however, was then and still is too high: cybernetic homeostasis reduces a real purpose to "an accompanying tune" that has no causal power. Most importantly, we *do not need* such a reduction: the phenomenological account that interprets all living beings as purposeful entities permanently caught in a dialectic of needful freedom let us grasp that primordial living beings already contained—just as the seed contains the tree— albeit *in nuce,* the ultimate level—namely, consciousness in its full-fledged form. If we admit a continuous gradation of purposes from high-level goals, to urges, all the way down to the most basic physiological functions, then it becomes meaningful and non-contradictory to speak of the "immanent, if entirely unconscious and involuntary purpose of digestion and its apparatus in the totality of the living body, and to speak of *life* as the *end-purpose* of just this body" (Jonas, 1984, p. 75, emphasis added).

The conclusion appears inescapable: anyone who accepts the Jonasian account of the intrinsic teleology of nature will have to regard a homeostatic explanation as unnecessary and even harmful. *Unnecessary,* because Jonas removes the psycho-physical problem that motivated homeostatic explanations; *harmful,* because by virtue of its reductionist strategy, homeostasis makes subjective teleological experience a redundant epiphenomenon (an "accompanying tune"). The reincarnated homeostat has thus become a profoundly contradictory contraption—a Jonasian homeostatic device.

Could perhaps Jonas's critique of homeostasis be incorrect? Could he have perhaps misread Ashby's interpretation of the living? Neither option is plausible. The core of Jonas's critique is the incompatibility between the subjective experience of teleology and the homeostatic "quiescence" that Ashby assumed as the supreme "goal" of the organism. As I pointed out above, Jonas is absolutely right in identifying the a-teleological vantage point (the "generalized homeostasis thesis") as the deep *philosophical* core of Ashby's conception of the living. All other theses—double-feedback loops (ultrastability), randomized step-wise functions, low connectance, and so forth—are *technical* details that Ashby needs mainly to articulate a plausible framework for his overall philosophical position. Jonas does not discuss the technical side of Ashby's work, and rightfully so: after all, Jonas is indeed constructing an argument between two opposing philosophical viewpoints. Once we grant to Jonas the right to part Ashby's overall philosophical position from the technical details accompanying it, we need also to acknowledge that Jonas's conclusions are correct: conscious living beings most of the time perceive themselves as being engaged in a multifarious series of goal-driven activities. If we, like Jonas, grant epistemological (and ontological) primacy to first-person experience, then the contradiction is inevitable. From Jonas's point of view, we cannot but reach the necessary conclusion that Ashby's "generalized homeostasis" is a philosophically uninteresting explanation of life and mind. It exemplifies just another radical, behavioristically-minded, and physicalist reduction of the mental (the phenomenal experience of conscious life) to the physical (or better to the physiological). It amounts to yet another version of a rat-in-the-maze psychology *à la* Clark Hull.

If no reconciliation between Jonas and Ashby is feasible at the philosophical level, then we could perhaps excise the more troublesome general thesis and focus instead on Ashby's technical contributions. As an analogue strategy we may think of, Habermas's well-known view of Gadamer as "urbanizing" Heidegger comes to mind (1985). According to Habermas, Gadamer, with the ambition to make Heidegger's work palatable again for general philosophical consumption, removed the extremes of his speculation about time and will that had brought his master so dangerously close to the abyss. I see a similar strategy implicitly at work also in the new Homeostat. A slight terminological variance masks it only partially. In his technical and non-technical work, whenever Di Paolo refers to "homeostasis" as the missing ingredient that lets us implement in technical artifacts Jonas's philosophical vision, what Di Paolo really means is "ultrastability." This crucial slippage is worth repeating: what the organism needs, Di Paolo argues, is "the capacity to regulate itself with respect to the boundaries of its own viability." This is an accurate and concise definition of Ashbian ultrastability. Di Paolo, however, first redefines it as adaptivity and then equates it with homeostasis. Then only one of the two forms of interaction between the original Homeostat and the environment possibly qualifies as "homeostatic": the higher level loop that checks the value of the essential variable. On the contrary, Ashby considered the homeostatic principle as governing *both* Homeostat's feedback loops (that is, the *Env-R* and the *Env-S-R* loops of figure 1.1 above), because both loops aim at keeping the organism in a state of equilibrium. The second-order mechanism, ultrastability, comes into play as soon as the first-order feedback fails to achieve it. In other words, Ashby's homeostat was built around a much stronger claim than Di Paolo's: ultrastability was merely the technical *means* that organisms and machines deploy to achieve their final end. As Jonas saw, that end is just homeostatic equilibrium.

Di Paolo's technical implementation of the simulated robots whose behaviors he studies reflects his overall approach to Ashbian generalized homeostasis. In his original paper (2000), the robot's controller is a small CTRNN network whose neurons' weights, biases, and time constants we have optimized through the application of a genetic algorithm. By modeling behavior along the dynamicist lines that Ashby championed, Di Paolo's setup is broadly consonant with Ashby's general approach. To this setup, Di Paolo added an additional element: the weight of each connection between neurons may change during the robot's operation according to a couple of parameters. The first parameter is genetically determined, and specifies how much and in which direction the synaptic weight may change (its range being $[-0.9, +0.9]$), whereas the second parameter is a plasticity function that depends directly on the actual firing rate of the neuron. Its value is minimal when the firing range is in the middle of the possible values, and maximal when the neuron is close to one of the extremes. If we equate a neuron's firing rate with the actual value of a Homeostat's unit, and if we assume that each neuron's value represents an Ashbian "essential variable," then the plasticity function of Di Paolo's system is equivalent to the dial *D* in figure 1.1 controlling the application of the parameter-switching mechanism *S*. It follows that in the course of the robot's behavior, the combination of the plasticity functions, of the synaptic plasticity levels, and of the weight change rules replicate Ashby's second-order feedback loop, whereas the genetically determined weights, the time delay constants, and the biases take care of his first-level loop. By using the second-level loop, and despite the swapping of their sensory inputs, some of the robots are still capable of carrying out a phototactic task.[34]

All the remaining aspects of the Homeostat are missing, however. At the technical level, Di Paolo makes no use of the two aspects that Ashby considered crucial to his model: the random stepwise functions and the low connectance among units. Likewise, he makes no reference to generalized homeostasis and he draws no link between the value of a neuron and some essential property of the organism. Again, Di Paolo's original model selectively recovers some of the elements of Ashby's theory, but arguably not the salient ones. It recovers the double-feedback thesis (number 3), but it rejects or leaves untouched the overall lack of goal that is characteristic of the Homeostat (theses 1 and 2). It also rejects randomness and step-mechanisms thesis (number 4), and the all important consequences of *dispersion*, or low-level connectivity (multi-stable systems, thesis 6) and deep embeddedness (communication through environment, thesis 7). In later experiments, by adding an internal (simulated) battery that powers the motors, Di Paolo (2003) explored the possibility of reintroducing some of the more genuine Ashbian elements. The battery can only charge up by approaching a light source, thus functioning as an Ashbian essential variable. Di Paolo also introduced a modified version of stepwise changes and, instead of tying the plasticity function to the firing rate of the neuron, he linked it to the level of the battery. Although the robot demonstrates a capacity to adapt to the different perturbations that occur during the phototactic tasks it performs, it also exhibits the same problems that plagued the original Homeostat—namely, no guaranteed time to adaption and no guaranteed preservation of successful adaptions—in short, no learning. As a consequence, Di Paolo did not pursue this line of inquiry any further.

Yet, even this, more urbane, Di Paolo's version of Ashby's creature may have got uncomfortably close to the precarious position that its less proper predecessor had occupied. On the one hand, Di Paolo's creature may be still too close to Ashby's lack of teleology (the *-stasis* part of homeostasis), so that some researchers have suggested to move towards a more dynamical approach. Ralf Der, J. Michael Herrmann, and Michael Holicki (2001; 2004), for instance, have proposed to replace homeo*stasis* with homeo*kinesis*. On the other hand, Di Paolo's creature may have strayed too far from Ashby's (alleged) strict adherence to the naturalistic canon, so that some other researchers have pushed for an approach where mechanisms of neural plasticity follow more closely the neurobiological findings of Turrigiano and others researchers I discussed above (Williams and Noble, 2007).

In conclusion, what do we make of the incompatibility that Jonas highlights between the incontrovertible phenomenal experience of purpose-driven self-identity (the teleology of the living) and Ashby's theory of ultimate quiescence (the generalized homeostasis thesis)? I have examined two possible solutions: the first one amounts to a simple rejection of one of the two horns of the dilemma. Neither alternative appears satisfactory, although for different reasons: rejecting homeostasis-generated ultrastability would leave us with the account that Di Paolo rightly stigmatized—in his critique of Maturana and Varela's autopoiesis—as insufficient to realize the teleology of the living of Jonas. Conversely, accepting Ashby's generalized homeostasis would force us to reject the validity of first-person experience of purposes and lead us to the narrowly conceived homeostatic account that Jonas's himself correctly criticized for its reductionist outcome. Di Paolo's move of grafting a selected part of Ashby's theory upon the trunk of Jonas's teleological account is an example of second solution. By renouncing generalized homeostasis and sticking to the technical device of ultrastability as a means to goal generation, Di Paolo tries to compromise between the incompatible positions defended by Jonas and Ashby. The compromise, however, is profoundly unstable. Di Paolo designed his robots to rely only on the technical mechanism of ultrastability, while avoiding the other Ashbian theses. In so doing, however, he put himself in a theoretically uncomfortable position. If ultrastability is just a proxy for full-blown Ashbian homeostasis (which future versions of his creatures will perhaps implement), then he would run afoul of Jonas's criticism: the so-called goals that his robots produce are nothing more than the "subjective representation of indifference." If, on the other hand, Di Paolo has no intention to accept generalized homeostasis and will continue to confine his creature to an "urbanized" version of Ashby's theory, then the empirical support he can provide to Jonas's philosophical outlook becomes really limited, ultrastability being just a technical device that fall far short of Ashby's theory.

Room for another option, however, may well be available. Jonas's starting point (an echo of Kant's similar argument in the *Critique of Judgment*) is that a physicalist interpretation of life, which neutralizes the indubitable and phenomenological experience of self-identity, ends up reducing the identity of the subject to a mystery. Physicalism (and generalized homeostasis is for Jonas an uninteresting form of physicalism) became acceptable only because the first and more preferable choice—the option in favor of life's intrinsic teleology—appeared to be incompatible with the natural sciences. The argument relies entirely on the epistemological and ontological primacy of the experience of purpose-driven self-identity, an experience that is grounded in the awareness of a constant process: "the organism's elemental urge [is] thus precariously to continue itself." But how reliable is the experience of *that* process? How primary is it? One obvious alternative to the dilemma Jonas highlights would be for instance to hold that not one but two series of processes are occurring at once: a fully homeostatic series whose only goal is, in Jonasian terms, quiescence, or full equilibrium, and another series that is fully purpose-driven. We would then have to show that a causal link exists between the two series. If we were able to do so, then a more originary and yet partially (or completely) unexperienced homeostatic process would produce, hence logically depend upon, the first-person experience of purpose-driven self-identity. The primary process that seeks a quiescent "goal" would generate the appearances of self-identity and purposiveness. Dubbing them "appearances" does not imply that they have no reality—instead, they represent genuine but derivative phenomena. To put it differently, the aspect of Jonas's critique of homeostasis that I propose to reject is the (onto-)logical primacy of phenomenal experience, not its phenomenological reality, nor, a fortiori, the possibility of causal efficacy. Such a position would deny to first-person experience the privilege of being a foundational property, but without coincidentally reducing it, as Jonas fears, to the mere epiphenomenon of an "accompanying tune." In other words: the physicalist strategy Jonas complains about in his critique of homeostasis is *eliminativist*. He regards the conscious purposes as really non-existing entities that play no causal role of any sorts in the explanation of behavior. The framework I am sketching here is not eliminativist at all: just the opposite. It does not deny, in those beings in which such an experience is present, the existence of consciously experienced purposes—-it just denies their foundational status. Those conscious experiences still play their all too crucial—though epistemologically reduced—causal role as the subordinate means to fundamentally homeostatic ends.

Jonas's emphasis on an essential teleology of living beings, which Varela and Weber together with Di Paolo accepted and pushed even further, is logically incompatible with my suggested refusal to grant ontological primacy to first-person experience of purposiveness. Will this imply that my rejection of Jonasian teleology will reduce all living beings and their behavior to the results that physical causes necessitate? Not necessarily.

Weber and Varela (2002), in the text that forms the basis for the rediscovery of Jonas's teleology and with regards to the theory of autopoiesis, rely on Kant's exhaustive discussion of the scientific use, in biology, of teleological judgments. In particular, Weber and Varela rely on the dense and penetrating opening paragraph of the third section of the "Analytic of Teleological Judgment." Toward the end of that long passage, Kant writes: "Since reason must be able to cognize the necessity in every form of a natural product if it would understand the conditions connected with its generation, the *contingency* of their form with respect to all empirical laws of nature in relation to reason is itself a ground for regarding their causality as if it were possible only through reason" (Kant, 2000, p. 242=AK V, pp. 369–370). If we are unable to account for the particular form that living beings take on the basis of physical laws, such form would be, with respect to those living beings, contingent. Yet, since we cannot really think of living beings as *merely contingent*, we must think them as teleological. Kant (not Weber and Varela, though) will later declare that only the regulative use of teleological judgments is allowed in biology: we can study living organisms *as if* their internal organization were set up to achieve goals. However, we cannot make direct use of end-reaching strategies within our biological theories. As far as the organization of life is concerned, the authors I have been discussing appear to follow the basic structure of an argument that Kant had already laid out. Only three alternative ways to conceptualize living beings are at our disposal: (1) living beings are the *necessary product* of empirical physical laws; (2) they are the *teleological source* of themselves; (3) they are the *contingent*, grossly undetermined, *result* of physical laws. Weber, Varela, and Jonas argue for the second alternative that they deem superior to the first one. New defenders of the homeostat—such as Di Paolo—add that only the introduction of the mechanisms of ultrastability will let the second alternative prevail. The claim I have been arguing in this chapter, on the contrary, is that Ashby's understanding of living beings is not only incompatible with Kant's alternatives number one and two, but that Ashby has built with his Homeostat and has put forward with his generalized homeostasis theory a sustained and all too fascinating case in favor of alternative number three. Because all life forms (life itself) turn out to be made out of contingent, dispersed, embedded, equilibrium-seeking beings that may (or may not), among the many devices they use to achieve quiescence, include the generation of goals and their related consciousness, Ashby's generalized homeostasis theory is a protracted argument in favor of the absolute contingency of all living beings.[35]

Is Ashby's general theory, like his device, a *unicum* whose articulation into a general framework would require us to start again from his pronouncedly technical work? My opinion is that this is not the case at all: if we go back to the first half of the twentieth century, we find at least another example of a theory that was developed by extending the principle of homeostasis beyond its original physiological confines. My previous mention of a possible resolution of the conflict between Ashby and Jonas in terms of primary and secondary processes may have given away where I think we could look. In Sigmund Freud's original work—including his neurophysiological speculations—and in some of its present-day reenactments, I surmise we may find an instance of a deeper consistency with Ashby than we have found in Jonas's. Indeed, we may find the framework that can possibly accommodate a homeostatic two-level theory of living beings.[36]

3. Conclusion

In this chapter I have tried to show that the Homeostat, Ashby's weird and historically contingent contraption, embodies a thesis that is at odds with most, if not all, the frameworks concerned with cognitive and biological interpretations that were prevalent in the 20th century, such as: the strict behaviorist interpretation that reduces cognition to the necessary stimulus-response behavior of the physiological apparatus; the cognitivist interpretation that separates the mental from the physiological; and the more recent enactive interpretation that reinserts cognitive functions on their physiological substrate on the basis of a Jonas-inspired account of the intrinsic teleological character of life. I have also argued, albeit all too briefly, that Ashby's generalized homeostasis thesis, plus his ancillary theses on randomness, dispersion, and deep embeddedness, are potentially most at ease within a theoretical framework that presently enjoys little or no intellectual currency—the Freudian one, for short.

If my analysis of the Homeostat is correct, two radically different conclusions are possible. On the one hand, we may say that Ashby's original homeostat (generalized homeostasis plus dispersion, randomness, and deep embeddedness) was *rightfully allowed to die*. A curiosity of mere historical value, it is deeply incompatible with contemporary major conceptual frameworks in (the philosophy of) biology and the philosophy of mind; hence we should let it rest where Ashby abandoned it when he left Urbana-Champaign—in the basement of 20th century cybernetics. On the other hand, the ideas the Homeostat embodied may be an alternative capable of capturing and integrating conceptions of life and cognition that come from different fields and that include: the re-evaluation of the role of contingent change in biology (Stephen Jay Gould's reframing of evolutionary theory, for instance); Freud's homeostatic theory of (un-)consciousness, in its original and in its more recent versions; and a few contemporary philosophical discussions on new ways to articulate the concept of action (from Heidegger's—Jonas's teacher—articulation of *Gelassenheit* to the more recent treatments of the concept of *desoeuvrement*).

Which of the two conclusions is more warranted? We could say, in a Kuhnian fashion, that because the choice between the two options depends largely upon a judgment concerning the relative usefulness and theoretical validity of the several theoretical frameworks I have analyzed, it is, on purely theoretical grounds, undecidable. We could say that it amounts to a perspectival decision that depends on an overall assessment of several research-programs. I would like to suggest—following the general methodology Ashby practiced and Di Paolo and the evolutionary roboticists so convincingly repeated—that returning to the original Homeostat, this time around armed with our current and powerful technology, could earn us some insight on the respective merits of those two alternatives. By way of computer simulation, we could try for example to resurrect not just the Homeostat's mind, but even its body, albeit in a virtual form. In other words, we could return to the original implement, reconstruct it, repeat Ashby's experiments, stretch them along the many different lines of thought that were already present in his writings and that also more current research suggests. In short, we could exploit our current technical prowess to connect a Homeostat to an outside environment (real or simulated), and to evaluate the results and the insights that come from it with regards to the two options I mentioned above. I am convinced that only then we will be able to decide, once and for all, whether the Homeostat, weird and fascinating as it is, deserves a second and perhaps more fruitful life.

NOTES

1. Boden, 2006, v. 2, pp. 1325–1327, 1448–1450. Boden's term is "Homeostasis in C(ontinuous) T(ime) R(ecurrent) N(etworks)." Her main reference is to Ezequiel Di Paolo's and Randall Beer's work, which I discuss below.

2. I should point out that Ashby's description of the Homeostat, although extensive, is not detailed enough to allow the reader to produce an exact replica of it (or a simulation thereof). The most crucial component of Ashby's device—the mechanism implementing the random search through a step-wise function—seriously lacks specification. The scarce literature on Homeostat simulators (for instance, Eldridge, 2002) noticed the issue, and I have confirmed it too in my own work (see also Andrew Pickering's remarks, 2010, fn. 17, p. 424, on Ashby's confusing discussion of number of possible states the homeostat could be in). The lack of details forbidding a duplication of Ashby's results, reinforces the thesis I defended above: in spite of the widely shared admiration the Homeostat received, it failed to exert a direct influence over the field.

3. As Andrew Pickering's analysis of Ashby's journal entries shows (2010, especially pp. 122–131), in the 1950s Ashby worked tirelessly on the DAMS system but his efforts, in spite of several prototypes and partially encouraging results, eventually failed. Still optimistic about the final outcome, Ashby described one of the early prototypes in an article that the French journal *Thalès* (1951) published, The second printing of *Design for a Brain*'s first edition (1954) still contained a reference to this article that the author instead removed from the second (1960) edition. After 1954, only Ashby's private journals will mention DAMS.

4. The label is Ashby's. It provides the title of section 5/6 of the second edition of *Design of a Brain*. Ashby used it often to refer to his overall view of the brain as regulator.

5. Wiener, 1948, p. 135. The text remains unchanged in the second edition (1961), published after Wiener had met Ashby personally and was acquainted with his work.

6. See Wiener and Schadé, 1963, p. 1, now in Wiener, 1985, vol. IV, p. 400. Ashby comments briefly on the few papers on homeostasis in this volume, praising Wiener for advocating the importance of homeostasis to the physiologists (*ibid.*, vol. IV, p. 407). Ashby does not mention, however, that his own view of homeostasis was far more general than theirs, as I will discuss below, and, by implication, far more general than Wiener's.

7. See Ashby, 1953; Ashby, 1952a; Ashby, 1956a; Ashby, 1947; Speidel, 1942. As I will discuss below, recent revivals of Ashby's paradigm have not kept either approach.

8. See Walter, 1961, p. 111. Walter concluded his review of Ashby's Homeostat by saying that "judging *Machina sopora* entirely by its behavior, the naturalist would classify

it as a plant." (112) Given Claude Bernard's classification of life in three distinct forms (which I will discuss below) and his consequent assignment of imperfect homeostasis to plants, Walter's comment appears to point to a fundamental misunderstanding of Ashby's project.

9. Cannon (1929) coined the term, although Claude Bernard had introduced the fundamental idea about 50 years earlier (1966 [1878-1879]). Next, Lawrence Joseph Henderson in the United States and John Scott Haldane in Britain (Henderson, 1928; Haldane, 1917 and especially Haldane, 1922) popularized in the English speaking world. The title of Cannon's *The Wisdom of the Body* pays homage to William Starling's homonym lecture, which had dealt with the self-compensating mechanisms of the heart (Starling, 1923). Starling was instrumental in the identification of hormones as the main agents involved in the body's self-regulating processes. As Pickering (2010, fn. 12, pp. 422–424) makes clear, Ashby elaborated his theory well before he became familiar with Cannon's work, although he only started to use the term "homeostasis" after reading it. See also Cooper, 2008 and Virtanen, 1960. Langley, 1973 contains a collection of sources.

10. Bernard, 1974, p. 84. Bernard goes on to say that "les éléments histologiques sont donc tous de véritables organismes élémentaires aquatiques…, parce que l'eau est indispénsable aux réactions chimiques, ainsi qu'à la maniféstation des proprietés de la matière vivante" (*ibid.*).

11. Keeping with the prevailing British orthodoxy of the time, I should perhaps call it, more than Kantian, Hegelian. Incidentally, Hegel's main translator in English was Haldane's sister, Elisabeth Sanderson Haldane.

12. Nicholas Mrosovsky (1990) formulates the hypothesis that the solar periodicity that the 24-hour cycle expresses provides the base curve around which the homeostatic mechanism acts. He then generalizes his claim to admit that a second-order (or even an $n+1$ order) homeostatic regulatory process may target a first order (respectively, n order) mechanism. Jay Schulkin (2003), following Sterling and Eyer, 1988's original introduction, discusses the concept of "allostasis" concerning a more complex regulatory mechanism that embeds positive-feedback and anticipatory cycles within negative feedback cycles. I should note that a return to Cannon's original formulation often frames these revisions to the classic theory of biological homeostasis.

13. Cannon, 1939[1932], p. 302. See Barcroft, 1932; G. P. Smith, 2008; Holmes, 1969, for Barcroft's contribution.

14. Descartes is often far from the stereotypical version of Cartesianism that we sometimes hear professed nowadays. A passage of the Sixth Meditation, for example, may sound like a proto-cybernetic declaration: Descartes equates bodily affections with useful or dangerous relationships between organism and environment and states that "I also perceived by my senses that this body was situated among many other bodies which could affect it in various favorable or unfavorable ways; and I gauged the favorable affects by a sensation of pleasure, and the unfavorable ones by a sensation of pain" (1984, p. 52 = AT VII:74).

15. See Galison, 1994, reprinted in this volume.

16. Ashby, 1981b, pp. 117–118. Ashby insists that he limits his investigations—like most scientific inquiries—to state-determined systems whose current state, not their previous history, completely determines their future behavior. This definition may be misleading, though: a system's past history may influence the behavior of a state-determined system as long as its state records some aspect of that history. So state-determined system are not "out of time." On the contrary, state-determined systems—as dynamic systems—are fully embedded in time, but only to the extent that past interactions alter their internal states. The decision of whether a system is state-determined or not is more epistemological than ontological, for the observer's descriptive framework, rather than the system's intrinsic properties, often decides it. Ashby's definition of a state-determined and non-state determined system is roughly equivalent to the distinction between autonomous versus non autonomous systems of differential equations. We can convert a non-autonomous system of n first order differential equations into an autonomous system of $n + 1$ equations. Thus, Ashby, 1960, 19/13, p.248 describes a system (a gently undulating putting-green-like slope with balls rolling down) that is non-state determined if the observer only knows about the balls' positions on the slope, whereas it is completely state-determined if the observer knows information about the balls' momenta. Only quantum systems—Ashby notes—would qualify as truly non state-determined.

17. R. E. Williams, 1961, references a more detailed mathematical discussion of Jenny and of stability, which he anticipates as forthcoming (Miller and Haroules, 1962). Williams concludes his brief communication with a set of specific recommendations for the optimal use of Jenny. The first and arguably most important one reads: "Describe the system operation by a set of equations obtained through a theoretical study (ref. 3) [i.e., *ibid.*], or by a more empirical study of the machine." I have been unable to track down this publication and I am not sure it ever saw the light of day. Since the Cambridge US Air Force Laboratory no longer exists, it will be difficult to uncover these details anytime soon. Similar considerations apply to the "empirical study of the machine."

18. See Capehart, 1967. More recently, Alice Eldridge (2002) followed a similar, batch-oriented, and non-interactive approach. She dropped Capehart's effort to simulate the Homeostat at the physical level through the numeric solution of differential equations that model continuous phenomena, and opted instead for a discrete time digital approach. As Asaro reports (see below, pages 112 ff.), Turing himself suggested a simulation of the Homeostat on the ACE computer that he was building at the time, but the project came to naught. More importantly, in his second and very successful book, *An Introduction to Cybernetics* (1956), instead of the language of continuous functions prevalent in *Design for a Brain*, Ashby adopted a discrete time formalism.

19. See Pickering, 2002, Gerovitch (in this volume). In spite of the different implications we respectively draw, my characterization of the Homeostat as a "unicum" within the standard canon of cybernetics and Artificial Intelligence is not far from Pickering (2005)'s, who calls it a "monster."

20. Di Paolo, 2000b, who recently moved to the University of the Basque Country in San Sebastiàn, Spain, reports the original experiment; H. Williams, 2006, and Her-

rmann, Holicki, and Der, 2004, present work conducted along similar lines; see Harvey et al., 1997; Harvey et al., 2005, for a general account of the Sussex approach to evolutionary robotics.

21. Cabanac, 1996, p. 39. The earliest discussion is in Cabanac, 1971, but his position did not become widely known, especially in the English-speaking world, until much later. Cabanac, 1995, offers the most complete discussion of his interpretation of pleasure. An expansion to the origin and role of consciousness is in Cabanac, 1996.

22. Cabanac, 2002, p. 76. Here is full definition: "I accept, therefore, that all emotions possess a strong hedonic dimension, either positive (pleasure) or negative (displeasure). In all these cases emotion may be described as a 'mental model of some stimulus or event occurring either in the subject's environment, or purely mental'. What these 'emotions' have in common is that they are all intense mental events aroused by exposure of the subject to situations more or less related to motivation, either positive or negative but all resulting in a behavior oriented to, or away from, the stimulus. Without this dimension the mental experience does not qualify as emotion. It may be proposed, therefore, that this hedonic dimension is what pathognomonically defines emotion."

23. See Rabinowitch and Segev, 2008, for an (Ashby-independent) discussion of this possibility and for the suggestion of a potential resolution to the conflict with Hebb-like plasticity.

24. See Giorgio Sandri's chapter, especially pages 309ff. For the deep compatibility between Ashby's model of computation and the structuralist account of the evolution of social structures, see Patrice Maniglier's chapter below.

25. Several alternative research programs developed neural networks that proposed different architectures. The distinctions between architectures and approaches were at the time often fluid. Sometimes researchers switched from one paradigm to another that we would now consider incompatible with it. For instance, we usually associate McCulloch's name with the feed-forward, timeless (static) networks he first introduced along with Walter Pitts, in 1943, even though later on he showed more interest for recurrent networks.

26. These are not the only differences. Hopfield networks have another important limitation: they exclude self-loops and admit symmetric synaptic weights only, both conditions being mathematically necessary to guarantee stability for arbitrary large networks (Hopfield, 1982; Hopfield, 1984). Although the neural network and AI communities largely ignored it until the 1980s, Stephen Grossberg's work (1968; 1969) was much closer to Ashby, both in general inspiration and formal detail. See also Selfridge, 1959; Beer, 1990.

27. In its simplest formulation, Hebb's associative rule takes the form $\Delta w_{ji} = \eta y_i y_j$ where y stands for the activity level of a given neuron, w is the weight, and η is a synaptic-specific constant that indicates the rate of learning of that synapse. This formulation provides the basis for the first rule of homeostatic change in Di Paolo (2000b) (see below).

28. For a recent introduction, see Robbins and Aydede, 2008.

29. Di Paolo, 2000b; J. G. Taylor, 1962. Taylor's experiments built upon similar work that the Innsbruck's psychological school in the two preceding decades had carried out (Kohler, 1955; Werner and Wapner, 1955; Kohler, 1964). Although some of the even

earlier work on motor disruptions (for instance, Sperry, 1947) was instrumental to Ashby's own conception of ultrastability (Ashby, 1960, p. 104), Taylor was the first researcher to explicitly use Ashby's ultrastability theory as a theoretical framework.

30. An incomplete list includes Williams and Noble, 2007; H. Williams, 2006; Herrmann, Holicki, and Der, 2004, Der et al., forthcoming; Der, 2001; Ikegami and Suzuki, 2008; Iizuka and Di Paolo, 2007.

31. Weber and Varela (2002, p. 115), quoting the definition of "cellular identity" that Varela gave in 1997, p.75, which continues: "This identity amounts to self-produced coherence: the autopoietic mechanism will maintain itself as a distinct unity as long as its basic concatenation of processes is kept intact in the face of perturbations and will disappear when confronted with perturbations that go beyond a certain viable range which depends on the specific system considered."

32. Di Paolo, Rohde, and Jaegher, 2011. "Enaction" is the general approach to the study of cognition that Varela, Thompson, and Rosch (1991) have elaborated on the basis of autopoiesis.

33. Jonas, 1984, p. 60. A drive or a fundamental bodily need would, in Ashbian terms, become the defense of an essential variable whose value is about to become critical. Notice that J. G. Taylor (1962), in the work that inspired Di Paolo's new homeostat, explicitly affirms this equivalence. We should not interpret "drives" as a direct rendering of the Freudian *Triebe*. Instead, "drives" refer to the behaviorist "drive theory" that Hull (1943)—to whom Taylor dedicated his book, besides Ashby and Kohler—championed.

34. The CTRNN model that Di Paolo developed following Beer and Hopfield is close in spirit to the dynamical approach Ashby championed. It highly differs, however, on at least one account: the physical inspiration that underlies Ashby's mathematical model of the brain. Ashby thinks in mechanical terms, and the differential equations describing his model of the brain describe the motion of physical bodies in real space. Although aware of the electrical properties of neurons that researchers were systematically investigating at the time, Ashby indulges in mechanical metaphors. At one point, for instance, he compares the brain to a large collection of beads strung together on elastic strings (Ashby, 1947). On the contrary, Hopfield's model and CTRNN neurons are expressed in terms of equations that describe the behavior of electric *RC*-like circuits. The primacy of Ashby's physical reference to mechanics, not to electricity, attests that he reasoned in terms of forces that produce accelerations, not velocities. Different physical intuitions such as these morph into different mathematical models and, in Ashby's case, into higher order differentials. The difference is significant: a simulated Homeostat that a set of 1st order equations models will, in general, not exhibit the interesting dynamical properties that Ashby studied on his device. More specifically, it will not be able to carry out even the simplest among the experiments that demonstrate adaptivity—the first of the seven experiments that Ashby discussed in *Design for a Brain*, 8/4, pp. 104–105. In addition, other technical differences are present, though not as relevant. Di Paolo's setup is more complicated than Ashby's, because: (1) it includes 4 different parameters-switching rules (it would have 4 different kinds of "S" boxes, in our diagram (see fig. 1.1), with evolu-

tion (GA) determining which box belongs to which neuron); and, (2) because in Ashby's model a measure of synaptic plasticity is not present. For instance, Di Paolo's first rule is $\Delta w_{ij} = \delta \eta_{ij} p_j z_i z_j$, where p is the plasticity function, z is the firing rate of a neuron, and η is the genetically determined parameter that controls the level of synaptic plasticity between neurons i and j (δ is just a theoretically irrelevant damping factor). When the neuron's firing rate is out of its optimal range, p becomes non-zero and produces a change in the synaptic weight w_{ij} that is proportional (or anti-proportional, depending on η's sign) to the joint firing rates of i and j, thus potentiating or depressing, respectively, a Hebb-like form of learning between them. This rule is identical to the standard rule of Hebbian learning I mentioned above, plus a plasticity function that controls its activation.

35. Maniglier (below, pp. 167 and ff.) reaches a similar conclusion about the Structuralist theory of the emergence of cultural forms.

36. Even in cybernetics' high times, researchers did not discuss very often the similarities between Ashby's conception of homeostasis on the one hand, and on the other Freud's constancy principle regulating the operation of the drives (*Triebe*) within the organism. But the argument does occasionally surface. Wilkins, 1968, p. 7, for instance, refers explicitly to Freud's theory as parallel, albeit in the psychological domain, to Ashby's. An early and not entirely convincing discussion is in Walker, 1956. Karl Pribram (1962) was perhaps the strongest advocate of a biological interpretation of Freud's theory as largely compatible with Ashby's generalized homeostasis thesis. Pribram asserted that "the principle of inertia [in Freud's abandoned 1905 *Project for a Psychology* (1953)] is in many respect similar to what we today know as homeostasis," but, "it is homeostasis in its boldest form: an organism, when stimulated, attempts to get rid of that stimulation, i.e., to return to the unstimulated form" (Pribram, 1962, p. 444). See also Alexander, 1948, pp. 35–36 for an explicit homeostatic interpretation of Freud's principle of constancy that Ego-psychology's interpretation of psychoanalysis carried out. See also, for a Lacanian interpretation of psychic life, Chiesa, 2007, pp. 15–17. Lacan regards psychic life as the product of a homeostasis that in humans structurally fails because of the phenomenon of neo-natality. In "Jammed Machines and Contingently Fit Animals: Psychoanalysis's Biological Paradox" I have attempted to provide the outline of a possible integration between Freud's and Lacan's psychoanalyses on the one side, with Ashby's generalized homeostasis on the other.

Two

THE ONTOLOGY OF THE ENEMY: NORBERT WIENER AND THE CYBERNETIC VISION

Peter Galison*

1. The Enemy

"I ... hope you can find some corner of activity in which I may be of use during the emergency," the mathematician and physicist Norbert Wiener wrote the czar of American war research, Vannevar Bush, on 20 September 1940. Britain was under unrelenting aerial attack, and a Nazi invasion seemed imminent. Wiener scrambled across the disciplinary map to throw his weight behind a technological defense. He suggested procedures to improve Bush's computational device, the so-called differential analyzer, in ways that would facilitate faster design of war materiel from airplane wings to ballistic shells. More concretely, he reiterated a previous proposal that the Allies loft air-bursting containers of liquified ethylene, propane, or acetylene gases to engulf a wide volume of the sky in a prolonged detonation (Wiener, 1940a). That repelling the onslaught of bombers had pushed all scientific questions aside is hardly surprising. For the German Air Force had dubbed 13 August 1940 "The Day of the Eagle," and with it the Battle of Britain had begun with an assault of almost 1500 aircraft flown against British air stations and aircraft factories. During the following two weeks over a thousand Londoners had died under the rain of bombs, and September was worse. On 7 September alone, 448 civilians perished; on 15 September the Germans pitched 230 bombers and 700 fighters against London, Southampton, Bristol, Cardiff, Liverpool, and Manchester (Gilbert, 1989, pp. 117–125).

*I would like to thank Arnold Davidson, Steve Heims, Gerald Holton, P. Masani, Howard Stein, George Stocking, Sheldon White, and especially Caroline A. Jones for many helpful discussions. Masani generously allowed me to see copies of Wiener's wartime papers and reports. Originals of these and other technical reports and general project files are housed at the National Archives, Library of Congress, Washington, D.C. References to the Norbert Wiener Papers are to collection MC-22 in the Institute Archives and Special Collections of the Massachusetts Institute of Technology, Cambridge, Massachusetts. I would like to thank Elizabeth Andrews of the MIT Archives and Marjorie Ciarlante of the National Archives for facilitating access to archival materials. I would also like to thank my able research assistants, Richard Beyler, Pamela Butler, and Jamie Cohen-Cole, for their help and Jean Titilah and Ann Hobart for manuscript preparation. This work was supported in part by a grant from the Andrew W. Mellon Foundation.
This essay was originally published in *Critical Inquiry*, 21:1 (Autumn 1994), pp. 228–266.

Over the next few years, Wiener's attention focused increasingly on the problem of destroying enemy airplanes. His early efforts at computation and antiaircraft fire coalesced in a remarkably ambitious calculating device that he called the "antiaircraft (AA) predictor," designed to characterize an enemy pilot's zigzagging flight, anticipate his future position, and launch an antiaircraft shell to down his plane. But Wiener's electronic manipulation did not stop with halting Nazi air attacks. In the course of characterizing the enemy pilot's actions and designing a machine to forecast his future moves, Wiener's ambitions rose beyond the pilot, even beyond the World War. Step by step, Wiener came to see the predictor as a prototype not only of the mind of an inaccessible Axis opponent but of the Allied antiaircraft gunner as well, and then even more widely to include the vast array of human proprioceptive and electrophysiological feedback systems. The model then expanded to become a new science known after the war as "cybernetics," a science that would embrace intentionality, learning, and much else within the human mind. Finally, the AA predictor, along with its associated engineering notions of feedback systems and black boxes, became, for Wiener, the model for a cybernetic understanding of the universe itself. This chapter is an exploration of that expansion. In it, I will be backtracking from the widest ontological claims of cybernetics into a collocation of vacuum tubes, resistors, and condensers designed to replicate the intentions of a hidden enemy pilot.

Enemies were not all alike. In the killing frenzy of World War II, one version of the Enemy Other (not Wiener's) was barely human; to the Americans, British, and Australians, the Japanese soldiers were often thought of as lice, ants, or vermin to be eradicated. As General Sir Thomas Blarney told a unit in Port Moresby in 1942: "Beneath the thin veneer of a few generations of civilization [the Japanese soldier] is a subhuman beast, who has brought warfare back to the primeval, who fights by the jungle rule of tooth and claw, who must be beaten by the jungle rule of tooth and claw .... Kill him or he will kill you." A year later, Blarney insisted on the Buna battlefield that "fighting Japs is not like fighting normal human beings .... The Jap is a little barbarian .... We are not dealing with humans as we know them. We are dealing with something primitive. Our troops have the right view of the Japs. They regard them as vermin" (quoted in Dower, 1986, pp. 53, 71). These monstrous, racialized images of hate certainly presented one version of the World War II enemy, but it was by no means the only one.

Another and distinct Allied vision held the enemy to be not the racialized version of a dreaded opponent but rather the more anonymous target of air raids. This enemy's humanity was compromised not by being subhuman, vicious, abnormal, or primitive but by occupying physical and moral distance. Viewed from afar, from the icy heights of thirty thousand feet, a city in Germany looked small, and individual people appeared to be invisible, partially shorn of their likeness to the bomber. After opening a spate of airmen's letters, one British censor from the Air Ministry reported on 21 June 1942: "[The letters] illustrate the effect of airmen's remoteness from their attacks on human beings. Expressions of satisfaction that the Germans are having to undergo the punishment they have hitherto meted out to others are found in almost all letters, but there is an absence of vindictiveness or fanaticism in the phrases used."[1]

But there is yet another picture of the enemy that emerged during World War II, less well known but in many ways more powerful and enduring than either the racialized or the anonymous enemy. More active than the targeted, invisible inhabitants of a distant city and more rational than the hoardelike race enemy, this third version emerged as a cold-blooded, machinelike opponent. This was the enemy, not of bayonet struggles in the trenches, nor of architectural targets fixed through the prisms of a Norden gunsight. Rather, it was a mechanized Enemy Other, generated in the laboratory-based science wars of MIT and a myriad of universities around the United States and Britain, not to speak of the tens of laboratories in the countries of the Axis.

On the Allied side, three closely related sciences engaged this calculating Enemy Other: operations research, game theory, and cybernetics. Each had its own prototypical war problem. Operations research focused, for example, on maximizing efficiency in locating and destroying German U-boats in the North Atlantic and along the coast of the Americas.[2] Game theory, though it had mathematical roots in the interwar years, exploded into view with John von Neumann and Oskar Morgenstern's masterwork of 1944, *Theory of Games and Economic Behavior*. Strategists picked up the technique as a way of analyzing what two opposing forces ought to do when each expected the other to act in a maximally rational way but were ignorant both of the opponent's specific intentions and of the enemy's choice of where to bluff. Wiener, the spokesman and advocate of cybernetics, in a distinction of great importance to him, divided the devils facing us in two sorts. One was the "Manichean devil" "who is determined on victory and will use any trick of craftiness or dissimulation to obtain this victory." Wiener's rational Manichean devil could, for example, change strategy to outwit us. By contrast, the other, the "Augustinian devil" (and Wiener counted the forces of nature as such) was characterized by the "evil" of chance and disorder but could not change the rules.[3] Exemplary of the Manichean enemy, von Neumann's game theory postulated a logical but cunning opponent; it was designed precisely to analyze an antagonist who played against us and would bluff to win.

Building on Wiener's own usage of the term *Manichean* to designate the continuing struggle against an active oppositional intelligence, I will call this triad of wartime enterprises—operations research, cybernetics, and game theory—the Manichean sciences.[4] I choose the third war science, cybernetics, as an entry point to these machine-human systems.[5] Working with the Greek word for steersman, Wiener coined the term *cybernetics* in the summer of 1947 to designate what he hoped would be a new science of control mechanisms in which the exchange of information would play a central role (Wiener 1948, pp. 11–12; hereafter abbreviated

C). If antisubmarine warfare was the formative problem for operations research, antiaircraft fire control was the key to cybernetics.[6] Faced with the problem of hitting fast maneuverable bombers with ground-based artillery, Wiener brought to bear his own established interest in feedback mechanisms, communication technology, and nonlinear processes.

I will argue that the system of weaponry and people that Wiener had in mind was predicated on a picture of a particular kind of enemy. On the mechanized battlefield, the enemy was neither invisible nor irrational; this was an enemy at home in the world of strategy, tactics, and maneuver, all the while thoroughly inaccessible to us, separated by a gulf of distance, speed, and metal. It was a vision in which the enemy pilot was so merged with machinery that (his) human-nonhuman status was blurred. In fighting this cybernetic enemy, Wiener and his team began to conceive of the Allied antiaircraft operators as resembling the foe, and it was a short step from this elision of the human and the nonhuman in the ally to a blurring of the human-machine boundary in general. The servomechanical enemy became, in the cybernetic vision of the 1940s, the prototype for human physiology and, ultimately, for all of human nature. Then, in a final move of totalization, Wiener vaulted cybernetics to a philosophy of nature, in which nature itself became an unknowable but passive opponent—the Augustinian devil.

Cybernetics no longer appears as a futuristic bandwagon or as a rising worldview that will leave mere mechanism in the dustbin of history, but it has much to tell us about the nature of the sciences in the mid-twentieth century and, as I will speculate, about postmodern theory in the late twentieth century.

2. The Calculating Enemy

Norbert Wiener's upbringing resembled none so much as John Stuart Mill's. His father, Leo Wiener, was an erudite and driven Harvard Slavicist who was determined that his son know languages, mathematics, and the sciences long before he was old enough to attend grade school. Graduated from Tufts University at the age of fourteen, and armed with a Harvard Ph.D. at eighteen, the younger Wiener found his own style of work in physics-based mathematics. By the late 1930s, Wiener was in his forties and a major figure at MIT, where he had contributed to quantum theory, ballistics, the theory of integration, and communication technology. Like so much else in his life, he came to war preparations early (for an American). Already in February of 1940, he joined a subcommittee under the direction of the Princeton mathematician Marston Morse to consider how the American Mathematical Society might contribute to a national emergency "which we hope will never arise" (Jackson, 1940). Needless to say, it did. Beginning on 10 July, German bombing attacks on British convoys grew significantly in number, ushering in the Battle of Britain. From 10 to 20 July the *Kanalkampf* as the Germans dubbed it, was prosecuted with the goal of depleting the British supply of fighters, to be followed by an all-out assault on the fighter bases in Britain itself. It was this spate of Nazi bombing that precipitated Wiener's primary war work. By 23 July, Wiener had received notice that the armed forces had in hand his suggestion about the use of incendiary bombs and his reiterated desire to participate in the war effort (Thomas-Stahle, 1940). But the main line of Wiener's military work centered on a general theoretical and practical inquiry into the possibility of radically improving antiaircraft technology. At root, Wiener's idea was to use electrical networks to determine, several seconds in advance, where an attacking plane would be and to use that knowledge to direct artillery fire. In an early simulated run of the AA predictor (November 1940), Wiener wanted to see how his machine would fare in four cases: a straight-line bomber trajectory, another having twice the slope of the first, a third with a parabolic slope, and a fourth with a curve following the integrated area of a semicircle. Since the circuit itself did not exist in wire and tubes, Wiener concocted a virtual mechanical equivalent and "tested" all four cases in simulated form on Vannevar Bush's differential analyzer (Caldwell, 1940; Wiener, 1940b).

Partly as a result of an extended exchange and collaboration with Bush and partly as a result of his own early circuit designs, Wiener over the next few months came to insist that the wartime mathematician had to stray far from the "purity" of the prewar years. In January 1941, Wiener hired Julian Bigelow, an MIT-trained electrical engineer, to work with him in circuits design (Wiener, 1942c). As a result of his commitment to the nitty-gritty of weapons work, Wiener had left the etherial domain of Tauberian theorems of mathematical analysis for the breadboard wiring schemes of the engineer long before American troops were attacked at Pearl Harbor. Writing to Morse back in Princeton in March 1941, Wiener used Bush's partial differential equation machine as an example of just how far mathematics would have to go if it were to advance. It could not remain "exclusively in the hands of mathematicians." He would need someone already versed in computing technique, someone, say, from Remington-Rand or IBM. This man (his term) would have to be familiar with vacuum tube work, and he would without question have to know his way around communication engineering techniques such as telephone technology.

> If I could not find these talents joined together in a single man, I would be forced to assemble a team of people each with particular talents in one field and a general knowledge of the others. In this team I would probably be the only mathematician thus the project as a whole would concern other engineering groups as much as the American Mathematical Society and it would be necessary for me to cross ordinary professional lines. (Wiener, 1941)

No one who did not already have a feeling for engineering, at least by taking on the construction of radio sets as a hobby, could even hope to participate. "There is nothing in a drawing in abstract algebra or topology ... which would prepare one in any way to cooperate in engineering design" (Wiener, 1941). Though Wiener worked his illustrative example from the Bush differential analyzer, he assured Morse that the same was true for projects from cryptography to ballistics. To be useful to the war effort, it was science itself that would have to change, becoming both materially grounded and squarely directed into the world of weapons. For Wiener, Morse, and their colleagues, science was at war, even if the country was not.

The scale of Wiener's work was not large. It was, in fact, infinitesimal compared with the scope of work getting underway at the Radiation Laboratory or at Los Alamos. Wiener's little group had, as its first 1941 budget (submitted in 1940), a paltry request for $2325, with $1200 devoted to circuit building, three man-months for differential analyzer studies at $450, and the remainder going to labor overhead (see Caldwell, 1940). Scale, however, can be measured in other ways. As the AA predictor came to fruition, Wiener came to see it as the articulated prototype for a new understanding of the human-machine relation, one that made soldier, calculator, and firepower into a single integrated system. His two thousand-odd dollars would be conceptually stretched to blanket the earth.

Preliminary circuit diagrams indicated that the AA predictor would be vulnerable to two kinds of sudden movements: irregularities introduced as the operator of the crosshair telescope cranked his gun to follow the plane and irregularities injected as the pilot zigzagged to escape. Both would have to be filtered to gain a smoother curve that the predicting circuit could handle. In the design of the filters, as in many aspects of the project, the underlying mathematical or calculational methods Wiener wanted to use for the AA predictor carried over from earlier studies of servomechanisms, that is, of feedback devices such as thermostats or self-guided torpedoes. Indeed, the only real difference between the two types of feedback problems was that in the AA predictor there was a longer time lag between action and effect: the shell took several seconds to reach its target. One striking indicator of the congruence of technological practice is that Wiener specifically asked for an enforcement of the same patent clause for the AA predictor as was used in the earlier servomechanism program (Figs 2.1 and 2.2).[7]

As Wiener and Bigelow gave form to the hardware in the summer of 1941 (Figs 2.3 and 2.4), they desperately needed realistic information on the character of input data:

> We realized that the "randomness" or irregularity of an airplane's path is introduced by the pilot; that in attempting to force his dynamic craft to execute a useful manoeuver, such as straight-line flight or 180 degree turn, the pilot *behaves like a servo-mechanism,* attempting to overcome the intrinsic lag due to the dynamics of his plane as a physical system, in response to a stimulus which increases in intensity with the degree to which he has failed to accomplish his task. A further factor of importance was that

> the pilot's kinaesthetic reaction to the motion of the plane is quite different from that which his other senses would normally lead him to expect, so that for precision flying, he must disassociate his kinaesthetic from his visual sense. (Wiener, 1942c, p. 6, emphasis added)

Here was a problem simultaneously physical and physiological: the pilot, flying amidst the explosion of flak, the turbulence of air, and the sweep of searchlights, trying to guide an airplane to a target. As Wiener saw it, humans acting under stress tend to perform repetitively and therefore predictably.

To recreate this tense concatenation of human and machine, Wiener, Bigelow, and Paul Mooney, an accomplished technician, began a series of experiments before the end of 1941 to simulate the data input of an enemy plane that would enter the AA predictor. Bigelow, who was an active pilot, took special responsibility for creating a mechanical apparatus "designed to have the 'feel' of an actual [airplane] control" (Stibitz, 1941). On a laboratory wall, a light-spot projector shot an intense white spot that followed a smooth but irregular back-and-forth motion, careening its way from wall to wall every fifteen seconds. At the same time, an operator—simulating the pilot—was given a deliberately sluggish control stick that guided the position and motion of a second colored light spot. The "pilot's" task was to guide the colored spot onto the "target," mimicking (as one contemporary witness put it) "a plane which is dodging, but flying a mission, i.e. the pilot is holding a general course, but with large swings away from the course" (Stibitz, 1942, hereafter abbreviated *D*). Quite deliberately, the experimenters made the task exceedingly difficult by racing the target across the wall at high speed and by inserting a mechanical lag into the control stick. This, they hoped, would create precisely the disassociation between kinaesthetic sense and visual information that the pilot had to face in the theater of war and would therefore lead to the same sort of feedback difficulties. Meanwhile, the position of the operator's light signal went down on tape alongside the position of the guiding light spot. The fluctuating difference between the "intended" position and the actual position of the operator's light dot provided "a way to duplicate .... the properties of the type of irregular motion of an airplane in flight" (Wiener, 1956, p. 251). These data would program the AA predictor (see Wiener, 1942c).

In particular, the pseudo pilots' "nervous reactions" exhibited two crucial features. First, there was no particular correlation among the recorded fluctuations of different operators; second, there was a high degree of autocorrelation between an operator's earlier and subsequent performances. More specifically, Wiener chose the following definition of prediction. Imagine a number of flight paths (ten, for example) that all coincide for a given segment of their trajectory but may differ after a given time, t. Now pick a point in space where we expect a plane to be at, say, $t+2$ seconds. For any such predicted point we can calculate the square of the difference between the predicted point and the actual position of the first plane at

$t+2$ seconds, and we can do the same for the other nine planes. The point for which the sum of squared errors is minimized is what Wiener calls the best prediction.[8] It turned out that prediction worked rather badly for one operator based on another operator's data, but any given operator was enormously self-consistent. "This suggests the use of such apparatus in the diagnosis of individual differences in reflex behavior, and of pathological conditions affecting the reflex arc. Many other extensions of these ideas will suggest themselves to the physiologist, the neuropathologist, and the expert in aptitude tests" (Wiener, 1942c, p. 7). More to the point, it suggested that a more refined AA predictor would use a pilot's own characteristic flight patterns to calculate his particular future moves and to kill him.[9]

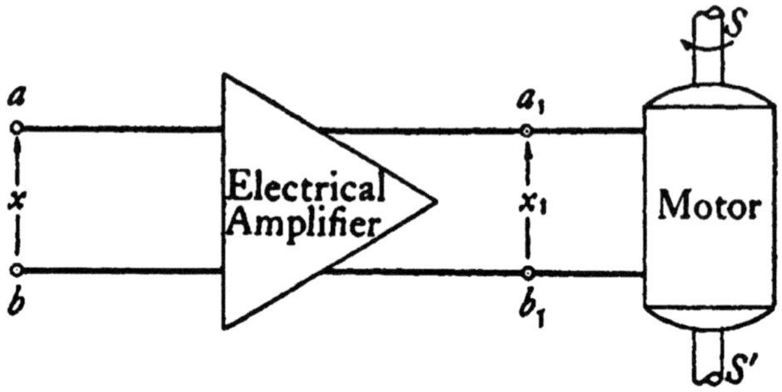

FIGURE 2.1: Sensor that actuates a motor. From MacColl (1945, p. 5).

The core lesson that Wiener drew from his antiaircraft work was that the conceptualization of the pilot *and* gunner as servomechanisms within a single system was essential and irreducible. As Wiener put it a decade or so after the war, we might succeed in eliminating this or that human feature in a weapons system, but the enemy's human behavior would not go away:

> It does not seem even remotely possible to eliminate the human element as far as it shows itself in enemy behavior. Therefore, in order to obtain as complete a mathematical treatment as possible of the over-all control problem, it is necessary to assimilate the different parts of the system to a single basis, either human or mechanical. Since our understanding of the mechanical elements of gun pointing appeared to us to be far ahead of our psychological understanding, we chose to try to find a mechanical analogue of the gun pointer and the airplane pilot. (Wiener, 1956, pp. 251–252)

Servomechanical theory would become the measure of man.

As the key discipline, servomechanical theory had a great deal to offer, and not just to laboratories devoted to enemy fire control. At MIT's huge and rapidly growing Radiation Laboratory, Wiener thought it obvious that suppressing

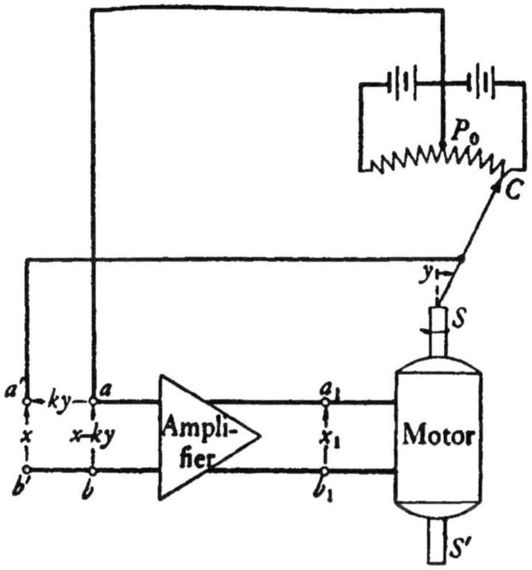

FIGURE 2.2: Sensor with feedback. According to most writers at the time, including MacColl and Wiener, a servomechanism was *defined* by the inclusion of feedback. From MacColl (1945, p. 6).

noise and conveying information should be the central electronic mission. Here his views collided with those of an extraordinary collection of "fundamental" physicists—the likes of Lee DuBridge, M. G. White, N. F. Ramsey, I. I. Rabi, Julian Schwinger, Edward Purcell, and many others. On the evening of 21 March 1942, Wiener in frustration submitted his resignation from the laboratory, cut his relations with it, and handed in his identification badge. To E. L. Bowles, director of the Rad Lab, Wiener complained the next day:

> New members of the staff of your Laboratory are recruited from the theoretical physicists or mathematicians of the country, or indeed anywhere except from among the ranks of communication engineers in the strictest and narrowest sense of the term .... It [noise suppression, operational notation, and circuit theory] is not something which a quantum physicist has any reason to know the slightest thing about, and to turn such an individual loose in your laboratory without special training, no matter what a big shot he may be in his own subject, is like ordering a corn-doctor to amputate a leg. Better three weeks delay while the big shot is learning his new trade than three months of puerilities and blunders. (Wiener, 1942a)

What the "big shots" lacked, Wiener contended, was a deep understanding of, for example, Brownian motion and generalized harmonic analysis. These were areas that Wiener had contributed to in fundamental ways; having found himself shunted

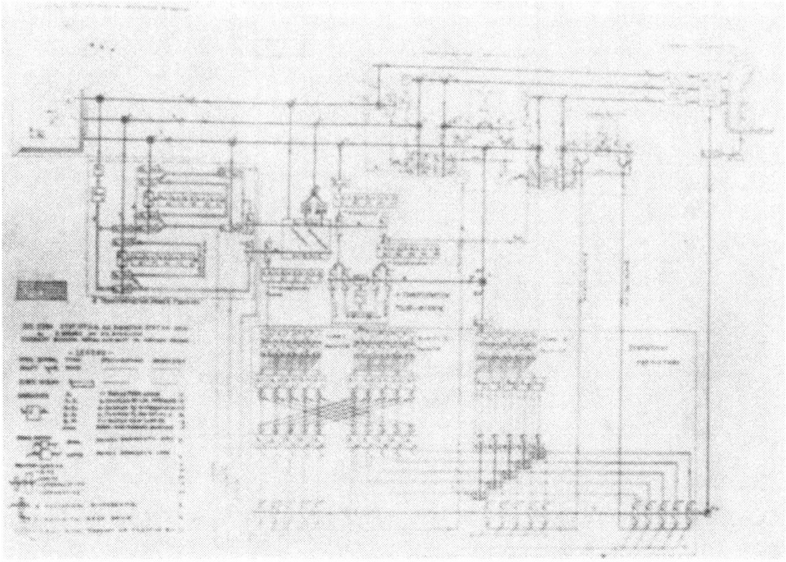

FIGURE 2.3: Circuit for Wiener's antiaircraft predictor. From Norbert Wiener (1942).

off to trivial problem solving, he was furious. Without the requisite communications knowledge, Wiener prophesied, "the military efforts of the Laboratory will be about at a good boy-scout level" (Wiener, 1942a).

Pressure and frustration began to overcome Wiener. He was working frantically, often with the powerful stimulation of Benzedrine (Wiener, 1956, p. 249). The day after his incensed letter of 22 March 1942 to Bowles, J. C. Boyce, professor of physics at MIT and technical aide to the National Defense Research Committee (NDRC), reported to Warren Weaver on Wiener's condition:

> He seems in an unusually bad nervous state the last few days, and I have been trying to get him to take a few days' rest. He had an unfortunate clash with the cleared patent attorney whom M.I.T. had asked to study some of his ideas on circuit theory, and at the same time he felt that the Radiation Laboratory was unappreciative of certain suggestions he had made to them on filter design. As a result of his state, Bigelow seems somewhat distracted, but I hope before very long this part of the zoo will be quiet again. (Boyce, 1942)

Weaver replied the next day, after seeing Wiener pacing furiously up and down a room, "perspiring profusely," and apologizing for being unable to transform an integral into a more easily calculable, rapidly converging series that the great statistician Jerzy Neyman could use. "Upon inquiry," Weaver concluded, "it turned

FIGURE 2.4: The antiaircraft predictor. From Wiener (*ibid.*).

out that [Wiener] had not been doing any of the things we particularly wanted him to do and that his busyness consisted of 'holding myself in readiness in case other jobs turned up'" (Weaver, 1942). If Wiener wasn't computing a faster-converging integral as quickly as Weaver wanted, he was already beginning to explore how the feedback mechanisms of his servomechanical theory might reshape rather distant fields. To J. B. S. Haldane, on 22 June 1942, Wiener put it this way:

> Behaviorism as we all know is an established method of biological and psychological study but I have nowhere seen an adequate attempt to analyze the intrinsic possibilities of types of behavior. This has become necessary to me in connection with the design of apparatus to accomplish specific purposes in the way of the repetition and modification of time patterns. (Wiener, 1942b)

Unmentioned was the content of these behaviorist studies. For security reasons Wiener would not reveal that the time-pattern behaviors were the pilot's evasive maneuvers and the test procedures Wiener employed to reproduce these patterns from the responses of test-subjects in the safety of the laboratory.[10]

The examination of an apparatus "from this point of view" is, Wiener told Haldane, a fundamental component of communication engineering, where the *function* of an instrument between four terminals is specified before anyone takes up the actual constitution of the apparatus in the box. He reported that this "black-box" vision of the nervous system had already generated information on a priori types of behavior, and it was clear that up to that point "no behaviorist ha[d] ever really understood the possibilities of behavior" (Wiener, 1942b). Whether his remarks were a spontaneous expression of excitement over the new results or a cryptic declaration of a priority claim, Wiener clearly saw the AA predictor, even before it was ready to shoot down a plane, as the prototype of a new behaviorist understanding of the nervous system itself. By the time Wiener wrote Haldane, he was in the final stages of preparing the machine for its great unveiling.

For a brief and shining moment, it seemed that the AA predictor would, in fact, foretell the future like a crystal ball and down enemy planes with ruthless efficiency. On 1 July 1942, G. R. Stibitz, Wiener's NDRC section chairman, visited Wiener's laboratory and registered his astonishment in his working diary:

> Most of the day is spent with Wiener, Bigelow, and Mooney. It simply must be agreed that, taking into account the character of the input data, their statistical predictor accomplishes miracles. Whether this is a useful miracle or a useless miracle, W[arren] W[eaver] is not yet convinced. The fact that predictions can at present be made only for a maximum of 2 seconds is a very serious limitation .... For a 1-second lead the behavior of their instrument is positively uncanny.[11] WW threatens to bring along a hack saw on the next visit and cut through the legs of the table to see if they do not have some hidden wires somewhere. [*D*]

These numbers were more impressive than they might at first seem, since the Wiener-Bigelow scheme compressed time by a factor of four to five, making a two-second prediction the equivalent of ten seconds in the real world. Since an antiaircraft shell took about twenty seconds to reach a bomber at altitude, the predictor seemed well on its way to success.

Even in the midst of a war project that did not yet approach field capability, Wiener clearly had already begun to reflect on the broader ramifications of this species of machine. On the same day he saw the predictor demonstrated, 1 July 1942, Stibitz recorded Wiener's wider ambition for the device:

> W[iener] points out that their equipment is probably one of the closest mechanical approaches ever made to physiological behavior. Parenthetically, the Wiener predictor is based on good behavioristic ideas, since it tries to predict the future actions of an organism not by studying the structure of the organism but by studying the past behavior of the organism. [D][12]

To get at the future behavior of the bomber-organism, Wiener and Bigelow made a tour that summer (1942) of the various installations charged with precisely measuring the flight of a plane. At Princeton and Tufts, they consulted on errors in tracking procedures; at Langley Field, experts offered them data on the regularities and irregularities of airplane motion; at the Aberdeen Proving Ground, at the Frankford Arsenal in Philadelphia, and at the Foxboro Instrument Company, additional information came pouring their way. But it was at the Anti-Aircraft Board at Camp Davis, North Carolina, that the two prognosticators received their most precious documents: tracking data on two test flights—the so-called flights 303 and 304—at one-second intervals (see Wiener, 1942d). These two trajectories through the sky were crucial because they gave, for the first time, realistic data that could be used as input to, and a test on the output of, the prognosticating machine.

Over the next five months, Wiener worked to reproduce these data—to little avail. By December 1942, it was all too clear that, however clever the general statistical analysis had been, it was barely able to compete with two simpler, geometrical prediction machines designed by Hendrik Bode. The first simply extrapolated the future from the derivative of the plane's trajectory, calculated at a fixed initial point. The second Bode method continuously recomputed its prediction on the basis of a trajectory derivative computed ten seconds back from the plane's current position. In December 1942 and January 1943, Wiener compiled the following chart for Weaver (Wiener, 1943a):

Track	(1) Bode	(2) 10 Sec. Bode	(3) Statistical
303	6 hits	22 hits	23 hits
304	35 hits	55 hits	49 hits

Bode, from Bell Laboratories, had developed a geometrical fire-control predictor that had the virtue of being based on already-existing technology and the vice of not taking into account the random fluctuations and irregular trajectories of the bombers.

Quite clearly, Wiener's own method (statistical) was barely better than the ten-second Bode method for track 303 and inferior to the ten- second Bode for track 304. In light of this manifest inadequacy, Wiener judged the only hope for the method to lie in a vastly increased statistical base involving the calculation of tens, if not hundreds, of tracks. Since this would tie up the computing facilities of the country, and because the likelihood of improvement struck him as "too distant to be significant in the present war," Wiener hesitated to recommend further research until after the end of the war (*ibid.*). What went wrong? Wiener speculated:

> To what extent the negative result of this investigation is due to bad tracking, to what extent to the restriction of the useable past [flight path] to 10 seconds, and to what extent to the fact that the enemy plane has a very considerable chance to change its flight pattern, whether voluntary or involuntary, in the

> twenty seconds of projectile flight, is not yet fully clear. (Wiener, 1942d, p. 7)

It may have been "not yet fully clear," but Wiener was "convinced" that it was the enemy's capacity to maneuver rather than anything else that would save him from inevitable destruction at the mechanical hands of the predictor. Failure came hard, for Wiener was frustrated by the predictor's weakness: "I still wish that I had been able to produce something to kill a few of the enemy instead merely of showing how not to try to kill them" (Wiener, 1943b).

3. From AA Predictor to Human Nature

What Wiener was willing to do, even in the worst days of war, was to turn to psychological and philosophical implications of the predictor. In their 1943 article "Behavior, Purpose and Teleology," Wiener and Bigelow collaborated with the cardiologist Arturo Rosenblueth, then visiting Harvard Medical School, to present a new, behaviorist description of the very concept of purpose. Aside from the pure satisfaction of classification, the authors were pleased to single out the class of predictive behavior because "it suggests the possibility of systematizing increasingly more complex tests of the behavior of organisms" (Rosenblueth, Wiener, and Bigelow, 1943, p. 22). Of particular importance, they contended that their classification rehabilitated "purpose" and "teleology" by bringing them under the aegis of a "uniform behavioristic analysis" that was equally applicable to living organisms and machines.

Where Darwin had assiduously tracked the similarities between human and animal in order to blur the boundary between them, Wiener's efforts were devoted to effacing the distinction between human and machine. Darwin's dog suffered remorse; Wiener's AA predictor had foresight. Indeed, over the course of the war, Wiener reported in 1945, men had grown ever more accustomed to attributing animation to servomechanical systems:

> The semi-humorous superstition of the gremlin among the aviators was probably due, as much as anything else, to the habit of dealing with a machine with a large number of built-in feedbacks which might be interpreted as friendly or hostile. For example the wings of an airplane are deliberately built in such a manner as to stabilize the plane, and this stabilization, which is of the nature of a feedback ... may easily be felt as a personality to be antagonized when the plane is forced into unusual maneuvers.[13]

Our consciousness of will in another person, Wiener argued, is just that sense of encountering a self-maintaining mechanism aiding or opposing our actions. By providing such a self-stabilizing resistance, the airplane acts as if it had purpose, in short, as if it were inhabited by a gremlin.

Within the rubric of "purposeful behavior," then, Wiener and his collaborators Bigelow and Rosenblueth allowed for those acts that do not involve feedback while the process is underway (such as a frog that shoots its tongue out towards a fly) and those (such as a self-guided missile or torpedo) that gather information and use that information to correct themselves en route. But beyond any particular features of humans or machines lay Wiener's deep-seated commitment to a behaviorist vision of both. His was not a claim that no criteria differentiated humans and machines. Quite obviously there was no machine that could (as yet) write a Sanskrit-Mandarin dictionary; and, similarly, no living organism rolled on wheels. But it was the behaviorist impulse to focus on broad classes of actions, and to do so on the basis of the input and output he knew so well from communication technology, that led Wiener to his blurring of the man-machine boundary. *Black boxes,* as Wiener used the term, meant a unit designed to perform a function before one knew how it functioned; *white boxes* designated that one also specified the inner mechanism. In this language, the more sophisticated feedback mechanism of the AA predictor opened a new universe of black boxes to the engineer—and to the philosopher.[14]

Behaviorists took note. The eminent Harvard psychologist and historian of psychology Edwin Boring found Wiener's suggestion that all functions of the brain might be duplicated by electrical systems "very attractive." Having had a chance to contemplate this circuit-reductionist program, Boring reckoned in a 13 November 1944 letter to Wiener that he could provide "a pretty complete list of psychological functions" in his spare time, all psychological facts being in principle expressible in terms of stimulus and response. "A symbolic process" would be "a delayed, adequately differential reaction"; "introspection" would be a reaction to a reaction, and Wiener's task, should he decide to accept Boring's challenge, would be to transfer these stimulus-response pairs and respond with his own matching catalogue that would give the "same specificity of 'output' to 'input.'" With fourteen psychological properties on the list already and others like "Generalization" and "Abstraction" to be added, Boring assured Wiener that a paper with these electrical designs would greatly benefit "us operationally-minded psychologists." "Is it a go?" Boring queried. "I do not know that you can [do it], but I should be betting on you" (Boring, 1944). Black-box engineering now had something more complex than electrical amplification as its functional goal: to re-create the mind itself.

Within a few weeks, Wiener's ambition left behind even the human mind. Collaborating with Howard Aiken, one of the pioneers in computer technology, and with von Neumann, the supremely versatile mathematician then at work on the computer, Wiener sent out a restricted letter on 4 December 1944 to a collection of seemingly unrelated experts:

> A group of people interested in communication engineering, the engineering

> of computing machines, the engineering of control devices, the mathematics of time series in statistics, and the communication and control aspects of the nervous system, has come to a tentative conclusion that the relations between these fields of research have developed to a degree of intimacy that makes a get-together meeting ... highly desirable. (Aiken, von Neumann, and Wiener, 1944a)

Because many of the relevant developments were directly tied to the war effort, Wiener asserted, the assembly would necessarily be nonpublic. It was a new vision of the world that was to emerge from this secret confluence of war sciences, one that would embrace matters of "engineering, physical, and even economic and social interest." (*ibid.* See also Wiener, 1944). Wiener, Aiken, and von Neumann named the group the "Teleological Society."[15]

The first meeting of the Teleological Society took place on 6–7 January 1945, and Wiener was delighted with its outcome. Rafael Lorente de No and Warren McCulloch, both physiologists specializing in the functional organization of the central nervous system, presented their work on the organization of the brain. "In the end," Wiener gushed to Rosenblueth, "we were all convinced that the subject embracing both the engineering and neurology aspects is essentially one" (Wiener, 1945c). It was time, Wiener contended, to turn separate avocations into an integrated, permanent research program, one that would be backed by many sources. These included the Rockefeller Foundation, with Weaver's support, along with "mysterious words from von Neumann concerning ... some thirty mega-bucks" from which powerful resources could be "siphoned" (*ibid.*).

For his part, von Neumann used the meeting to set up a division of labor: Wiener and Walter Pitts (a logic student of Rudolf Carnap who had used logic to analyze the switching properties of neurons) would cover filtering and prediction problems such as the prototypical AA predictor; the mathematical statistician W. Edwards Deming (who would become a principal advisor to Japan during its postwar economic miracle), von Neumann, and several others would cover the application of fast, mechanized computing methods to statistical problems; the application to differential equations (astronomy, hydrodynamics, ballistics, and so on) would come from Aiken, H. H. Goldstine, and von Neumann; and the neurological features would go to de No, McCulloch, and Pitts (John von Neumann, 1945b). This was fine with Wiener, though he found von Neumann's sketch to lack the crucial transition from the computing machine to the control machine. "The issues that come up here are those of transfer from continued data to counted data; of the final transition from counted data to the motion of a shaft effector; and the sensing of the motion of the effector by feedback or other quasi-proprioceptor apparatus." Such a feedback system, which Wiener had stressed from his earliest work on servomechanisms, continued to occupy a central place in his thinking. For it was this same proprioceptive process that occurred in mechanical controls, organic controls, and in hybrid mechanico-organic systems. Von Neumann ceded the point (Wiener, 1945d; see John von Neumann, 1945a).

Despite the fantastic array of supporters that Wiener's approach elicited, there was resistance. In 1950 Richard Taylor, a young philosopher from Brown University, asked with incredulity if Wiener and his collaborators could seriously be proposing a definition of purposefulness that was built purely on the culmination of a sequence of events. Rosenblueth, Wiener, and Bigelow's definition was this:

> The term purposeful is meant to denote that the act or behavior may be interpreted as directed to the attainment of a goal—i.e., to a final condition in which the behaving object reaches a definite correlation in time or in space with respect to another object or event. Purposeless behavior then is that which is not interpreted as directed to a goal. (Rosenblueth, Wiener, and Bigelow, 1943, p. 18)

To Taylor this definition was both so all-encompassing as to rule out nothing and so devoid of content that it had no overlap with any common meaning of the term. Let a clock run for many years only to break down at midnight on New Year's Eve. What rules this out as an instance of purposefulness? A brick tumbles off a building, killing a passer-by. Is this, too, to be considered as purposeful? For Taylor, the utter arbitrariness imposed by the clause "may be interpreted" makes it all too easy to allow these tumbling bricks and failing clocks to be counted as purposeful, voiding the term of any similarity with our usual understanding (R. Taylor, 1950a, p. 311).

Of course Taylor recognized that Wiener and Bigelow wanted to lay special stress on self-regulating machines. This did not move him. Could the authors really be claiming that a roulette wheel—by construction a purposeless device—could be rendered into a purposeful machine by the addition of a lead weight on its perimeter? Even the guided missile, that paragon of purpose, is hardly to be philosophically distinguished from non-self-regulating devices. Consider a missile following mechanically along a taut wire attached to a target. Such a mechanism might be less etherial than radar guidance but would hardly be distinct insofar as it could be considered self-regulating. "The expression 'target-seeking missile' is," Taylor concluded, "metaphorical." Wiener and Bigelow might choose to redefine the very concept of purpose, but their discovery would amount to no more than the redefinition of the plus sign with that of multiplication: "entirely correct, but scarcely significant" (*ibid.*, pp. 316, 317).

In a joint postwar response to Taylor, Wiener had no apologies for classifying a crooked roulette wheel as purposeful. But he and Rosenblueth (who collaborated on the riposte) reemphasized that the weighted wheel and the magnetic compass differ from the servomechanisms of guided missiles and AA predictors because the former are passive whereas the latter are active. In the laboratory, Wiener and

Rosenblueth insisted, the physics and engineering practices behind self-regulating systems are utterly different from that of bricks and clocks; the former are governed by time-reversible causal stories whereas the latter are unidirectional in time. (The AA predictor, for example, makes its statistical forecast on the basis of the *history* of the pilot's past performance. See Rosenblueth and Wiener, 1950, p. 328).

Offering a cornucopia of war-related electromechanical feedback systems —guided missiles, target-sensing torpedoes, and radar trackers among them— Wiener and Rosenblueth saw pragmatically defined novelty where Taylor saw none.[16] But beyond the issue of novelty, Taylor had emphasized what was fundamentally an objection to the behaviorist input-output analysis that underlay the Wiener program. And here the two sides found no meeting of minds. Explaining that they did not care whether, in the abstract, machines "are or can be like men," Rosenblueth and Wiener insisted that the question was "irrelevant" for scientific objectives:

> We believe that men and other animals are like machines from the scientific standpoint because we believe that the only fruitful methods for the study of human and animal behavior are the methods applicable to the behavior of mechanical objects as well. Thus, our main reason for selecting the terms in question was to emphasize that, *as objects of scientific enquiry, humans do not differ from machines*.[17]

We should read this last remark critically and historically. In 1941 and 1942, it had made sense to Wiener and his collaborators to view humans, qua pilots and gunners, as undifferentiated from the bombers and antiaircraft units in which they fought. Seen as man-machine *enemies*, from a military perspective "humans do not differ from machines." It was then a short step from viewing the enemy as a cybernetic entity to seeing the quasi-automated Allied aircraft gunner the same way. What had begun in the Manichean field of science-assisted warfare had now been decontextualized. By 1950, Wiener had globalized his claim: under the gaze of scientific inquiry, human intentionality did not differ from the self-regulation of machines, full stop.

Taylor rejoined: How could Wiener and Rosenblueth base a notion of purposefulness on *observable* behavior alone (Taylor's emphasis), ignoring the blatant distinction between the various intentions behind the observation that a car is following a man? Is the driver trying to run the man down? Making a joke? Trying to frighten him? Or simply veering to rid his car of a pesky bee? Taylor concedes that observation may well be our best or even only evidence, but surely we want to distinguish between the definition of purpose and the evidence we may or may not have to ascertain what that purpose is. Further complicating the purely behaviorist notion are cases in which the goal, as a distinct physical entity, does not even exist: knights seek the Holy Grail, alchemists pursue the philosopher's stone, and people stumble around in the dark looking for matches that are not there. Intention, like desire, is something that is just as real as more tangible acts. Taylor protested: We should not abandon concepts simply because they are not operationally useful to science (see R. Taylor, 1950b).

While Wiener let the debate end at this point, the fundamental conflict remained unresolved. With the mathematics of cybernetic feedback systems, the formalism of game theory, and the flow charts of operations analysis, the Manichean sciences had, in a sense, reached the apotheosis of behaviorism, as Boring had hoped. In fact, in February 1945, despite a recent and dramatic collapse due to an ulcer, Boring tried to deliver on the side of the bargain he had proposed to Wiener three months earlier. Riveted by the progress of Wiener's interdisciplinary research, Boring attested: "What I had done before [his collapse] is to make out a list of what I thought all the functions of the brain are, putting them in positivistic reaction terms of the organism, terms which could be translated into in-put, out-put and adjustment of a mysterious box with binding posts and knobs on it." His conclusion: "That's about what a person is, really" (Boring, 1945). W. Ross Ashby writing from England had similar panegyrics for Wiener's black-box program:

> When I consider how the psychologists have been trying to solve exactly this problem for decades (if not for centuries), the black box being the brain, and when I think how little attention they have given to the principles involved, my opinion of psychologists falls to a new low. The trouble with the psychologist is that he is too proud to learn to walk before he tries to run. So today he lies on his back, foolishly waving his legs, and pretending to be a ballet-dancer, when in reality he hasn't yet learned how to crawl. For this reason I regard it as highly complimentary when I say that your study of the "black box" problem is a first step towards a scientific psychology![18]

Wiener, unlike Boring, thought he could actually make the hardware that would put the specific black-speckled boxes on the table. Such a radical position necessarily left unsatisfied those like Taylor who could not abide the elimination of inner states of human intention, desire, pleasure, and pain in favor of purely observable manifestations. But with the power of wartime materiel and the glittering promise of future industrial riches, it was clearly not Taylor's view that prevailed.

4. The Philosophy of Nature and the Delivery of Cannon Fire

If humans do not differ from machines from the "scientific standpoint," it is because the scientific standpoint of the 1940s was one of men-machines at war. The man-airplane-radar-predictor-artillery system is a closed one in which it appeared possible to replace men by machines and machines by men. To an antiaircraft operator, the enemy really does act like an autocorrelated servomechanism. What is astonishing is the globalization of this technological *aperçu* into a new age for humanity and a general philosophy of human action. In 1947, as Wiener reflected on the events of the war, he divided the thoughts of the ages into three epochs. A first era was characterized by the clockmakers, surveyors, and planetary astronomers. Their science was one of prediction by laws and their economy that of the merchant. Boats sailed across seas based on the clocks and astronomical calculation of longitude; this was, as Wiener put it, the "engineering of the mercantilist" (*C*, p. 38). As the seventeenth and eighteenth centuries drew to a close, Wiener asserted, a new day dawned in which clocks gave way to the steam engine as the symbol and real center of technological work. Huygens and Newton ceded their place to Rumford, Carnot, and Joule, and it was the manufacturer not the trader who embodied the new culture. Finally, for Wiener, the present age, ushered in by the vast array of electromechanical devices of the war, was the age of information and control. If these developments reached back to Kelvin and Gauss, they found their real form (and interpreters) only in the laboratories and factories of radar and its associated systems. This age, our age, was that of the servomechanism.

As Wiener argued, each age engendered its own simulacrum of humanity—clockmakers of the eighteenth century made their pirouetting mechanical figures, steam engineers of the nineteenth glorified their engines as versions of the body. Our age? We make computers to calculate differential equations, open doors with photocells, and, not surprisingly, "the present automaton … points guns to the place at which a radar beam picks up an airplane" (*C*, p. 40). In a sweeping totalization Wiener had, within two years of the end of the war, elevated his AA predictor to the symbol for a new age of man. Whether or not we accept Wiener's techno-periodization of the history of humanity, there seems little doubt that he and many of his contemporaries saw themselves as standing at a historical and philosophical watershed in which the Manichean sciences would undergird the cybernetic age.

To a certain extent, Wiener's hopes and fears for cybernetic technologies were in place before Hiroshima and Nagasaki, but they were multiplied one hundredfold by the August 1945 nuclear bombing of Japan. In the weeks following the atomic blasts, Wiener was too distracted even to respond to a letter from his friend and collaborator, the philosopher Giorgio de Santillana. Finally, in October 1945, Wiener put pen to paper:

> Ever since the atomic bomb fell I have been recovering from an acute attack of conscience as one of the scientists who has been doing war work and who has seen his war work a[s] part of a larger body which is being used in a way of which I do not approve and over which I have absolutely no control. I think the omens for a third world war are black and I have no intention of letting my services be used in such a conflict. I have seriously considered the possibility of giving up my scientific productive effort because I know

> no way to publish without letting my inventions go to the wrong hands. (Wiener, 1945a)

In short, almost telegraphic prose, Wiener reported to de Santillana on the full range of his cybernetic work, ranging from wave filters and predictors, computing machines, automatic control of assembly lines, and control of chemical plants to random nets of switching devices, quantum theory, and cardiac fibrillations. But, agonizing over the possible uses of his research, he halted his letter with these "sketchy" descriptions in part because of a claimed modesty and in part because "in these troubled times I do not feel any too certain that I shall continue in science indefinitely. I do not know how to publish work without making it available for the strongest hands and I do not like the strongest hands of the present time. I feel it most intensely personally and in particular what I have seen in looking upon a world completely inadequate to receive the atomic bomb" (*ibid.*). Two days later, he drafted a letter to the president of MIT, Karl T. Compton, in which he rehearsed his fears about the scientist's loss of "control" over the civil and military uses of science, concluding that he "[intended] to leave scientific work completely and finally. I shall try to find some way of living on my farm in the country. I am not too sanguine of success, but I see no other course which accords with my conscience" (Wiener, 1945b). In the years that followed, Wiener repeatedly stressed the power of cybernetics to save, enslave, or destroy humanity. Already built into the AA predictor and its progeny was a set of cultural meanings not easily shed.

Nineteen forty-seven closed with Wiener still at MIT, despite his moral discomfort with the technical possibilities of cybernetics. In handing over the technology to what he called "the world of Belsen and Hiroshima," he could only hope to "confine our personal efforts [in cybernetics] to those fields, such as physiology and psychology, most remote from war and exploitation" (*C*, p. 28). Paradoxically, during the war Wiener had extended the cybernetic vision beyond its narrow applications because of the weakness of the AA predictor; now that he associated cybernetics with the power of cataclysmic weapons, he tried to push cybernetics away from the military arena because of its deadly efficacy. Either way, for Wiener and many colleagues, the association of cybernetics with its wartime origin was forcefully and deeply inscribed in the cultural meaning of the new science and its machines.

The Josiah Macy, Jr. Foundation opened the cybernetic age for the social sciences. On 8 and 9 March 1946, the foundation gathered psychologists and anthropologists to meet with mathematicians and physicists on the general subject of circular causal systems. Gregory Bateson, already persuaded of the importance of the new ideas, led the contingent of non-physical scientists and helped organize the second such meeting, "Teleological Mechanisms in Society," on 20 September 1946, and a third, "Feedback Mechanisms and Circular Causal Systems in Biology and the Social Sciences." Those invited included Paul Lazarsfeld, Margaret Mead, and F. S. C. Northrop, among many others. Backed strongly by Bateson and enthusiastically led by Wiener, von Neumann, McCulloch, and de No, the group's intense discussions brought systems, information theory, and feedback mechanisms onto the center stage of sociology, psychology, and anthropology.[19] Northrop later acknowledged the impact of servomechanical theory as "of revolutionary significance for natural science, moral as well as natural philosophy, and for one's theory of the normative factor in law, politics, religion, and the social sciences" (Northrop, 1949, p. 414). To Bateson, the new vocabulary of communication theory and cybernetics presented a turning point in his work; his biographer David Lipset called it a "theoretical conversion" in which his older terms, such as *schismogenesis,* were reworked into the language of the purposeful machine: "regenerative feedback" (Lipset, 1980, p. 182. See also Heims, 1977).

While reaching out to the social sciences, Wiener also wanted to raise the Manichean sciences to a more abstract philosophy. At least since the early 1930s, Wiener had held a deep interest in Leibnizian philosophy. He extolled Leibniz's philosophically open mind (as opposed to the Newtonians' dogmatism), he celebrated Leibniz's commitment to relativity, to the quantum mechanical-like identity of indiscernibles, even to the idea of monadic self-containment (by analogy to certain higher-dimensional theories of the electron. See Wiener, 1932). But in the years after the war, Wiener saw more in Leibniz. He extracted an overarching philosophical umbrella that covered and combined cybernetics *and* operations research.

Both cybernetics and operations research, he told the Operations Research Society in 1953, were grounded in a modern parallel to Leibnizian monads. Leibniz's own conception of monads are, Wiener assures us, far too anthropomorphic. It was a world picture in which "monads [were] quasi-souls whose activity was confined to the mirroring of the universe of the monads themselves." Cybernetics provided "a similar world-picture": nodes of communication interact by the exchange of orders or commands. According to the cyberneticist, the world is nothing more than the mutual internal relations of these incoming and outgoing messages—ultimately cybernetics carries, on Wiener's own account, a "quasi-solipsistic" vision of the universe. Taken in its epistemological function, cybernetics can be either observational (purely incoming messages) or experimental (incoming and outgoing messages). At the same time, Wiener wanted to make plain that while epistemology may well capture the knowledge-gathering function of the science, cybernetics will not rest there: "messages may be sent for the purpose of exploring the universe, but they may also be sent with the intention of *controlling* the universe." Precisely because Wiener wanted to accentuate the dual aspect of information, he distinguished between messages that could be sent "in the indicative and the imperative mood" (Wiener, 1953a, pp. 2, 3).

As the windowless monads suggest, and as Wiener's own proclamation of quasi-solipsism made explicit, the cybernetic philosophy was premised on the opacity of the Other. We are truly, in this view of the world, like black boxes with inputs and outputs and no access to our or anyone else's inner life. This same opacity prevails in von Neumann's game theory, where the opponent acts according to certain universal maximization principles but where the thought process that eventuates in any given move is hidden from us. Although in his later life Wiener came to reject von Neumann's game theory as containing an inadequate psychological basis (Heims, 1980, pp. 305–310), in the years directly after the war, he sympathized with the project, even identifying it as being of the same "spirit" as cybernetics.[20]

The impact of the Manichean sciences not only on computation and automata theory but also on the social sciences should not be underestimated. For Mead, Northrop, and Bateson, the impact of Wiener's models of feedback and homeostasis became essential components of their analyses. Even *Time* saluted Wiener in 1950 as one of the leaders of the new "computermen" who were blurring the boundaries between the wet sciences of the brain, psychological properties, and the machine (caricatured in fig. 5). Given such adulation, it is perhaps not too surprising to find many social scientists identifying themselves with the new sciences emerging from the war. The social scientists' fascination with systems in the 1940s and 1950s may have roots in older turn-of-the-century networks of telephony and power. Recent fascination with information-based feedback systems, however, tracks its roots more proximately—to the radar and tracking systems of World War II.

There is a more contemporary phase in this continuing history of the Manichean sciences. More startling than the continuity between wartime and postwar systems theory is the role Wiener's cybernetics played and continues to play in postmodernist discourse. Beginning with Jean-François Lyotard, whose *The Postmodern Condition* (1979) is often counted among the founding documents of postmodernism, we can trace a continued role for cybernetic theories. Lyotard himself rather nervously contended that his social analysis, produced for the government of Quebec, departed radically from cybernetics. On the contrary, I want to argue that the link between the two is profound and the continuity nearly complete.

First, Lyotard asserted that, contra cybernetics, "messages have quite different forms and effects depending on whether they are, for example, denotatives, prescriptives, evaluatives, performatives, etc." Here at least two of Lyotard's categories (denotative and prescriptive) directly parallel Wiener's distinction between the indicative and imperative moods of messages. Secondly, Lyotard contended that "a cybernetic machine does indeed run on information, but the goals programmed into it [leave no way] to correct in the course of its functioning ... its own performance." Nothing could be further from the mark. This self-correction is *exactly* what Wiener's machines did. Indeed, in every piece of his writing on

FIGURE 2.5: The cybernetic entity. *Time* featured this wonderfully literal representation of the ultimate cyborg. Note its three best features: military threads, feedback as it watches its output, and biomechanical form. Cover, *Time*, 23 Jan. 1950.

cybernetics, from the first technical exposition of the AA predictor before Pearl Harbor up through his essays of the 1960s on science and society, Wiener put feedback in the foreground, returning again and again to the torpedo, the guided missile, and his antiaircraft director. Moreover, even in the predictor *both* the performance *and* the rules governing performance were corrected "in the course of its functioning." Third, Lyotard found the "trivial cybernetic version of information" to miss the decisively important "agonistic aspect of society." On his postmodern vision of social relations, Lyotard saw each "player" as undergoing a "displacement" as he sent and received messages. "Moves" and "countermoves" characterize Lyotard's world, as he insisted that what we need for postmodern understanding is an "agonistic" theory of communication and a theory of games (Lyotard, 1984, p. 16).

But it was on the agonistic field that Wiener, von Neumann, and the operational analysts were most at home. Formally, militarily, and philosophically, theirs was a universe of confrontation between opponents: Allies to Axis, monad to monad, message to message, and mechanized "man" to servomechanical enemy. The opposition between Lyotardian postmodernism and the "trivial" cybernetic vision (presumably of Wiener) is unsustainable (*ibid.*, p. 16).

From this continuity between cybernetic and Lyotardian postmodernist social relations, two things might follow. We could conclude that Wiener and his allies were postmodernists *avant la lettre*. Or, as I incline to believe, it might be the other way around: we track Lyotard's postmodernist and game-theoretical worldview back deep into the heart of the Manichean sciences. As we study the development of postwar science, then, it seems to me of utmost importance not to seize uncritically the central metaphors of operational analysis, game theory, and cybernetics and make them our own while claiming all the while a new "postmodern" periodization.

Donna Haraway invoked cybernetics in a more subtle, yet still conflicted *postmodern* way. In "A Cyborg Manifesto: Science, Technology, and Socialist-Feminism in the Late Twentieth Century" (1979), she used the term *cybernetics* to characterize post-World War II biological sciences in terms, and with a periodization, that Wiener would have recognized. Before the war (according to Haraway) biological discourse had been organized around the organism viewed through the categories of medicine and the clinic. These included intelligence testing, human relations, physiology, and racial hygiene. After the war, the new sciences of information- and control-dominated systems reshaped biology, including sociobiology. This new, more cybernetic biology emphasized communication and feedback. For Haraway, E. O. Wilson's work typified the latter set of developments with his stress on information transfer among insects, including efficiency, noise, and capacity (Haraway, 1991c, pp. 44–45). In her view, cybernetics, although often used to sanction the status quo, is ultimately far more open to a new and more liberating vision of the biological sciences than the psychobiological and organic functionalist theories that preceded it. The cybernetic biological view (sociobiology) is, in Haraway's view, *less* open to racism or sexism because in cybernetics the organic body is depicted as an engineering entity, always modifiable, and never defined essentially (Haraway, 1991c, p. 67).

Haraway opened "A Cyborg Manifesto" (1985) with a partial, ambivalent continuation of these Wienerian themes: "A cyborg is a cybernetic organism, a hybrid of machine and organism, a creature of social reality as well as a creature of fiction" (Haraway, 1991a, p. 149). I say the continuation is partial and ambivalent because the cultural meaning she struggled to ascribe to the communication and information technologies is utterly different from the cultural meanings that emerged from cybernetics.[21] Haraway alluded to the "cyborg orgy" that she saw "coded by C^3I, command-control-communication-intelligence, an $84 billion item in 1984's US defence budget." Just this cyborg root in military feedback systems is, she allowed, the "main trouble" with cyborgs: "But illegitimate offspring are often exceedingly unfaithful to their origins. Their fathers, after all, are inessential" (*ibid.*, pp. 150–151). Can the cybernetic vision be so easily detached from its military historical origins and present location? After all, the very notion of a cyborg issued from an Air Force contractor's extension of Wiener's ideas.[22] I would argue that the associations of cybernetics (and the cyborg) with weapons, oppositional tactics, and the black-box conception of human nature do not so simply melt away.

For the classic cyberneticists (exemplified by Wiener, Rosenblueth, McCulloch, and their colleagues), the blurred boundary between human and machine opened an infinity of possibilities; Haraway, like Wiener, stressed the possibility that machines could be open-ended, nondedicated in their function, and able to reproduce, learn, and interconnect with the human. But Wiener, unlike Haraway, saw power and control as absolutely central to the very definition of cybernetics, for better or worse. Indeed, by the end of his life, as if to push this theme to its theological *Endstation,* Wiener had come to see the human-machine relation as a model, if not an incarnation of the bond between God and "man." The paradoxes of religion ("Can God create a rock too great for him to move?") reemerged as questions about the cyberneticist and his offspring ("Can a human create an entity that can beat him at chess?"). On the last lines of the last page of his last book, Wiener put it this way: "Since I have insisted upon discussing creative activity under one heading, and in not parceling it out into separate pieces belonging to God, to man, and to the machine, I do not consider that I have taken more than an author's normal liberty in calling this book

GOD AND GOLEM, Inc." (Wiener, 1964, p. 95)

We who make cyborgs are, in the end, like gods.

Haraway, by contrast, took the variability, the unfixed nature, of the cyborg as grounds for the *partiality*, not the *omnipotence*, of what is human. As she put it, we are ourselves already in so many respects cyborgs—through our reproductive technologies, our psychopharmacologies, our prostheses (mechanical and computational)—that we can no longer put any stock in essentialist definitions of the classic dichotomies of mind and body, animal and human, organism and machine, public and private, nature and culture, men and women, primitive and civilized (Haraway, 1991a, p. 163). I understand her project to resonate with the more critical branch of postmodern theory: a refusal to espouse a nostalgia for a "natural" or "feminine" world that preexisted technology and a concomitant move to use (rather than simply shun) the built world of technology and science. Postmodernism holds cybernetics in an uneasy embrace. As a postmodernist challenge to a fixed human, racial, or gendered nature, the cyborg presents an alternative, a way out. But (as Wiener and Lyotard attest in different ways) the successes of cybernetics in blurring the human and nonhuman have been most striking in the agonistic field, if not the battlefield itself; the choice between fighting Augustinian and Manichean enemies, as Wiener pointed out, is merely one of tactics. In choosing the cyborg to lead the flight from modernism, one risks reducing the picture of human capacities to one of tactical moves and countermoves in a metaphorical extension of automatic airwar.

Whether we accept or reject the ontology of the Manichean sciences, in discussing the technologies of cybernetics we find ourselves in the grip of a powerful set of cultural meanings. By this, I do not mean that feedback systems were born (so to speak) with a full complement of symbolic associations. As with any set of artifacts, it is possible to trace back fragments of servomechanisms, game theory, and operational reasoning long before 1940. One can cite, as Wiener often did, fragments by James Clerk Maxwell, Leibniz, and many others who attended to issues of self-regulation, interconnection, and communication. Wiener, for example, knew perfectly well that the nineteenth century had a well-developed theory of the steam engine governor, and by the 1920s electrical analogues in the form of voltage regulators were legion.

As Otto Mayr has so exhaustively demonstrated, pre-twentieth century feedback devices were culturally located quite differently from systems discussed here. Known in the golden age of Islam, feedback mechanisms—especially liquid-level regulators—flourished in antiquity. Then, from the Middle Ages through the baroque period, the technology vanished almost completely in Europe. Clocks, not self-regulating machines, held pride of place. Timekeeping machines served as a cultural symbol of authority; these were the mechanisms that appeared everywhere, celebrated from literature and poetry to philosophy and political theory. According to Mayr, "the authoritarian conception of order was directly and patently shaped by society's experience with the mechanical clock."[23] When the feedback device came back into European favor in the seventeenth and especially in the eighteenth century, it did so not on the Continent but in the British Isles, a manifestation, in Mayr's view, of the "liberal attitude" that at one and the same time shaped the "socio-intellectual" and the "technological" sides of culture. Regulating devices, especially as popularized by Watt's incorporation of the governor into the steam engine in the 1780s, were celebrated alongside political rhetoric of "dynamic equilibrium," "self-regulation," "checks and balances," and "supply and demand" (*A*, p. xviii).

But there are differences, crucial differences, between these devices and the wartime work at MIT or Los Alamos. First, as Mayr points out, there was no early modern entity coextensive with the abstraction *feedback systems*. There were fluid regulators, steam-engine regulators, and windmill governors aplenty—but no notion of these constituting a collocation defined by an abstract causal loop.[24] Second, neither the political theorists nor the inventors of the seventeenth and eighteenth centuries made explicit any link between the two domains of self-regulation rhetoric. Consequently, while Mayr conclusively demonstrates the coperiodization of feedback talk in technology and politics, historiographically the bond between them remains that of an acausal Zeitgeist, albeit one located (spatially and temporally) with the onset of "liberal attitudes" and a "liberal conception of order." He concludes: "About the details of the causal nexus between the advent of the liberal conception of order and the rise of self-regulating mechanisms in technology we are reduced, at this point, to speculation" (*A*, p. 199). The links holding the disciplines of cybernetics together need not be so speculative.

What we have seen in Wiener's cybernetics is the establishment of a field of meanings grounded not through Zeitgeist but explicitly in the experiences of war. For however far telephone relaying technology or A. N. Kolmogoroff's statistics had come before the war, it was the mass development and deployment of guided missiles, torpedoes, and antiaircraft fire that centralized the technology to scientists and engineers. To the thousands of servicemen who used and faced this new generation of weapons, the "human" character of self-regulating machines seemed all too human. After all, trying to shoot down a Junkers JU 88 heading for London or a V-1 buzz bomb doing the same thing was not all that different. A skipper trying to dodge a self-guided torpedo could be excused for referring to the device as "trying" to kill him, as could the pilot ascribing airfoil self-adjustment to the work of "gremlins." And in the specific case of Wiener, Bigelow, Weaver, and their colleagues, it is perhaps understandable that the pilot of an enemy plane could be said to "behave like a servo-mechanism." While prewar behaviorists might have cautioned against the ascription of internal states, war made it impossible; reading the hidden enemy meant reading his actions. In the mechanized battlefield, in those life-and-death confrontations with an enshrouded enemy, the identity of intention and self-correction was sustainable, reasonable, even "obvious."

Face to face with another person, with no way to avoid the full depth and ambivalence of human interaction, feedback may seem (as it did to Taylor) to be a ludicrously simplistic representation of intentionality. But to Stibitz and Weaver, as they stood in Wiener's MIT laboratory that July day in 1942, the system of simulated pilot and AA predictor was positively "uncanny" in its capacity to predict a pilot's next move. World War II elevated the stakes of understanding the enemy's intention to survival itself; it stripped human behavior to moves of pursuit, escape, and deception; and it introduced a new class of self-regulating weapons. It is in this specific context that the identity of intention and self-correction was forged.

Symbols matter: it counted for a great deal in the reception of cybernetics that its war applications were lethal, or potentially so. After all, as Wiener himself recognized, much of the theory of servomechanisms (time series autocorrelation, for example) had been developed before the war in nonmilitary contexts, and a great deal more had come from the "pure" statistical investigations of Kolmogoroff. Would cybernetics, information theory, and "systems thinking" have proved such a central and enduring metaphor without combat? Would the pervasive postwar ontology of the enemy have had such a runaway success without the seduction of victorious military power? I doubt it. Without the specific lived horror of V-2s raining down on London, the air-dropped torpedoes ruthlessly diving for Japanese warships in the South Pacific, or the fire-controlling AA predictor that promised to stem the enemy barrage, it is hard to imagine the new technologies appearing so much *like* human pilots and gunners that the two could be conflated. In World War II, the mechanized soldier faced his opponent as a machine, and machines manifested themselves as people. After the AA predictor but before Nagasaki, Wiener responded to the newfound moral, political, and industrial power of automatic control like this: "It ... occurred to me that we were here [with self-regulating machines] in the presence of another social potentiality of unheard-of importance for good and for evil" (*C*, p. 27). After the bomb, as we have seen, Wiener's association of danger and moral weight with cybernetics grew even stronger. War gave the new cybernetic technologies a role to play in the Manichean drama of the world. Mere governors, thermostats, and voltage regulators could not usher in a cybernetic age—weapons could.

In general, the cultural meaning of concepts or practices, I would argue, is indissolubly tied to their genealogy. To understand the specific cultural meaning of the cybernetic devices is necessarily to track them back to the wartime vision of the pilot-as-servomechanism. In the air-ground battle, it was a short step for Wiener and Bigelow to take the pilot-as-servomechanism directly over into the AA gunner-as-servomechanism and thence to the operation of the heart and proprioceptive senses. From the body, it was us more generally—we humans—whose intentions could be seen as none other than self-correcting black-boxed entities and finally nature itself that came to be seen as a correlated and characteristic set of input and output signals.

If this cybernetic conception seems to differ from more familiar conceptions of the Other, it should. The cybernetic Enemy Other has little to do with the racialized Other so horrifyingly invoked by Blarney, and examined, for example, by Edward Said in *Orientalism* (1978). There is no sense in which Wiener sees the German bomber pilot as a racially lesser being. Nor is the German pilot an Other in being simply invisible. Finally, I take it to go without need of much elaboration that the servomechanical pilot is not Emmanuel Levinas's Other, where the recognition of the ineradicable humanity outside of oneself is the fundamental move in the establishment of an ethical philosophy (Levinas, 1989). No, Wiener's conception of the Enemy Other is more like his depiction of the game players in von Neumann's theory: "perfectly intelligent, perfectly ruthless operators" (*C*, p. 159). This is a theoretical representation in which information, statistics, and strategies are applied to moves and countermoves in a world of opposing but fundamentally *like* forces.

Surprisingly, the cybernetic Other is not negatively contrasted with us, nor are we the model upon which the Other is empathetically formed; our understanding of the cybernetic Enemy Other becomes the basis on which we understand ourselves. Wiener's image of the human and natural world is, in the end, a globalized, even metaphysical, extension of the epochal struggle between the implacable enemy from the sky and the Allies' calculating AA predictor that did battle from the ground. It is an image of human relations thoroughly grounded in the design and manufacture of wartime servomechanisms and extended, in the ultimate generalization, to a universe of black-box monads.

In principle, can the cultural meanings of feedback systems be disassociated from the origins of the technology? Of course. After all, early modern "self-regulation" rhetoric spanned steam-engine governors and liberal political economy; feedback rhetoric in 1946 bound together very different meanings. Another instance of shifting cultural meanings comes from flying itself. During the 1920s, the airplane and "airmindedness" came, in America, to stand for a concatenation of individual freedom, religious renewal, and the inauguration of a new, modern, international era. In Germany, at nearly the same time, opposite associations prevailed: gliding and then powered flight carried with them an inextricably nationalist association; airmindedness meant a self-conscious revolt against Versailles and a constant reassertion of community over individualism.[25] Of interest is not the mere identification of associations, but the cultural historical account of their assembly, persistence, and deconstruction.

Cultural meaning is neither aleatory nor eternal. We are not free by fiat alone to dismiss the chain of associations that was forged over decades in the laboratory, on the battlefield, in the social sciences, and in the philosophy of cybernetics. At the same time, it would clearly be erroneous to view cybernetics as a logically impelled set of beliefs. Nothing in the feedback device implies a representation of human beings as behavioristic black boxes; nothing in the mathematics entails by deduction alone a universe reducible to Wiener's monadic input-output analysis.

What we *do* have to acknowledge is the power of a half-century in which these and other associations have been reinstantiated at every turn, in which opposition is seen to lie at the core of every human contact with the outside world.

As a twenty-year old, Wiener published his first philosophical paper opposing the notion of a "highest good" and underlining the irreducible internal and external conflicts that fixed our notion of morality. When disputes could not be settled by reason, young Wiener wrote, "the conflict can be settled, if at all, only by the suppression by brute force of the disputant or disputants on one side" (Wiener, 1914; rpt. Wiener, 1985, v. 4, p. 49). Differing from Hobbes (Wiener did not believe that people were fundamentally selfish), he nonetheless saw morality as a conflict not resolved in the distant past but as continuing into the here and now. Such a relentless struggle continued in the cybernetic *Weltanschauung*, though it took a new, scientific, and more subtle form, embracing not only morality but our relation with the world itself. Wiener queries whether the world is an active (Manichean) opponent or merely a passive (Augustinian) antagonist, the only difference being that the "Manichean devil" used tricks, craftiness, and dissimulation against us, while the "Augustinian devil" did not change methods: "The difference between these two sorts of demons will make itself apparent in the tactics to be used against them" (*HU*, p. 35). To Wiener, the essential and unrelieved reality of the world was that the individual lived in isolation, struggling (searching for tactics) to create order out of chaos. Science itself, as it faced nature, was such a battle: " The scientist," he declared late in life, "is always working to discover the order and organization of the universe, and is thus playing a game against the arch enemy, disorganization. Is this devil Manichean or Augustinian? Is it a contrary force opposed to order or is it the very absence of order itself?" (*HU*, p. 35). Cybernetics, that science-as-steersman, made an angel of control and a devil of disorder.

But perhaps disorganization, noise, and uncontrollability are not the greatest disasters to befall us. Perhaps our calamities are built largely from our efforts at superorganization, silence, and control.

NOTES

1. Quoted in Hastings, 1979, p. 146. Other recent literature on the moral, political, and military aspects of bombing civilians includes Sherry, 1987, which argues that the slide towards civilian bombing led inexorably to the use of atomic weapons against civilian targets, and Crane, 1993, esp. pp. 53–59, which tracks aviators' attitudes in the field as differing according to their specific tasks: piloting, navigating, or bombing. Distancing, however, had to be maintained, often under immense stress. One pilot (who was killed in

November 1944) wrote shortly before his death: "The whole idea was to blow up just as much Germany tomorrow as possible. From way up high, it wouldn't mean a thing to me. I wouldn't know if any women or little kids got in the way." Such remarks were followed in the very next sentence by doubts: "I'd thought about it before, but that night it was close. The more I thought of it, the uglier it seemed" (Stiles, 1947, p. 21).

2. On the origins of operations analysis, see Fortun and Schweber, 1993; and Rider, 1988; 1992; 1994.

3. Wiener, 1954, pp. 34–35; hereafter abbreviated *HU*. Elsewhere in the book Wiener writes that irrationality in human behavior (Freud) is of a piece with the chance element in the physical world (Willard Gibbs) and with the statistical system of reasoning itself (Henri Lebesgue): "This random element, this organic incompleteness, is one which without too violent a figure of speech we may consider evil; the negative evil which St. Augustine characterizes as incompleteness, rather than the positive malicious evil of the Manicheans" (*HU*, p. 11). Wiener emphasizes over and again that the evil opposition of nature is not active: "I have already pointed out that the devil whom the scientist is fighting is the devil of confusion, not of willful malice. The view that nature reveals an entropic tendency is Augustinian, not Manichean" (*HU*, p. 190).

4. It is part of my argument that the division between active and passive antagonists is not sustained in the cybernetic view of the world. In particular, within Wiener's vision of the world there is great continuity between the antientropic self-regulating strategy for controlling (or destroying) an enemy and controlling the built and found world around us. Indeed, Wiener himself contends that the Augustinian position easily breaks down into Manicheanism. See *HU*, p. 191.

5. On earlier human-machine links, see, for example, Rabinbach, 1990.

6. On Wiener, see the rich and provocative book by Heims, 1980. An extraordinary source for original writings by Wiener is the four-volume set of his collected works; there one finds the best technical summary and analysis of Wiener's antiaircraft predictor. See Masani and Phillips, 1985. I have also found very valuable the mathematically sophisticated treatment of Wiener's work on the AA predictor in Masani (1990, esp. pp. 184–193). Three authors writing recently on the links between cybernetics and the biological and psychological domains include Kay, 2000; Keller, 1996; and Edwards, 1996.

7. See Caldwell, 1940. A standard text on servomechanisms that Wiener approved of (and cited frequently) was MacColl, 1945. MacColl clearly identified servomechanisms with slaved systems that used feedback; see pp. 1–9.

8. This method of prediction (least squares) was mathematically felicitous but controversial. Stibitz, for example, argued that a pilot heading into a burst of AA fire would turn right or left, but the Wiener Predictor would lay the next burst dead ahead, and he argued, "is almost certain to miss." Duncan Stewart (for whom Stibitz worked at OSRD) was not at all confident that the enemy would behave in a consistent way. Would he see the burst? Would he fly on to complete a bomb run? "Present predictors, I fear, are useful largely because the psychological effect on the enemy serves to keep them far enough away to ... reduce the damage" (Stibitz, 1941).

9. Because Wiener's AA predictor based its algorithm for prediction on statistical input from the pilot's past performance, the device was a kind of learning machine. As such, it came to stand in for Wiener as the prototype of other game-playing and potentially war-fighting machines. See, for example, Wiener, 1948, pp. xi-xii. The central idea in Wiener's statistical approach is the quantification and characterization of noise. His analysis begins with the determination of the autocorrelation coefficients and an associated spectrum. Given a sequence of numbers, $a(n)$, where $a(n)$ is the value of a at time $t = n$, the autocorrelation coefficients $\phi(m)$ for lag m are defined as the mean value, taken over all time, of the product of one of these numbers with its mth neighbor. The spectrum is then defined as the Fourier series where the $\phi(m)$ are the autocorrelation coefficients of the original series. Wiener characterizes what it means for stability to exist for these time series and therefore establishes conditions under which predictions can be made in his crucial work: *The Extrapolation, Interpolation and Smoothing of Stationary Time Series with Engineering Applications*.

10. As he made clear to a conference organizer shortly afterwards, even alluding to the connection between statistics and prediction could be disastrous. When a joint meeting was planned in 1942 between the American Mathematical Society and the Institute of Mathematical Statistics, several papers were slated to discuss statistical prediction; Wiener shot off an urgent note to one of the organizers, J. R. Kline, contending that even titles might be "a tip-off" to the enemy on subjects "vital and secret in more ways, and vastly more important ways, than I have been able to tell you" (Wiener, 1942e).

11. Stibitz invokes the term *uncanny* at just the moment—1 July 1942—when Wiener's machine began predicting as if it were animated (whence Weaver's half-joking call for a saw). One is reminded here of Stanley Cavell's reflection that Freud, in his essay on the uncanny, may be protesting too much when he claims (no less than four times) that the animate/inanimate conflation does *not lie* behind the uncanny. For Cavell, the uncanny reflects precisely the philosophical anxiety exacerbated by the ambiguity created when it is unclear whether a mind or merely an inanimate object is at hand. The sentiment of uncanniness resulting from such an ambiguity is therefore tied to the philosophical problem of other minds. According to Cavell, this philosophical difficulty (surrounding the existence of other minds) is part of, not subordinate to, the psychology of uncanniness. See Cavell, 1988. I would add this: each generation has its own conception of what constitutes a mind. Wiener's notion circulates around feedback, control, and the capacity to predict. Since characterizations of mind change, a philosophical-historical account of the uncanny would necessarily pass through many epochs.

12. *Behaviorism* as used by Wiener, Stibitz, Boring, and others encompassed a field and spirit of inquiry far wider than a behaviorism defined as a lineal descent from J. B. Watson to B. F. Skinner. For more on the scope of behaviorism, see Woodworth and Sheehan, 1964, pp. 43–92, 111–213; Heidbreder, 1933, pp. 234–286.

13. Wiener, 1945e, pp. 14–15. In particular, the wings of an airplane rise from the fuselage upward towards the wingtips (this rise is known as the dihedral). When the plane banks (while maintaining direction), the plane side-slips towards the lower wing. Since

the lower wing is now positioned more nearly parallel to the ground, the lower wing encounters the relative wind strongly while the upper wing, now tilted more nearly perpendicular to the ground, encounters the relative wind more weakly. This raises the lower wing, righting the airplane. See, for example, the popular 1944 flight instruction book by Langewiesche and Collins, 1972, (1944), especially the subsection "What the Airplane Wants to Do," pp. 125–127, which addresses the dihedral.

14. See Rosenblueth, Wiener, and Bigelow, 1943, pp. 23–24. The term *black box,* commonly used at the MIT Radiation Laboratory during the war, became popular through the use of common black-speckled boxes to encase radar electrical equipment such as amplifiers, receivers, filters, and so on. Wiener himself referred during the war to "boxes" with unspecified interiors, as in his 1942 letter to Haldane, cited above. After the war, Wiener elaborated on the notion of a black box, contrasting it with a "white box" in the sense invoked here. See, for example, *C,* pp. xi and 180 and Wiener, 1961b. On the black box as part of "radar philosophy" at the Rad Lab, see Galison, 1997.

15. Wiener, Aiken, and von Neumann identified the common center of their interests to revolve around intention: "Teleology is the study of purpose of conduct, and it seems that a large part of our interests are devoted on the one hand to the study of how purpose is realized in human and animal conduct and on the other hand how purpose can be imitated by mechanical and electrical means." Their intention was to found a society, a journal, a patent and support mechanism, a means of popularization, and, finally, a protective net to guard against "dangerous and sensational publicity" (Aiken, von Neumann, and Wiener, 1944b).

16. See R. Taylor, 1950b, p. 328. Taylor did concede to Wiener and Rosenblueth that his Newtonian scientific conception of causes and categories needed updating in light of recent developments in physics; see p. 332.

17. Rosenblueth and Wiener, 1950, p. 195, emphasis added. The problem of integrating intentionality into the broad project of behaviorism had vexed psychologists for decades, and proffered solutions to the difficulty were numerous. One of the most famous of the attempts was by Tolman, 1932, but his approach was by no means universally accepted.

18. Ashby, 1951a. Ashby went on to write a well-known text, *An Introduction to Cybernetics.*

19. On the Macy's meetings, see Bateson, 1946; McCulloch, 1946; *Conference on Feedback Mechanisms and Circular Causal Systems in Biology and the Social Sciences.* Heims has a good discussion of the tenor of these meetings in *John von Neumann and Norbert Wiener: From Mathematics to the Technologies of Life and Death*, pp. 201–207 and a wide-ranging study of the impact of cybernetics on the social sciences in *Constructing a Social Science for Postwar America: The Cybernetics Group, 1946–1953.*

20. "[Morgenstern's] very important joint book on games with Dr. von Neumann ... represents a most interesting study of social organization from the point of view of methods closely related to, although distinct from, the subject matter of cybernetics" (*C,* pp. 18–19).

21. On "cultural meaning," see Galison, 1993.

22. The term *cyborg* itself was, as Matthew Price has shown, first used by Air Force contractors, in 1960, in the context of speculative research on biochemical means for extending the capability of astronauts. One line of inquiry was the search for drugs that would alter osmotic pressures within the body to allow unprotected "walks" in space. See Matthew Price, "'Man Must First Conceive'—A Critical Philology of the Cyborg," unpublished manuscript. This links rather closely on the one hand with Wiener (whom the Air Force contractors cite) and on the other with the bionic implants required by the space pilot in the science fiction representations of cyborgs cited by Haraway in (1991, p. 179).

23. Mayr, 1986, p. 197; hereafter abbreviated *A*. See also Mayr, 1970.

24. "The abstract concept of the closed causal loop which provides the common basis for all the regulating mechanisms discussed in this study, and which is expressed most compellingly by the graphic symbols of the block diagram, is an achievement of the 20th century" (*ibid.*, p. 129). While the nineteenth century was rife with studies of speed regulation and its theoretical representation in terms of differential equations, the general feedback problem was posed only with developments in electrical self-regulation, especially through the work of Karl Küpfmüller (1928), Harry Nyquist (1932), Harold L. Hazen (1934), and Harold S. Black (1934), and assembled into treatise form by Rudolf C. Oldenbourg and Hans Sartorius (1944) and Le Roy A. MacColl (1945). See Mayr, 1970, p. 132.

25. For the American history of the popular culture of flying, see Corn, 1983; on the German cultural history of aviation, see Fritzsche, 1992.

Three

COMPUTERS AS MODELS OF THE MIND: ON SIMULATIONS, BRAINS, AND THE DESIGN OF COMPUTERS

Peter Asaro

After the war, together with a small group of selected engineers and mathematicians, Johnny built, at the Institute for Advanced Study, an experimental electronic calculator, popularly known as Joniac, *which eventually became the pilot model for similar machines all over the country. Some of the basic principles developed in the* Joniac *are used even today in the fastest and most modern calculators. To design the machine, Johnny and his co-workers tried to imitate some of the known operations of the live brain. This is the aspect which led him to study neurology, to seek out men in the fields of neurology and psychiatry, to attend meetings on these subjects, and, eventually, to give lectures to groups on the possibilities of copying an extremely simplified model of the living brain for man-made machines.*

Klara von Neumann, wife of John von Neumann (John von Neumann, 1958, p. viii)

Turing knew perfectly well what the job he had to do, which was to manufacture or design a machine that would do the complicated sort of mathematics that had to be done in the Mathematical Division of NPL. *But he had all sorts of interesting things that he liked to do: for example, he was really quite obsessed with knowing how the human brain worked and the possible correspondence with what he was doing on computers .... Turing thought that the machine should be made quite simple, and at the same time should make everything possible that could be done. His particular purpose was to permit the writing of programs that modify programs, not in the simple way now common but rather in the way that people think.*

Ted Newman, colleague of Alan Turing (1994, p. 12)

1. Introduction

The purpose of this chapter is to clarify some of the important senses in which we might consider the relationship between the brain and the computer as one of "modeling." I will also consider the meaning of "simulation" in the relationships between models, computers, and brains. While a fairly broad literature on models and simulations in science has emerged, it has primarily focused on the physical sciences, not the mind and brain. And while the cognitive sciences have often invoked concepts of modeling and simulation, they have been frustratingly inconsistent in their use of these terms, and the implicit relations to their scientific roles. My approach is to consider the early convolution of brain models and computational models in cybernetics, with the aim of clarifying their significance for more current debates in the cognitive sciences. I believe that clarifying the historical senses in which the brain and the computer worked as models of each other in the historical period prior to the birth of AI and cognitive science is a crucial task for an archeology of AI and the history of cognitive science. "Model" is a challenging concept, in part because it is a noun and a verb, an object and a practice, and because it takes many prepositional phrases. Let *X* be the model, *Y* the thing modeled, and *Z* the agent doing the modeling. We may distinguish the following different meanings of "model": *X* models *Y*, *X* is a model of *Y*, *X* is a model for *Y*, *Y* is modeled on *X*, *Y* is modeled after *X*, and even *Y* models for *Z* (it would be more proper to include "for *Z*" in each of the prepositional phrases, but I have omitted it when the context is unambiguous). We could say that the brain was a *model for the structure* of computer (or the computer was *modeled on* the brain) in the sense that the designers of the early computers, such as John von Neumann, treated the biological brain like an artist's model, and crafted the computer in its image. Once completed, we might think of the artist's sculpture as a sort of model of its subject—and so we might think of the computer as a *model of* the brain.

This would seem to be the case for some electronic devices that were *models of the performance and behavior* of the brain, such as W. Ross Ashby's Homeostats, which intended to be models of the adaptive properties of the brain, or W. Grey Walter's Tortoises, which intended to be models of the dynamic drives of the living brain. In yet another sense, researchers such as Alan Turing, saw the first computers as *models of the fundamental structure* of the brain in the sense that the digital computer was an engineered device that worked on the same principles as the brain and as such it could be used to test theories about how the higher level functions of the brain might operate. For each of these cases, the modeling—the construction of each device—was unique and complex and involved further concrete and abstract models of the brain and its behavior. I have considered the nature of working models that the Homeostats and Tortoises represent elsewhere (see Asaro, 2006). In this chapter I will add to theoretical and working models another class—the "simulation"—which is a term we use in nearly as many ways as "model." Many authors understand "simulation" as a synonymous of "model," while others see simulations as "working models," because their behavior can become the subject of empirical inquiry. More precisely, "simulation" generally refers to a special class of models whose use defines them—specifically computational models that we use to approximate the behavior of a system of equations that are too difficult to solve by analytic techniques.[1] Because of the power of the computer, simulations have come to be one of the most significant kinds of models in scientific practice. Yet they occupy a strange place *vis à vis* working models like the Homeostat and the Tortoise. In fact, we might see computational simulations as challenging the distinction between theoretical and working models. On the one hand, simulations appear to be something like automated theories. On the other hand, simulations are the most abundant form of models that *do* something—working models. This duality of character leads to the power of simulations and to the difficulty of properly understanding their role in science.

I begin with a review of some crucial aspects of the work of John von Neumann and Alan Turing, who were involved in the theorization of mathematical computation and the design of the first general-purpose stored-program computers. They also theorized the ways in which the computer could simulate the brain and mind. This examination reveals that the design of the first general-purpose computers drew heavily upon the McCulloch and Pitts (1943) neuron model and other aspects of neurophysiology. The brain, or more precisely the theoretical models of neurons, *were* the models of the first computers. We also have a sense of how the human practices of performing mathematical computations shaped the mathematical theory of computation. We do not often discuss these basic forms of modeling, yet I believe they have had a significant impact on our theoretical intuitions regarding the more complex forms of modeling of the computer after the brain. This was especially true early on, when we used to refer to computers as "giant brains" (Berkeley, 1949).

The computer is a unique artifact among the tools of modeling for several reasons that deserve special attention. The very notion of a symbolic simulation is tied up with ideas about the nature of computation and the technological performance of working computers. These ideas, and the conceptualization and design of the first computers, are intimately related to models of the brain and mind in multiple and complex ways. Part of my purpose in this chapter is to make these multiple and complex relations visible as instances of the different types of scientific modeling and models. In considering the computer and the brain, we can find instances of each type of model we have just considered—theoretical and working models—as well as the simulations that I will discuss shortly.

Before embarking on our historical journey to the electronic brains of the 1940s, I would like to consider the view of the mind that has dominated cognitive psychology since then. Scholars have dubbed "computationalism" the philosophical view that the mind *is* a computer. The preface to a recent book on Computationalism summarizes the view:

> Are minds computers?... Computationalism—the view that mental states are computational states—is based on the conviction that there are program

> descriptions of mental processes and that, at least in principle, it is possible for computers, that is, machines of a particular kind, to possess mentality. (Scheutz, 2002, p. ix)

Despite many different formulations, Computationalism is rooted in the basic analogy between mental states and computational states—the different formulations are the result of different definitions and conceptions of these two types of states. Most often, we define computational states in terms of the stored-program digital serial computer, or the mathematical theory of computation. Two of the mathematicians most responsible for the development of the mathematical theory of computation, and who had the greatest influence on the design of the first such computers—Alan Turing and John von Neumann—were also building synthetic minds. As I will show in this chapter, while they shared a common understanding of computation, they differed in their views of how a computer might simulate a brain or the mind. Still different from theirs was W. Ross Ashby's attempt to model mental phenomena. Ashby saw "information processing" as central to the adaptive mind, but not strictly as a form of computation. Hence, he built analog simulations (working models) of the brain, whereas Turing and von Neumann were seeking different kinds of symbolic simulations (theoretical models).

First, a clarification of the distinction between simulations as working models and simulations as automated theoretical models is in order. An examination of the early history of the computer and of the attempts to simulate the brain with these early machines will demonstrate the usefulness of distinguishing the different types of models and modeling. In so doing, I will consider the automata theory that von Neumann developed as he was designing the EDVAC, arguably the first stored-program computer. I will also consider his use of the McCulloch and Pitts neuron model in this design, and his later thoughts on simulating the brain on the computer. I will then consider Turing's preoccupation with the universal (Turing) machine as a model of the mind and brain, which will lead me to Turing's suggestion to Ashby to simulate the Homeostat on his ACE computer, and to a direct comparison of these two types of simulations. I will then conclude with the implications of the view of models and simulations for Computationalism that we have developed over time.

2. Analog and Symbolic Simulations

Making a careful distinction between theoretical models and working models has an advantage. While the notion of theoretical models has its roots in the normative philosophy of science, theoretical and working models are useful to a descriptive naturalistic study of scientific practice. We have to make the crucial distinction between, on the one hand, working models—which do something in the world and have a material agency that is independent of humans—and, on the other hand, theoretical models, which—while sometimes endowed with agency of a normative and disciplinary sort (Pickering, 1995, p. 120)—are not completely independent of humans, but require an agent willing to conform to their normative rules and disciplined constraints. Let us consider more carefully the distinction between theoretical and working models, and how they relate to the current philosophical discussion of simulations.

In his recent work on computational simulations, Winsberg (2001; 2003) considers three traditional attempts to account for computational simulations in physics: as metaphors, as experiments, and a third middle mode. The view of simulations as metaphors, while perhaps never expressed as such, nonetheless holds that simulations are just brute-force number crunching procedures, that we use when analytic techniques are impossible—in other words, a degenerate form of theorizing. The view that simulations are experiments, and the computer is an experimental target, holds that there is some mimetic relation between the simulation and the simulated, such that the simulation can mimic the real and act as a stand-in. The third mode holds that simulations are an entirely different kind of thing, lying somewhere between theorizing and experiment.

Winsberg is careful to distinguish between simulations in which analytic solutions produce closed form expressions, and simulations that use numerical methods producing a "big pile of numbers" that require the usual tools of experimental practice to analyze: visualization, statistics, data mining, and so forth (2003, p. 111). The difference is not merely one of the mimetic qualities of the mathematics, but of the practices that scientists use to engage and work with the models. The numerical methods are more like experiments than theories, because the same practices used to study them are used in experimental investigations.

As I have mentioned above when discussing working models, we employ these synthetic numerical simulations to generate, or synthesize, phenomena that we want to investigate, explore, and experiment on. Unlike the theoretical models or the simulations based on tractable equations that we deduce from theory—which lend themselves to straightforward mathematical analysis—these synthetic numerical simulations are themselves objects of empirical study. In these analytic simulations we do not need to employ the data analysis techniques of experimental practice to discern the structures and patterns in the phenomena—these models only produce data as an instantiation of the theory for a given case, and we can easily derive the desired results from the equations. This distinction also corresponds nicely to an analytic/synthetic distinction. Here the analytic simulations engage in deriving local models from general theories in a formal way, whereas the synthetic simulations seek to fill in the gaps of missing theory and data by generating something new that we can manipulate, experiment on, and use to generate data on which to devise and test theories.

Much of the philosophical literature on simulations appears to Winsberg to be hung up on visualization as a key aspect of what makes simulations interesting. The notion of the mimetic qualities in the simulation's representations (Hughes,

1997) captures this. Like isomorphism, mimetic relations aim to provide an objective way of expressing the precise relations between the real and the simulated systems, but mimetic relations have the added requirement of preserving the graphical aspects of the original:

> The extensive use of realistic images in simulation is a stepping stone that simulationists use in order to make inferences from their data. It is also a tool they use in order to draw comparisons between simulation results and real systems; a move that is part of the process of sanctioning of their results. It is the drawing of inferences and sanctioning of results that give rise to the interesting philosophical connections between simulation and experimental practice. (Winsberg, 2003, p. 113)

The graphical aspects of such models can provide a means to the employment of visual laboratory practices, but it is not the only such aspect, and it is not essential. As Winsberg notes, we can derive visualizations from many mathematical models, and this appears tangential to what makes them good models. Further, if we put too much faith in the power of the mimetic features, we see simulations as purely and truly experimental, taking literally the notion of the "numerical experiment" and interpreting the computer simulation as a stand-in for the real phenomena. This begs the question of just how well a simulation mimics the real phenomena (2003, p. 115). It matters not just that it "stands in" in some respects, but in which respects, to what extent, and to what precision and accuracy. If we are interested in the epistemic status of models and simulations, these questions are of great importance, and we cannot take for granted that because we can perform experiments on simulations, they must have the same epistemic, much less metaphysical, status as real experiments. Still, they must not completely lack epistemic status either, or be metaphysically weak or unreal.

Viewing simulation as a unique new mode of scientific practice between theorizing and experimenting, Winsberg asserts that this is merely a good place to start thinking about simulation, not an explanation of it:

> What is of interest philosophically is to understand (a) how it is that what is at root a theoretical enterprise, takes on characteristics of experimentation, (b) what those characteristics are—at the abstract, reconstructed level, (c) what consequences there are of such a hybrid for our understanding of the nature of modeling, theorizing, and experimenting, and (d) how simulation produces knowledge and what kind of knowledge that is. (2003, p. 118)

In staking out this middle ground for simulation, Winsberg is careful to note that the techniques and practices of modeling can carry their epistemic credentials, independently of the theory that produces them:

> [T]he credibility of [a] model comes not only from the credentials supplied to it by the governing theory, but also from the antecedently established credentials of the model building techniques developed over an extended

> tradition of employment. That is what I mean when I say that these techniques have their own life; they carry with them their own history of prior successes and accomplishments, and, when properly used, they can bring to the table independent warrant for belief in the models that they are used to build. (2003, p. 122)

It appears that for Winsberg the epistemic basis for simulations comes from theory and from the practices of modeling. More importantly, he also holds that those simulations are autonomous (or semi-autonomous, as he says) from theory only to the extent that a tradition of practice roots these independent epistemic foundations. I agree with this approach, and only wish to add that working models are also produced in and support the development of a tradition of modeling practices.

Before discussing the early attempts at using the computer to simulate the mind and brain, I want to consider one more position on simulation that considers more explicitly how a model represents the system it models. During the Connectionist debates of the early 1990s, a recurring argumentative theme occurred, which the distinction between analog and digital computation motivated. Perhaps we made too much of this distinction, or rather, we failed to fully recognize its real nature. The work by Russell Trenholme (1994) is an exception. Developing a line of thought connecting the ideas of Kenneth J. W. Craik, Norbert Wiener, Philip Johnson-Laird, and Rodney Brooks, Trenholme presents a view of simulation that I believe is compatible with the idea of a working model. I hope that applying his view to the work of von Neumann and Turing on the early computers will provide a new perspective on how we variously conceived the computer as a simulation of the brain and mind.

In short, while all synthetic simulations are working models in the generic sense of being automatic, just as all models are "representational" in a generic sense, the nature of the scientifically relevant aspect of their being models can differ in significant ways. An analog simulation directly models a natural system, while a symbolic simulation employs an intermediate symbolic system and is thus an indirect model. This level of indirection is highly significant to the extent that it shifts the relevant aspect of a simulation from the realm of working models, epistemic artifacts, and material agency, to the realm of theoretical models and normative disciplinary agency. Trenholme calls a naturalistic analog simulation what I have been calling a working model, and he calls a symbolic simulation a special kind of theoretical model, one whose disciplinary agency has been transferred from human hands and minds to automatic computations.

Trenholme's argument begins with a clarification. While much of the debate at the time was couched in terms of "analog *vs.* digital" computation, Trenholme is careful to point out that the real issue is "analog *vs.* symbolic" representation. As I will show in the discussion of von Neumann's automata theory below, "analog *vs.*

digital" is a matter of how computers represent *numbers*. It says nothing, however, about what or how those numbers come to represent or simulate anything else beyond themselves. The real issue is whether numbers are an essential part of the simulation at all. The "analog *vs.* symbolic" dichotomy intends to capture the notion that analog simulations do not require or depend upon symbolic representations. Instead, they serve as simulations, primarily or completely, in virtue of their causal structure. Trenholme calls these "naturalistic analog simulations" in order to distinguish them from the more careless definitions of analog computation.

Trenholme goes on to argue that naturalistic analog simulations are not representational in the same sense that symbolic simulations are representational.[2] The involvement of an additional "mapping," in the sense of multiple levels of representation, motivates the difference. In short, analog simulations relate the causal structure of a natural phenomenon to the causal structure of the simulation by isomorphism or by the looser "similarity" relation (along with a probabilistic causal theory). We derive, or map, a symbolic simulation, on the other hand, from a formal theory of a natural phenomena, and we then map the result into a computational simulation of the formal theory. Missing from the symbolic simulation is the obvious sense of an isomorphism between causal structures. Symbolic simulations are instead representational in the sense of semantic relations like denotation. Analog simulations lack this purely representational layer.

Symbolic simulation is a two-stage affair: first comes the mapping of inference structure of the theory onto hardware states, which defines symbolic computation; second comes the mapping of inference structure of the theory onto hardware states, which (under appropriate conditions) qualifies the processing as a symbolic simulation. In contrast, a single mapping from causal relations among elements of the simulation to causal relations among elements of the simulated phenomenon defines analog simulation (Trenholme, 1994, p. 119).

While a symbolic simulation depends on causal structures at *some* level, the symbolic level is crucial to its successful performance as a simulation. Similarly, while an analog simulation depends on representation at *some* level, it is the causal structure, or material agency, that is operative in its performing as a simulation. Both types of simulation entail a generic sense of representation: the simulation is in some sense a representation of the system that is simulated, but the kind of representation at stake is not necessarily symbolic representation.

For instance, the mercury barometer "represents" air pressure without employing symbolic systems *in its performance*—the symbols only come at the end, when we read the result off from the markings on the device. Instead of depending upon the symbolic relations to achieve a well-behaved simulation, analog simulations depend upon causal structures. These causal structures are the material agency of working models. On the one hand we need to constrain these causal structures in specific ways so as to constitute a valid simulation (they are disciplined material agencies), while on the other they, to the extent that causal structures engage the potentially infinite causal relations of the world and can always exhibit new emergent properties, are always unconstrained and open-ended. In analog simulations, an implicit respect is in order for this dual aspect of material agency, which is largely missing in symbolic representations that attempt to completely constrain the behavior of the physical system to produce only the precisely intended symbolic system. When a symbolic simulation behaves irregularly, its output is meaningless.

Symbolic representation is a quite specific form of representation in which symbols represent via denotation and reference. Symbols alone do not have a causal structure, and in order to base an automatic simulation on them, we ought to have a computer, which Newell and Simon call a "physical symbol-system":

> *The Physical Symbol System Hypothesis.* A physical symbol system has the necessary and sufficient means for general intelligent action. By "necessary" we mean that any system that exhibits general intelligence will prove upon analysis to be a physical symbol system. By "sufficient" we mean that any physical symbol system of sufficient size can be organized further to exhibit general intelligence (Newell and Simon, 1981[1976], p. 41)

Simulations based upon symbolic representations are so closely tied to the development of the modern computer that we often just call them "computer simulations." Other expressions evoking notions of modeling and simulation are in use, but they confuse the issue:

> Certain ways of speaking—for example, saying that a theory is "modeled on a computer"—risk conflating the two mappings, thus blurring the distinction between analog and symbolic simulation, as does the notion of representation when applied to analog simulations .... Thus we may say that a user (or an observer) recognizes an analog simulation as such when properties in the analog device are held to *represent* the corresponding properties (those that play the corresponding causal role under the causal-structural isomorphism). In the case of symbolic simulation, once the user identifies the phenomenon to be simulated, the term of the theory coded into the computer may be held to *represent* the relevant features of the phenomenon under its standard (intended) interpretation; here the notion of *representation* stands for language-world semantical relations such as *reference* and *denotation*. An obvious distinction can be made between these semantical relations and the notion of isomorphism of causal structure used in characterizing naturalistic analog simulation. (Trenholme, 1994, p. 119)

Once the distinction between analog and symbolic representation is clear, we can begin to see how this argument relates to the larger questions concerning Computationalism.[3]

There are at least two significantly different ways to elaborate the isomorphism between a simulation and the system or process it simulates.[4] One is the simple input/output correspondence, or what Ashby first called the "black box" simulation. Under this view, one system is a black box simulation of another if it produces similar outputs under similar inputs. In the case of analog simulations, the inputs and outputs are causal, in symbolic simulations they are symbolic. The matter is more complicated in the second type of isomorphism in which the *internal processes* of the simulation and the system or process that are being simulated have to correspond. In this form of simulation not only do the input/output relations matter, but also the internal states and processes of the simulation. Such cases make the difference between analog and symbolic simulations clearer.[5] While symbolic simulations must realize symbolic processes, these are independent from their physical realization. We generally refer to this feature as "multiple realizability," meaning that different causal systems can realize the same symbolic process. For analog simulations, which lack an independent symbolic level, the causal processes that realize the simulation must be isomorphic to the system being simulated for it to count as a simulation.

Let us turn from this abstract discussion of simulation to some more concrete historical examples. We will see in the ideas of von Neumann and Turing two different kinds of symbolic simulation. For von Neumann, we ought to use the computer to simulate a *physical model* of the brain, while for Turing the computer ought to simulate the same essential algorithm, or program, that the brain itself simulates.

3. Analogies, Digits, and Numbers

The distinction between analog and digital is due to John von Neumann (Aspray, 1990). His work on the first stored-program computers and views on the relationship between the computer and the brain presents a complex history involving numerous layers of analogy, modeling, and simulation. A complete account of the development of his thought is well beyond the scope of the present chapter, but it is worthwhile to consider how the view of working models and simulations would apply to the different positions he stated. Clearly, his design of the first computer memory drew heavily upon his review of research in neurophysiology, and quite explicitly upon the work of Warren McCulloch and Walter Pitts (1943) and their conception of logical neural networks. This much is clear from his 1945 EDVAC design proposal (1993). The story becomes more complex when we consider his 1948 theory of automata (1951), his 1946 letter to Norbert Wiener about simulating the brain (1997[1946]), and his later reflections on the relationship between the brain and the computer for his posthumously published Silliman lectures (1958). In this section I will briefly review some of von Neumann's thoughts on the issues involved in thinking about the computer as being modeled after the brain, and his approach to simulating the brain with a computer.

The key to understanding the development of von Neumann's ideas on the relationship between the brain and the computer is to keep in mind that for him the computer is a representational system of a specific type. In his automata theory, the computer is fundamentally an automatic system for representing and manipulating *numbers*. Numbers are not the same thing as quantities or numerals—they are abstract mathematical entities. Quantities and numerals are concrete ways of representing numbers for practical purposes. To automate mathematics we need to develop a physical system that can represent numbers; before we consider the technological possibilities, the choice between quantities and numerals is an open question.

On 20 September 1948, von Neumann presented a theory of automata at the *Hixon Symposium on Cerebral Mechanisms in Behavior* (1951). In his presentation, "The General and Logical Theory of Automata," von Neumann spelled out the need he saw for a rigorous theory of computation, and outlined a formal theory of *automata* (axiomatically idealized computational mechanisms). He began by distinguishing two general classes of automata by their mode of representing numbers. The class of automata built on the "analogy principle" represents numbers by analogy. They use some physical quantities in the same way a thermometer represents temperature by the height of its mercury column. If, for example, we are representing two numbers by the electrical currents in two circuits, we can add the numbers by combining the circuits appropriately and we will register the result as the total current output from the combined circuit. We can do all the basic arithmetic operations ($+, -, *, \div$) in roughly this way by using the currents in a circuit and providing the appropriate configurations of the circuit's relays and switches. Automata built on the "digital principle," instead, do not represent numbers as physical quantities, but as aggregates of numerical digits in the manner human beings typically do when they write them down on paper or count on their fingers (the etymological origin of "digit" itself). Such an automata might have a dial with ten positions representing the range from 0 to 9, or a series of such dials for the ones column, tens column, hundreds column, and so on, of the decimal numbering system. The digital representation that nearly all modern computers use is a binary system where wires, which carry electrical currents of two sufficiently distinct magnitudes, employ a set of canonical circuits to perform mathematical and logical operations on the binary representations.

The two types of automata that von Neumann described correspond to Trenholme's (1994) "analog *vs.* digital" distinction, though Trenholme, distinguishing between types of simulation, criticizes it as spurious. In each case, we seek to represent numbers and, unless we want to "simulate" pure mathematics, both are forms of symbolic simulations and both are caught up in a double-mapping. Even if we use an analog computer, such as a differential analyzer, we still need a theory about how the causal structure of the analog computer realizes the desired mathematical calculations, as well as a theory about how those calculations relate to the natural phenomena in question. A differential analyzer is not a simulation, for instance, of some hydrodynamic system, but an automated device for solving a set of equations derived from a theory of that hydrodynamic system.

The main consequence of the different kinds of numerical representation in automata is the practical realization of them when building electronic computers. Though analog computers have the advantage of arbitrarily high precision, the constant introduction of noise into their circuits make them deeply susceptible to errors. When dividing, as when dividing "5" by "7," we cannot express the result in a finite number of digits and so the number of digits that a computer can represent at one time will limit a digital computer's answer, while an analog computer can represent the answer to an unlimited precision.[6] The down side is that this is only an ideal; in actuality to get analog circuits to perform calculations with much accuracy was quite difficult, because slight fluctuations in the physical quantities, due to sources outside the computation, build up as noise in the system. Consequently, while our analog division of "5" by "7" might be perfectly precise in theory, even low probabilities of very small disturbances in the current of the circuit will limit the actual precision enormously. Von Neumann insisted that to maintain a reliable signal-to-noise ratio that prevents the analog computer from becoming more precise than a few decimal places is a difficult engineering problem, and that it just so happens that the same problem does not arise in digital machines—which have the further advantage that we can increase precision indefinitely and economically by increasing the number of digits represented by duplicating components.

While we often overlook this point, von Neumann claims that, in spite of their great practical differences, he sees no essential mathematical difference between the two kinds of automata. Any kind of calculation that we can, in principle, do on an analog automata we can also do on a digital automata, and *vice versa*. Von Neumann argues that organisms, natural automata, are *mixed* automata utilizing both principles for different particular functions, and that high speed computers are strictly digital machines. He is careful to point out that trying to base a theory of mind on the kinds of computations achievable in digital computers should not affect its applicability to the living brain since research has shown that neuronal activity is digital and, even if it were not, an appropriately constructed analog machine could realize the theory anyway. In other words, a computational simulation of the brain will be a symbolic simulation, as long as that simulation depends on the computation of theoretical models, regardless of whether we use an analog or digital computer.

4. Modeling the Computer on the Brain

> *These repeated excursions into biological information processing and the interdisciplinary study of cybernetics have been ignored in previous accounts of von Neumann's computing, yet they clearly shaped his ideas.* (Aspray, 1990, p. 189)

This section will digress from my discussion of simulations, but it will still be relevant, I hope, to the extent that it will consider the practice of "modeling" in the development of the computer as a machine. In particular, I will tell the story of how the computer was modeled on the brain, or at least upon some specific theories of how the brain and its neurons work. The stored-program computer required something more than numerical representation to become a technological reality. Namely, the stored-program computer would need to represent its *instructions* as numbers, which it could then store in memory, retrieve, execute, alter, and so on. In other words, some numbers are *representations to the computer* of instructions that the computer must interpret and perform. This great leap in design turned mere calculators into universal computers and also sowed a deep confusion regarding the representational character of computations. The connections between computational memory and neural structures in the brain soon fertilized this early kernel.Von Neumann's description of the EDVAC design made clear that in the design of the first computer memory he looked to the brain and to McCulloch and Pitts neuron for inspiration. This led to a peculiar kind of analogy between the computer and the mind/brain that the proponents of Computationalism would soon embrace, but which von Neumann himself would ultimately reject.

While researchers had constructed different electromechanical calculators before, during, and after World War II, the principal technological impediment to creating a general purpose computer (a working Universal Turing Machine) was with how to give it a memory.[7] The problem was one of both organization and technological realization. We generally hold that the first machine to solve the memory problem was von Neumann's Institute for Advanced Study machine. Von Neumann solved the problem by introducing a memory unit, or "organ" as he called it, as a central part of the computer architecture. The representation, or storage, of logical instructions or "codes," as well as numbers, distinguished the memory organ from numerical counters. The computer could, by following the instructions in memory, calculate partial solutions, store the intermediate results, reconfigure itself to perform a new function, and resume the calculation with the instructions that the memory stored as numbers (Goldstine and von Neumann, 1987[1946]). The advantages over having to reconfigure the machine by hand for each portion of a calculation, or having to build a computer so complicated that all the necessary steps in a computationally demanding problem had to be preprogrammed were obvious.

We often overlook the extent to which von Neumann drew upon McCulloch and Pitts neuron model and the neurological language he used to describe the new design. Instead, we see a clear case of modeling on—the transfer of a theory in one domain to the design of a technology in another domain. As the language he uses to describe the computer unmistakably manifests, von Neumann modeled his design after neurophysiological theory and McCulloch and Pitts neuron:

> The three specific parts CA [Central Arithmetical organ], CC [Central Control organ], and M [Memory organ] correspond to the *associative* neurons in the human nervous system. It remains to discuss the equivalents of the *sensory* or *afferent* and the *motor* or *efferent* neurons. These are the *input* and the *output* organs of the device, and we shall now consider them briefly. (1993, p. 20, emphasis in the original)

Most notably, von Neumann does not refer to the parts of the computer's architecture as "units," "components," "modules," or any other such connotatively neutral engineering terms. As a mathematician, he chose the semantically loaded term, taken from the description of biological systems, "organs." These organs explicitly *correspond to specific elements in the human brain*—a theme that his description of the machine carries out:

> It is worth mentioning, that the neurons of the higher animals are definitely elements in the above sense. They have all-or-none character, that is two states: Quiescent and excited. They fulfill the requirements of [section] 4.1 with an interesting variant: An excited neuron emits the standard stimulus along many lines (axons). Such a line can, however, be connected in two different ways to the next neuron: First: In an *excitatory synapses*, so that the stimulus causes the excitation of that neuron. Second: In an *inhibitory synapses*, so that the stimulus absolutely prevents the excitation of the neuron by any other (excitatory) synapses. The neuron also has a definite reaction time, between the reception of a stimulus and the emission of the stimuli caused by it, the *synaptic delay*. Following W. Pitts and W. S. MacCulloch [*sic*] ("A logical calculus of the ideas immanent in nervous activity," *Bull. Math. Biophysics*, vol. 5 [1943], pp. 115–133.) we ignore the more complicated aspects of neuron functioning: Thresholds, temporal summation, relative inhibition, changes of the threshold by after effects of stimulation beyond synaptic delay, *etc*. It is, however, convenient to consider occasionally neurons with fixed thresholds 2 and 3, that is neurons which can be excited only by (simultaneous) stimuli on 2 or 3 excitatory synapses (and none on an inhibitory synapses). It is easily seen, that these simplified neuron functions can be imitated by telegraph relays or by vacuum tubes. Although the nervous system is presumably asynchronous (for the synaptic

> delays), precise synaptic delays can be obtained by using synchronous setups. (*ibid.*, p. 24)

Clearly, von Neumann is making a serious consideration of the structure and function of biological neurons, albeit he idealizes them in the manner of McCulloch and Pitts neuron model. Here we see how he uses the functional identity of neuron activity and mathematical logic (a unification of neural mechanisms and Turing machines) as a modeling tool in the design of the electronic computer. He sees its usefulness as a model as metaphorical or analogical, to be sure, but this kind of model is powerful, because it influences design *decisions*, and not merely the construction of design *alternatives*. In that sense it is evaluative or normative:

> The analogs of human neurons, discussed in 4.2–4.3 and again referred to at the end of 5.1, seem to provide elements of just the kind postulated at the end of 6.1. We propose to use them accordingly for the purpose described there: as the constituent elements of the device, for the duration of the preliminary discussion. We must therefore give a precise account of the properties which we postulate for these elements. (*ibid.*, p. 30)

But he saw this model as not the only criteria that is operative in design decisions:

> At this point the following observation is necessary. In the human nervous system the conduction times along the lines (axons) can be longer than the synaptic delays, hence our above procedure of neglecting them aside of *t* would be unsound. In the actually intended vacuum tube interpretation, however, this procedure is justified: *t* is to be about a microsecond, an electro-magnetic impulse travels in this time 300 meters, and as the lines are likely to be short compared to this, the conduction times may indeed be neglected. (*ibid.*, p. 30)

McCulloch's and Pitts's neuron model evidently almost completely circumscribed von Neumann's conception of physical computation.

What conclusions can we draw from these observations about the significance of the McCulloch and Pitts neuron model on the design of the first computers? First, the computer was from its very conception a kind of model of the brain. Researchers in Artificial Intelligence or Cognitive Science did not "discover" any analogy between the mind and computer. It was always there. Instead, they reconfigured it in an effort to develop an analogy between computer programs and psychological theories. One way of approaching the question of Computationalism might be to ask whether the computer itself, regardless of the program or simulation it is performing, is a model of the mind, or a good model of the mind. It obviously could be, and has been, such a model. The question of whether the mind is a computer, or a computer is a mind, however, remains. But almost no one holds the view that every computer is a mind—even strong AI holds that we must properly program computers. Turing appears to have held the view that minds are Universal Computers, and perhaps also the converse, that Universal

Computers are at least capable of being minds. I will consider his views shortly. First I will consider whether the computer may somehow simulate the brain so efficiently that it becomes a mind.

5. Simulating the Brain on the Computer

I would like to consider at this point how von Neumann envisioned the computer as a simulation of the brain. Even though he had based his automata theory on a strict distinction between analog and digital numerical representation—and modeled the memory of the stored-program computer after McCulloch and Pitts's essentially digital model of neurons—in other writings he made clear that the brain itself was much more complicated than these theoretical idealizations let on. As a result, the simulation of the brain by a computer would be far more elusive.

On 29 November 1946, just after the second Macy Conference, von Neumann wrote a letter to the mathematician and cybernetician Norbert Wiener in which he assessed the situation regarding the theory of biological information processing, and the replication of the brain's abilities in a computer. The letter is interesting for several reasons, and warrants more thorough examination than our current purposes permit. Despite his early enthusiasm, von Neumann was perhaps the first to realize the limitations of computers for simulating the brain. The clearest articulation of the shortcomings of this approach comes in his letter to Wiener. Here von Neumann argued that directly modeling the physical structure of the brain will be so complicated as to be nearly hopeless, an idea he would express more explicitly in later work. He concluded from this situation that a much better approach would be to turn instead to detailed cytological work. Specifically, he proposed to understand simple organisms in complete atomic detail (literally), by starting with the study of bacteriophagic viruses.

In the letter, von Neumann intimated one of the primary reasons for the difficulty in making the analogy between brains and computers:

> Besides, the [brain] system is not even digital (i.e. neural): It is intimately connected to a very complex analogy; (i.e. humoral or hormonal) system, and almost every feedback loop goes through both sectors. If not through the "outside" world (i.e. the epidermis or within the digestive system) as well. (1997[1946], p. 507)

Two crucial points in this passage call for attention. First, von Neumann was keenly aware of the embedded and situated nature of the human brain, and that its information processing relies on feedback loops with its environment. Unfortunately, von Neumann did not direct his energies in pursuing this issue. The other point is that the supposedly neat distinction between analog and digital was not so clear in the living brain and von Neumann directed a great deal of his energy in this direction.

One significant consequence of McCulloch and Pitts neuron model was to establish the digital character of neuronal behavior in the brain as the character relevant for understanding its organization. While this move had great benefits in terms of formalizing neural networks, and thereby treating them using mathematical logic, the idealization was ultimately a gross simplification. Von Neumann stated the case most devastatingly in his Silliman manuscript. In one passage, he assailed the notion that it is safe to treat the mechanism of neuronal excitation and inhibition as a straightforward summation function:

> It may well be that certain nerve pulse combinations will stimulate a given neuron not simply by virtue of their number, but also by virtue of the spatial relations of the synapses on a single nerve cell, and the combinations of stimulations on these that are effective (that generate a response pulse in the last-mentioned neuron) are characterized not only by their number but also by their coverage of certain special regions on that neuron (on its body or its dendrite system, *cf.* above), by the spatial relations of such regions to each other, and by even more complicated quantitative and geometrical relationships that might be relevant. (1958, pp. 54–55)

While it might be tempting to treat things as if all inputs to a neuron were equal, in fact the complex three-dimensional geometry of the synaptic connections to the neuron's dendrites is relevant to the electro-chemical processes that trigger a pulse. Similarly, the ideal of the synchronous timing of neuronal activity, essential to McCulloch and Pitts's assertion that we can treat multi-layered networks as equivalent to logical propositions, was also untrue:

> On all these matters certain (more or less incomplete) bodies of observation exist, and they all indicate that the individual neuron may be—at least in suitable special situations—a much more complicated mechanism than the dogmatic description in terms of stimulus-response, following the simple patterns of elementary logical operations, can express. (1958, p. 56)

So while he found the suggestions of treating neurons as logical units performing summations useful for designing computer memory circuits, he was deeply disturbed by just how remote this idealization was from real brains when it came to building a simulation.

Not only were the neurons not performing the idealized functions that McCulloch and Pitts required, even von Neumann's own distinctions between analog and digital automata applied to the brain in complex, and by no means straightforward, ways:

> The observation I wish to make is this: processes which go through the nervous system may, as I pointed out before, change their character from digital to analog, and back to digital, etc., repeatedly. Nerve pulses, i.e. the digital part of the mechanism, may control a particular stage of such a process, e.g. the contraction of a specific muscle or the secretion of a

> specific chemical. This phenomenon is one belonging to the analog class, but it may be the origin of a train of nerve pulses which are due to its being sensed by suitable inner receptors. When such nerve pulses are being generated, we are back in the digital line of progression again. As mentioned above, such changes from a digital process to an analog one, and back again to a digital one, may alternate several times. Thus the nerve-pulse part of the system, which is digital, and the one involving chemical changes or mechanical distortions due to muscular contractions, which is of the analog type, may, by alternating with each other, give any particular process a mixed character. (1958, pp. 68–69)

Ultimately, these complexities led von Neumann to believe that the study of the brain would lead to a new mathematics.

Von Neumann begins the letter to Wiener with a devastating critique of his own work, as well as that of Wiener, Turing, McCulloch and Pitts, to formulate a substantive theory of information processing in the brain:

> Our thoughts—I mean yours and Pitts' and mine—were so mainly focused on the subject of neurology, and more specifically on the human nervous system and there primarily on the central nervous system. Thus, in trying to understand the function of automata and the general principles governing them, we selected for prompt action the most complicated object under the sun—literally .... The difficulties are almost too obvious to mention: They reside in the exceptional complexity of the human nervous system, and indeed of any nervous system. (John von Neumann, 1997[1946], pp. 506–507)

Evidently, von Neumann recognized that the lack of formalized scientific understanding of the operation of the neurons was a major hurdle to constructing synthetic brains with artificial automata. The consequence of his efforts in clarifying these ideas over the course of several years was one of exasperation:

> What seems worth emphasizing to me is, however, that after the great positive contribution of Turing-cum-Pitts-and-McCulloch is assimilated, the situation is rather worse than better than before. Indeed, these authors have demonstrated in absolute and hopeless generality, that anything and everything Brouwerian can be done by an appropriate mechanism and specifically by a neural mechanism—and that even one, definite mechanism can be "universal." (*ibid.*, p. 507)

Besides being perhaps the most devastating critique of Turing's project before the rise of AI, this passage clearly stated the paradoxical relationship between the universal computer and the synthetic brain, which I will consider in the next section. The critique was this: the universal computer, and its equation with the functioning of the brain is infinitely potent, yet flaccid. The "absolute and hopeless generality" of the theory is such that while it could replicate the processes of mind, if those are formally definable (Brouwerian), it gives absolutely no insight into the nature, structure, or organization of those processes. The universal computer can imitate any machine, and thus the brain-machine, but this tells us nothing about brains. Instead, a rigorous theory of the biological brain was in order:

> Inverting the argument: Nothing that we may know or learn about the functioning of the organism can give, without "microscopic," cytological work any clues regarding the further details of the neural mechanism. (*ibid.*, p. 507)

Ultimately, the study of neural networks in the abstract has left us with the fundamental problem of understanding the brain empirically. Our theoretical understanding of neuroscience limits the simulation of the brain on the computer—unsurprisingly so, I would add, when we recognize that von Neumann's approach to brain simulation was a form of symbolic simulation, and hence a theoretical simulation. His notion of a computer simulation of the brain completely depended on a theoretical model of the brain that a symbolic representation on a computer could map. We see a clear example of the alignment between the symbolic simulation and the theoretical model. I will now turn to a different conception of how the computer can simulate the brain, which is nonetheless a symbolic simulation based on a theoretical model.

6. Computer as Universal Modeling Machine

> *It is possible to invent a single machine which can be used to compute any computable sequence. (Turing, 1937, p. 127)*

After the examination of the complex relationship between the computer and the brain in von Neumann's work, I turn to the more straightforward, if more abstract, relationship between the computer and the mind in Alan Turing's work. Put simply, Turing was preoccupied with his idea of the Universal Machine (now called Universal Turing Machines or UTMS). The idea of the UTM that Turing described in (1965 [1936]) is of an abstract formal computer able to compute any computable function. Elsewhere he refers to the ability of the UTM to "model" any other machine. Our earlier distinction between analog and symbolic simulation circumscribes this way of talking. The UTM is a purely symbolic conception, and real working computers are only approximations of this mathematical formalism. Turing did, however, seek to design and build real computers that were based on this formalism, most notably the Advanced Computational Engine (ACE), and later the Mark I at Manchester (Carpenter et al., 1986).

In nearly every speech and paper related to computation that Turing produced from 1944 to 1950, he cited the similarity between his UTM and a programmable digital computer as being highly significant:

> Some years ago I was researching on what might now be described as an investigation of the theoretical possibilities and limitations of digital

> computing machines. I considered a type of machine which had a central mechanism, and an infinite memory which was contained on an infinite tape. This type of machine appeared to be sufficiently general. One of my conclusions was that the idea of a 'rule of thumb' process and a 'machine process' were synonymous. The expression 'machine process' of course means one which could be carried out by the type of machine I was considering. It was essential in these theoretical arguments that the memory should be infinite. It can easily be shown that otherwise the machine can only execute periodic operations. Machines such as the ACE may be regarded as practical versions of this same type of machine. There is at least a very close analogy. (1992, pp. 106–107)

Turing saw universality as the critical element of UTMs—their ability to model any other machine made them so remarkable. The mathematical theory of UTMs had to be the model for the design of computing machines —the only difference in Turing's mind between the mathematical formalism of UTMs and the physical computer was the finite size of the memory, and the additional physical limitations of the machine, such as time. Apart from these practical constraints, we might easily call Turing's vision of the computer a "universal modeling machine."

His focus on universal modeling was also central to his consideration of learning and intelligence, and how to model them on a computer. He cast the problem as being that of an unorganized system becoming organized. But instead of accepting a thermodynamic interpretation of organization, as the cyberneticians would, Turing's approach sought to show how unorganized systems could efficiently organize themselves into UTMs through reinforcement by pleasure and pain.

Turing applied the same conception of universal modeling machines to the human brain as he did to the design of computers:

> All of this suggests that the cortex of the infant is an unorganized machine, which can be organized by suitable interfering training. The organizing might result in the modification of the machine into a universal machine or something like it. This would mean that the adult will obey orders given in appropriate language, even if they were very complicated; he would have no common sense, and would obey the most ridiculous orders unflinchingly. When all his orders had been fulfilled he would sink into a comatose state or perhaps obey some standing order, such as eating. Creatures not unlike this can be found, but most people behave quite differently under many circumstances. However, the resemblance to the universal machine is still very great, and suggests to us that the step from the unorganized infant to a universal machine is one which should be understood. When this has been mastered we shall be in a much better position to consider how the

> organizing process might have been modified to produce a more normal type of mind. (1992, p. 16/120)

For Turing, it appears that the key to unlocking the secrets of the mind lay with the UTM. For him, the mind had the ability to model other machines, and was thereby a universal model of a particular sort, though perhaps not completely identical to the uncreative computer.

We can safely contrast this approach to intelligence and mind to that of W. Ross Ashby, who saw behavior not in terms of logical rule-following, but as trajectories in phase space. He sought to embody his ideas in machines that he could directly interact with, such as the Homeostat, and later the DAMS. Learning for Ashby was a never-ending dynamic process in which the goal was survival, and the environment was continually changing. Modeling was a consequence of this, not necessarily the cause. Turing saw learning as a search for a single stable goal—the organized universal computer. For Turing, it was the patterns of symbols in memory, the symbolic "instruction tables" which held the secrets of the mind, while for Ashby, it was the patterns of interactions and feedback loops between the system and its environment. In both cases, they could build a machine to embody their vision. However, for Turing this machine would epitomize the symbolic simulation of mind, while for Ashby the machine would epitomize the analog simulation of mental behavior. We can best understand this contrast by considering their letters to one another.

7. Simulating the Homeostat

When Ashby returned from his service with the Royal Medical Corps in India in the Spring of 1946, he had been concerned with the mechanisms of learning for over a decade and had begun thinking about constructing a machine to demonstrate the principles he had concluded to be essential to adaptation. During that year he filled his notebooks with mathematical formalisms for different stability-seeking systems. These eventually turn to diagrams of simple, and then more complex, electrical circuits to realize the behavior of a set of equations in a system that he could directly engage with manually (Ashby, 1948; Ashby, 1952b; Asaro, 2008).

In the fall of 1946, Ashby wrote to Turing about the adaptive machine Ashby was designing.[8] As with all other machines and processes, Turing believed that a programmable computer such as the ACE could model Ashby's Homeostat, as he explained in his reply to Ashby on 20 November 1946:

> The ACE will be used, as you suggest, in the first instance in an entirely disciplined manner, similar to the action of the lower centres, although the reflexes will be extremely complicated. The disciplined action carries with it the disagreeable feature, which you mentioned, that it will be entirely uncritical when anything goes wrong. It will also be necessarily devoid

> of anything that could be called originality. There is, however, no reason why the machine should always be used in such a manner: there is nothing in its construction which obliges us to do so. It would be quite possible for the machine to try out variations of behaviour and accept or reject them in the manner you describe and I have been hoping to make the machine do this. This is possible because, without altering the design of the machine itself, it can, in theory at any rate, be used as a model of any other machine, by making it remember a suitable set of instructions. The ACE is in fact, analogous to the 'universal machine' described in my paper on computable numbers. This theoretical possibility is attainable in practice, in all reasonable cases, at worst at the expense of operating slightly slower than a machine specially designed for the purpose in question. Thus, although the brain may in fact operate by changing its neuron circuits by the growth of axons and dendrites, we could nevertheless make a model, within the ACE, in which this possibility was allowed for, but in which the actual construction of the ACE did not alter, but only the remembered data, describing the mode of behaviour applicable at any time. I feel that you would be well advised to take advantage of this principle, and do your experiments on the ACE, instead of building a special machine. (1946)

This letter did not stop Ashby from pursuing his machine and, in fact, shortly after he received this letter he came up with one of the basic elements of the Homeostat circuit. The difference of perspective and approach between Ashby and Turing is now clearer. Turing's insistence, even while admitting the importance of learning to intelligent behavior in general, on the universal character of the ACE meant that there was no point in building a special machine of the type Ashby was proposing. A tension exists between the rigid determined behavior of the ACE and the adaptive behavior of the brain. Considering how the ACE could model the Homeostat will bring into focus the differences between analog and symbolic computation.

While Turing's letter seems to move casually from talking about the ACE modeling the brain to talking about the ACE modeling the Homeostat, we should pause to consider the difference. This difference is the essence of the analog and symbolic distinction: Turing did not believe that the unprogrammed computer, which in spite of some resemblance to the "lower centres" was too rigid and disciplined to be brain-like, was a model of the brain. Instead, the computer required the proper instruction tables, the program defining the machine of the brain. And while Turing had some idea that conditioned behavior might be the base of it, he had little idea how to program his machine to behave in this way. The ACE might be able to perform a symbolic simulation of the brain, but only if it receives the proper symbolic organization. Ashby's Homeostat instead, because of its causal and not its symbolic structure, was itself a model of the brain. Turing saw in the Homeostat the potential to model the brain indirectly. Because the Homeostat was a machine whose design we understood, we should be able to simulate it symbolically on the ACE. Because the Homeostat simulated the brain, the ACE's simulation of the Homeostat would be a simulation of the brain (once removed). The level of indirection in this case was the Homeostat itself as a theory of the brain's adaptive behavior. Any simulation that the ACE performed would only be as good as the theory upon which it relied. Simulations told us something about the Homeostat, and only indirectly something about the brain. But we may still wonder about the advantages to the directness we got from the analog simulation that the Homeostat provided. Several were the reasons why the analog Homeostat was a better model of the brain than the symbolic Homeostat that the ACE simulated. Ashby and Turing faced first practical issues, then methodological issues of scientific practice, and finally the issue of the effectiveness of the two simulations as a demonstration. I will now turn to these questions.

First is the question of the practicality of building each simulation. As it turned out, the Homeostat was far easier to build than the ACE. Pilot ACE would not become fully operational until May of 1950, two years *after* the Homeostat was demonstrated to the EEG Society in Bristol (von Neumann's machine was not fully operational until 1952, though this was due more to the perpetual redesign of its parts). In addition, the Homeostat was a far more economical solution than the ACE, even if it had only one application. Ashby was an intellectual entrepreneur who built a significant machine on his kitchen table out of surplus parts from the war, in a self-financed project. While this was true then, the opposite is true now, when computers sit on the majority of desks and finding the analog circuit components that Ashby used would be extremely difficult. Therein lies the true advantage of the symbolic computational simulation—because the computer potentially suits everyone's modeling needs, we can produce them cheaply in quantity and thus, despite their vastly more complex structure, they turn out to be cheaper and easier to build than specialized models. By 1946, Turing had this much down. The quality of the symbolic simulation when compared to the analog was less clear, though.

In his analysis of simulations (2003), Winsberg argues that often only the local outcomes of a specific situation, not the mathematical theory underlying the simulation, are in question. One of the great advantages of simulations for these cases is the ability to visualize the model, what R.I.G. Hughes (1997) calls their mimetic properties. This advantage is due to the ability of researchers to employ experimental practices (most notably observation) and laboratory techniques in understanding the local simulations, as opposed to depending upon mathematical practices and analytic techniques. Yet these mimetic properties are dependent upon the development of instrumentations within the simulation to make those properties visible. The symbolic processes themselves are rather opaque and not susceptible to direct observation, thus entailing more levels of indirection in the interpretation and visualization of data that simulations produce. From an epistemic perspective, we are getting further and further away from the phenomena in question. Consider the Homeostat and its ACE simulation. We can directly interact with and observe the Homeostat. But we must preprogram, any interactions with the ACE simulation, and we must interpret the results from the symbolic output of the machine. While today we could write a simple Java program with a colorful on-screen visualization of the Homeostat, this was hardly conceivable in 1946—which just proves the point that any such symbolic visualization introduces another layer of interpretation.

Other important practical issues of time, which the increasing sophistication of computing machines is similarly obscuring, are also present. First are the issues of mathematical complexity and computation time. To simulate the behavior of the Homeostat's circuits using the equations of electrical engineering, or physics, would probably have taken far too long for a machine like ACE to perform in real-time (something approximating the response time of a real Homeostat). The significance of having a real-time simulation relates to the mimetic property mentioned above. Namely, we can interact with a real-time system in ways that we must simulate in false-time systems. Science does not always need this, but it is extremely important for understanding certain phenomena. Real-time here applies to the observers' time frame and ability to act upon the system, receive feedback from that action, and evaluate the relationship of the two. This feedback loop must be real-time for the observer, even if the time-frame of the simulated phenomena is greatly sped-up (as in astronomical phenomena like the movement of the planets) or slowed-down (as in high-speed phenomena like protein-folding). The significance of this aspect of analog models becomes even more apparent when we consider the idea of a regulator as model.

8. Conclusions: Simulation and Computationalism

Von Neumann saw the power of mathematics to simulate detailed physical models, and ultimately to allow for the control of enormously complex systems involving many dynamically interdependent variables. One of the goals of the computer he built was to automate the mathematical calculations needed to simulate these systems so as to allow mathematicians to develop ever more complex and detailed simulations. The brain, too, was a natural phenomenon that could be so simulated, once we understood it scientifically. The difference between Turing and von Neumann was that von Neumann felt that Turing's universal machine concept added almost nothing to our understanding of the mind because it told us nothing about the brain. Without a clear understanding of the brain's operation, the computer would never embody an artificial mind.

Von Neumann ultimately came to see the task at hand to be one of modeling the *detailed microstructures of brain*. For him, the computer was not itself a brain, though his design for the EDVAC computer drew heavily on analogies to brain structures and neuronal functioning. The crucial aspects of the design of the computer were technical considerations of scalability, reliability, and numerical accuracy as computers were given more memory and faster processing speeds. He quickly realized that the scale of the simulations that would be needed for brains would require vast numbers of calculations, and even small or infrequent errors would result in critical failures. He also quickly became skeptical about the pace of progress in building simulations of the brain.

Turing's view of the computer was as the universal modeling machine. But rather than worry about modeling the brain in physical detail, he sought to simulate its behavior symbolically. Turing and von Neumann sought after symbolic simulations of the brain, but approached these in a remarkably different way. Ashby remained committed to analog simulations through most of the 1950s, but appeared to give up on them after his frustration with DAMS to accept that the increasing capabilities of digital computers better suited his needs.

Another way of thinking about the difference between analog and symbolic simulations is to note that the symbols of a simulation inside a computer are not connected to the world in the same way as the causal structures of an analog simulation are. For the Homeostat, these connections were highly relevant to Ashby's theoretical views of the mind.[9] Turing's claim that the distinctions are irrelevant thus bespeak a significant theoretical disparity between the two men. The best and simplest way of characterizing this difference is in terms of the relationship between a system and its environment, which for Turing is irrelevant, yet for Ashby is central. It is a distinction of theoretical and practical import. Theoretically, Ashby starts from the rich complexity of relationships between system and environment, while Turing considers the environment only in the simplest and most idealized terms possible. Practically, Ashby was interested in building a machine that an observer could directly interact with in real-time. Such a direct interaction was essential to his pedagogical and rhetorical aims, and to his experimentalist desires (see Asaro, 2006). For Turing, logical proof was the ultimate form of demonstration, experimentalism was not a priority, and the details of the physical machinery were inconsequential. These distinctions became even more pronounced over time. For even as Turing turned to deeper considerations of learning and adaptation, he became more frank about his reasons for avoiding issues of general intelligence, human behavior, and human-like bodies for computers (Turing, 1950; Turing, 1953; Teuscher, 2002).

NOTES

1. A classic example are the Monte Carlo simulations devised by John von Neumann, Stanislaw Ulam, and Enrico Fermi in the development of the atomic and hydrogen bombs.

2. I will call these "analog simulations," but will be careful to distinguish them from analog representations and analog computations, as is necessary.

3. For instance, the kind of argument made by Searle, 1980, in his famous Chinese-room thought experiment turns on just these semantic aspects of symbols. Searle aims to show that intentionality is an essential element of symbolic representation and that Computationalism, or at least that the strong program in AI fails to explain it.

4. I use isomorphism here not in its strict sense of correspondence, but in the looser sense of similarity proposed by Teller, 2001.

5. For black box simulations, we do not care about internal processes so it makes little difference whether these are achieved symbolically or causally, or even magically.

6. Computer programmers call this a "truncation" error—when the number of digits needed for an answer exceeds the number the machine can represent. In division, this generally results in a truncation or rounding-off of the decimal positions after the last available digit is used. Consider a machine multiplying two 10-digit numbers, and capable of only representing 10 digits at a time. The result of the multiplication will be at least 18 digits long, but the machine can only represent half of these, and hence can not represent a meaningful answer at all—or must convert it to scientific notation, truncate it, and thereby lose precision.

7. The early computing machines were highly specialized to perform a single class of functions or, like ENIAC, had to be programmed manually by arranging walls of dials and networks of patch cables in a fashion similar to hand-operated switchboards for telephones. For instance, Howard Aiken's Mark I Automatic Controlled Sequence Calculator, built for IBM in 1944, used 72 rotary counters for storing numbers mechanically, but had no means for storing the calculation instructions themselves—the "function" resided in the mechanical configuration of the device.

8. Unfortunately, this letter has been lost.

9. See Ashby, 1952b; Ashby, 1956a; Ashby, 1961; Ashby, 1962; Ashby, 1967; Ashby, 1968; Ashby, 1972; Asaro, 2008; Asaro, 2009.

II

AI'S PERIPHERIES

Four

AT THE PERIPHERY OF THE RISING EMPIRE: THE CASE OF ITALY (1945–1968)

Claudio Pogliano*

1. Cybernetic Disclosure

During the years of rising autarchy that the fascist government imposed on Italian culture, only a few people dared to disobey the prohibitions and opened up to foreign experiences. The *Studi per un nuovo razionalismo* [*Studies for a new rationalism*], which Ludovico Geymonat symbolically dated 25 April 1945, Liberation Day in Italy, collected all his works as follower of Logical positivism. Likewise, Antonio Banfi asked for the constitution of a new "systemic" that could counter the proclaimed radical crisis of knowledge and continuous changing of principles and methods. He also chose to translate and present the works of Alfred Whitehead, a philosopher still unknown in Italy, to counter the immobile and content-less Italian philosophy ("world with no more existence and value") with the dynamism of modern science's "Great Adventure" (Geymonat, 1945; Banfi, 1945a; Banfi, 1945b). In spite of their initial allegiance to Benedetto Croce's Neoidealism, some scholars did dare to declare the falsity of the Crocean distinction between *philosophical concepts* and *scientific pseudo-concepts*. They realized the importance of acting and planning, and of "getting involved with machines, all the machines that men built, from politics to workshops, from literature to sciences and to philosophy" (Balbo, 1966, p. 24). Thus wrote Felice Balbo, the communist and Catholic philosopher, in 1945. The following year, he explained his project thoroughly in a text he emblematically titled *Il laboratorio dell'uomo* [*The laboratory of man*]: the war had torn down a whole civil structure and our only chance was to convert to those machines that were "so much part of our contemporary living," namely to a technology that had to be improved against the prejudices of the Italian old, lazy, and romantic humanistic culture (*ibid.*, pp. 112–114).

The intellectuals gathered around the magazine *Il Politecnico* hailed Balbo's utopia. They too were prophets of a culture inclined to build a new Italy after Fascism. On 29 September 1945, the first issue of *Il Politecnico*, that Elio Vittorini directed and Einaudi published, was released. It would last only about two years,

* I would like to thank Joseph Raho for his careful linguistic revision.

but it became the paradigm of a haunted relationship between intellectuals and politics that depending on the point of view became a positive or negative model. The frequency of a scientific and technical "ingredient" in the fusion of themes and debates of *Il Politecnico* was remarkable, and Giulio Preti was mostly responsible for it. While examining the theories of Evolution and Relativity, Preti discussed forms and solutions to the crisis of modern science, warned about the risks that threatened Italian schools, and stigmatized the bad alternative between humanistic rhetoric and technophilia. Others wrote for Vittorini's periodical (a weekly at first, then a monthly) about the ambiguous results of nuclear physics so as to inform the public or to lament the Italian political narrow-mindedness that caused Italy's scant contribution to the progress of research.

In December 1945, an editorial condemned the self-assurance and ignorance with which teachers and supporters of Idealism had spoken of science, as if they were slaves of a philosophy that was incapable of becoming "master of folks." In Milan, enthusiasm greeted the birth of *Analisi,* a quarterly journal of "critique of science" that published five issues in two years (Poli, 1945). Its core objective was to be a venue for a free exchange of ideas between scholars interested in problems of general and specific methodology in the scientific and philosophical disciplines. Giuseppe Fachini, Livio Gratton, and Giulio Preti—a biologist, an astronomer, and a philosopher, respectively—were the main people in charge, and they recruited about thirty collaborators. In the first issue, *Analisi* received news about the foundation of two Centers for Methodology in Milan and Rome which, with Fachini and Silvio Ceccato as secretaries, would integrate scholars from the different fields of scientific and philosophical research (Quaranta, 1978, pp. 106–107; Cambi, 1983, pp. 33–39; Castellana, 1988–1989). The debate surrounding the concepts of causality and determinism, as well as the related question of whether the laws that the new physics formulated were "objective," stood out in its pages. The journal attempted to anchor biological investigations to the tested methods of physics and chemistry. It outlined the bases of the Operationalism of the Fifties, which some scholars viewed as a likely lifeline. Fachini would later consider that some essential features of the new discipline that Norbert Wiener in 1948 would call *cybernetics* were already present in those few issues of *Analisi* (Quaranta, 1978, p. 27).

Sigma (six issues starting in 1947) was another similar and ephemeral experience that intended to focus on the "continuous interaction among...different forms of knowledge" and on the necessity to submit their results to the control and verification that a unitary method provided. For over ten years in Europe and America, the idea of a unitary science "system" had been emerging; clearly, a new category of researchers devoted to a synthetic perspective had to flank the specialists. The editors of *Sigma*, Giuseppe Vaccarino and Vittorio Somenzi, were two young scholars from Sicily and Lombardy respectively, with degrees in industrial chemistry and theoretical physics. The first presented the dense *Elementi per una teoria della conoscenza* [*Elements for a theory of knowledge*], and the second, taking inspiration from intellectuals such as Sartre and Valéry, investigated the cognitive aspect of artistic activity.[1] They were ready to publish posthumous contributions by Eugenio Colorni and Italo Cotone, whom the fascists had killed in 1944, as well as some *Regulae* with which Ceccato began his long career as "technician among philosophers." The psychoanalyst Emilio Servadio made also a quick incursion, and Preti too, with a letter which, in spite of warning against the risks of Existentialism or mythical metaphysics, cheerfully welcomed the appearance of *Sigma* as "a first step out of the swamp." Another psychoanalyst, Enzo Morpurgo, had a good time writing dialogues about animals and machines that showed a curious "cybernetic" insight (Servadio, 1948; Preti, 1948; Morpurgo, 1948). The last issue explained the journal's closure with the need to concentrate efforts around a new magazine, *Methodos*, which I will discuss later.

The common horizon of that generation, determined to escape the hold of traditions perceived as suffocating and dogmatic, appears generically or vaguely "neopositivist." Attributing all the blame to the "dictatorship" of Croce and Gentile might have been a mirage but, as Viano pointed out, "even mirages matter" (1985, p. ix). The embrace of Positivism—a movement that was already on the wane abroad—was a tactic and momentary rite of passage that quickly wore out, leading to new trends. In 1953, when Francesco Barone, fearing the advent of a "barbaric anti-humanism," wanted to confront the logic of Neopositivism, he would historicize it, while at the same time granting it the merit of having spotlighted the relation between philosophy and the sciences (Barone, 1953, p. 395).

Within a few years, Italian culture experienced the failure of the political and cultural hopes that the anti-fascist Resistance had raised and felt the impact of Stalinism and the Cold War. First illusions were shattered and then a rigidly divided world ossified them. To align with one side or the other of this cruel battle appeared the only available option at a time of a supposedly self-evident showdown between good and evil. As the economic reconstruction of the country plodded through, from the political and social standpoint the Italian Fifties appeared sad and obscure. The suspicious and un-enlightened Italian philosophers did not realize that an epochal change was taking place in a different domain. Scientists and technicians, instead, were quicker to appreciate the new events. Between 1949 and 1950, Manlio Mandò entertained the specialized audience of *Periodico di matematiche* [*The periodical of mathematics*] on *Grandi calcolatrici moderne a successioni automatiche* [*Large modern automatic sequence computers*]. He started out, quite pedantically, by drawing a distinction between two categories—analog and digital—to highlight the new and significant qualities of the second. He went on to discuss the ASCC and ENIAC computers in detail. He then included a paragraph on machines with memory and time-delayed lines that addressed previous shortcomings such as the reduced, rapid memory and the long, industrious computations. We could conclude that these ongoing developments would have nurtured the future contribution of computing machines, with their versatility, flexibility, and speed, to the progress of mathematics that were soon to come. Such were the neat boundaries within which Mandò, prone to no philosophical extrapolation or temptation (1949; 1950), operated. Centers for this new method of numerical computation were already operating in the United States, Great Britain and France. As for Italy, he concluded:

> let us hope that some initiative will take place here as well, in order to give mathematicians, physicists, and technicians this powerful instrument of research and work. Hopefully, we will not witness an exodus of mathematicians toward the country of electronic computing machines as we did witness the exodus of physicists toward the land of "cyclotrons," "betatrons," etc. (1950, p. 97)

Again in 1949, writing in *Sapere*—the only popular science magazine published in Italy since the 1930s—Bruno de Finetti explained how "electronic computing machines"—whose incredible capacities the press was relentlessly flaunting—actually worked. They emanated, he believed, a sort of "fairy-like taste": machinery and devices (such as the brand new IBM computer) arranged along the walls of a room of thirty meters by eleven, with thousands of relay valves, and the ability to perform thousands of operations in a few seconds, and so on. De Finetti's purpose was explanatory; he did not want to shock his audience. He tried to catch the fleeting last phase of an evolution that had already begun, and which sought to make computation more rapid and automatic. He described the functions and structure of the "sequence controlled" machine installed in Madison Avenue, New York, and even dared a not quite correct prediction:

> Machines of such complexity could possibly never be mass produced and be at the disposal of many institutions. However, their computing power is such that even a few of them can carry out a great amount of calculations of different kinds. (1949, p. 41)

He presented the new statistical IBM 101 machine, which applied electronic organs to a push card system that at the end of nineteenth century Hermann Hollerith had planned to facilitate the data processing of the U.S. population census (1949), as a remarkable "improvement." While working as actuary at *Assicurazioni Generali*, charged with the task of mechanizing the main office in Trieste (*Atti Trieste,* 1987), de Finetti had acquired broad experience in punch card systems. In 1950, several mathematical meetings in Berkeley and Cambridge (Mass.) invited him, and he spent this time in a sort of "cybernetic pilgrimage" from machine to machine, (in Los Angeles, New York, Boston, Washington, Philadelphia, and Princeton) taking part in several conferences and spending time at the Watson Scientific Computing Laboratory at Columbia University. Once back in Italy, he again informed his colleagues about those "*Macchine 'che pensano' (e che fanno pensare)* [*Machines 'that think' (and make us think)*]."[2] Those who exalted the characteristics of the machines confronted against those who stressed that at any rate real people ordered their operations, thus highlighting that only human beings were capable of thinking. To de Finetti, the question, conceived as two theses propped one against the other, did not appear to be well formulated: a moral prejudice biased the metaphysical dilemma. Instead, he thought that a close examination of the given data, with patience and method, was in order. A comparative study on the "psychic and physiologic functions" of animals and computers was undoubtedly instructive for two reasons: first, to facilitate the planning of appropriate devices on the basis of knowledge of analogous organs; and second, to formulate a hypothesis or theory of organic functions and dysfunctions with the help of knowledge of artificial widgets. A spontaneous question arose, to which de Finetti tried to answer:

> So do all differences disappear? I do not say that. We *feel* it is not so. And yet, it is difficult to say where such difference begins without fearing a return to more backward positions.... A doubt persists that the difference may be perspectival, not substantial: when we think of a machine, we set ourselves at the level of the individuals who built it—we are aware of the fact that they know the trick and that the machine's activity is related to them; when we think of a man or an animal, we place ourselves on their same level, without appealing, in order to explain their behavior, to their constructor. Therefore—and this would be the problem—if we could only observe the behavior of the being we face, without understanding whether its constructor is a machine, an animal, or a human, which criterion could we apply to distinguish them? (1952, p. 15)

After an allusion to the "imitation game" that Alan Turing suggested in 1950, de Finetti devoted some pages to a description of the advantages of the binary system, memory systems, and general modalities of programming. In the second part of the article, or appendix, in order to give the reader some "support to the notions that the text exposed" he finally showed, with pictures, six examples of machines he had had the opportunity to study directly (1952, p. 14). Italy had always held a relevant position in the field of pure and applied mathematics, as the initiative of Unesco for an international center for mechanical computation in Rome attested. The CNR, (*Centro Nazionale delle Ricerche* [*National Research Center*]), where already in 1932 the *Istituto nazionale per le applicazioni del calcolo (Inac)* [*National Institute for Computing Applications*] operated under the direction of Mauro Picone, was to host it (little matters that a meeting of experts focusing on the purpose and organizational modalities of the new center, followed by a symposium on *Modern Computers* (1957), took place only in October 1956.) At that time, raising questions about the current and future Italian situation could hardly be avoided. De Finetti foresaw that even Italy, despite the huge cost of purchasing computers, would soon get out of the mythical phase. To those who feared that progress would quickly render obsolete a recently produced machine, he replied that by now research was beyond the initial stage and we therefore needed not to follow the example of the Pirandellian village of Milocca—a village that after the regular interruption of all projects had been deliberately left without electricity, "because of its awaiting new studies and new discoveries."

In 1951, Somenzi too left for the United States, on a Fulbright grant, to work at Harvard's Physics Department (that Percy Williams Bridgmandirected) and at the Philosophy Department (that Phillip Frankdirected). He later shared his crucial and even critical experience of Bridgman's Operationalism with the likes of Vaccarino and Ceccato, who in turn felt the need to use these new instruments to outline the beginnings of a *Scuola Operativa Italiana* [*Italian Operative School*] (Somenzi, 1987). In 1951, Somenzi published a popular science article on cybernetics in the weekly *Il Mondo* that Mario Pannunzio edited, stating:

> The development of computers based on the use of radio valves instead of rotating mechanisms brought to the discovery of strange analogies between them and the animal nervous system. Some resemblance should have been clear from the beginning, since machines have been studied under an anthropomorphic point of view to carry out tasks, which are normally a prerogative of the human brain. Therefore, we are less surprised that a "memory" or other human or animal faculties are proper to machines than the criteria their construction follows supply the basis for a satisfactory explanations of the self-regulation phenomena that all living creatures show in general. (1951)

The following year, Somenzi informed *Il Mondo*'s readers of the Unesco decision to make Rome the site of the *Centro internazionale per il calcolo* [*International computing center*] I mentioned above, thereby pointing to the theme of a technological gap that was bound to generate an intense discussion.

> Today, to the specialists of European countries that lack the United States' abundance of means the possibility of using their help in the most rational way, by concentrating it into shared equipment like it happens to researchers at Harvard University, is the top priority. (1952)

By the second year of the weekly journal (1950), Somenzi had published five articles on physics and philosophy. After a brief review of the path leading from the thematization of *feedback* as an active process in machines and organisms to the copyright of the term *cybernetics*, Somenzi highlighted the great potential for its extension. He discussed a variety of perspectives that could justify the enthusiasm—"which to an European cannot fail to appear excessive"—with which Wiener had launched himself into philosophical and sociological speculations.

After Liberation, the rank and file of the workers' movement and the Italian Left hailed the idea of development as progress: they defended scientific rationality because it purportedly represented the productive fields that, according to the revolutionary dialectic, the capitalist system had until then sacrificed and compressed, and because it would sooner or later throw into disarray the extant production relations. The relation between Marxism and scientific knowledge has arguably been for a long time idealized and altogether problematic. The model of the USSR, where a new human being appeared to be born thanks to an educational system based on science, dazzled researchers. In the West, on the contrary, doubting about such knowledge was "modern." Under Lenin's aegis, and at least until the end of the Fifties, the polemic idol of Italian Marxists remained the sort of "empiriocriticism" that they found in everyone—from Benedetto Croce to the language methodologists and analysts, to those who considered science a mere apparatus of rules and conventions. Marxists liturgically repeated that only dialectic Materialism would have supplied modern scientific research with a theoretical framework capable of expanding its growth.

Società, the theoretical journal of Italian Communism, disseminated and debated scientific themes. In 1953, Massimo Aloisi wrote a piece on cybernetics (or, in Latin, "gubernetica"). The biologist was interested in understanding whether cybernetics allowed us to get around (as Wiener had argued) the opposition between Mechanicism and Vitalism, thus overcoming the American enthusiasm—which jaded Europeans perceived as slightly childish—as well as the behaviorist and operative approaches he criticized. Aloisi granted that the new servomechanisms could shed light on some brain modalities, but he was also sure that the electronic brain could never explain the human brain "in its true essence." He thought this purpose could be reachable only once we understood birth conditions and the development of living matter: he viewed the physiological problem merging with the ontogenetic one. According to him, a mere analogy, whose use depended on the isomorphism among the models that human beings (Aloisi, 1953) build, was all that remained. In 1953, Bompiani translated and published the cybernetic "manifesto" that Wiener launched in 1948, extolling the new devices in a publisher's note as follows:

> "Cybernetics" is the word for the science of our time, the magic technology creating complicated control mechanisms or servomechanisms that realize in metal the equivalent of the circuitry of the nervous systems; mechanic super-brains that are even equipped with memory, like thermostats. Like the gyroscopic compass that regulates the automatic driving systems of ships, the bullet rockets that search for the target, the anti-aircraft pointing systems that set the cannon toward the point individuated by the radar, the ultra-quick computers solve the most difficult differential equations, shape the electronic prosthetic devices and the marvelous series of the most recent "robots." (Wiener, 1953b)

In his book, Wiener, who had certainly much contributed to the foundation of the new field of study, framed the scientific and technical aspects, as well as the "sociological" ones, in such a way that the publisher felt authorized to close his presentation with an uneasy question:

> The new mechanic slaves that cybernetics promises humanity will free human beings from unpleasant and slavish duties, but they are also bound to make the competition between machines and humans more dangerous. The first industrial revolution, the revolution of the "dark, satanic factories," produced a devaluation of human work force. Will the current revolution produce a devaluation of human brain power? (Wiener, 1953b)

Einaudi had preceded Bompiani by a few months, publishing *Introduzione alla cibernetica* [*Introduction to Cybernetics*], a timely translation of Wiener's *The Human Use of Human Beings*, with a foreword by Gino G. Sacerdote, then an engineer at the *Istituto Elettrotecnico Nazionale "Galileo Ferraris."* The volume claimed that cybernetics was a "new science" aimed at uniquely coordinating a scheme of different knowledge elements and at dominating broad horizons: as a matter of fact, it was a "great process of synthesis." From studying communication and information, it shifted to machines equipped with superior faculties such as memory, prediction, ideas' coordination, and learning—a move that could also potentially make the reader uneasy. Sacerdote defended cybernetics against the accusation of introducing mechanization into the study of human behavior arguing that, besides instinct, other, already existing human faculties no machine could ever possess and replace, existed. Assessing automata and providing a definition of their limits was, after all, the main target. We should embrace the hope that Wiener's book expressed, namely:

> that human beings, helped by the machine, could develop their best faculties, without becoming living machines themselves. And that all human beings could tend to a form of civilization that is not the civilization of ants. (Sacerdote, 1953, p. 10)

The *Giornate della scienza* [*Science Days*] that CNR promoted—a series of meetings in Milan, which took place in April 1954 and were devoted to electronics and television—had a section on cybernetics. Among others, the essays by Anna Cuzzer (1954) and Somenzi (1954–55) stood out. Scholars believe that the two-year period 1954–1955 marks the beginning of information technology in Italy, with almost simultaneous initiatives in Milan, Pisa, and Rome (Cuzzer, 1993). The Politecnico in Milan started a *Centro per i calcoli numerici* [*Center for Numerical Calculations*]. The center was equipped with a stored-program general-purpose computer, the CRC 102A, which the Computer Research Corporation of California had manufactured and which was purchased thanks to funding from the European Recovery Program. On 11 October 1954, the machine was unloaded at the port of Genoa and, after about ten days, was temporarily installed in Milan at the *Istituto di elettrotecnica generale* [*General Electrotechnical Institute*], waiting for the new Center to be inaugurated (Dadda, 1993; 2005). In Pisa, by contrast, the stimulus came from Enrico Fermi who, before dying, suggested to the Chancellor of the University to use the available funds to acquire "a research tool that could bring priceless advantages to all the sciences and related fields of research" (letter dated 11 August 1954) (Maestrini, 2002; Denoth, 2005). This is how the "legendary" history of the *Cep* (*Calcolatrice elettronica pisana* [*Pisan Computing Machine*]) began. The first Italian specimen of the new species became operative in 1961, after a long gestation under the guide of the physicist Marcello Conversi. In the meantime in Barbaricina, on the outskirts of Pisa, Olivetti had opened a laboratory for electronic research affiliated with the University and had appointed Mario Tchou—a young Chinese-Italian born and raised in Rome—as director. In the spring of 1957, Tchou recruited a group of brilliant engineers to build the *Macchina Zero* [*Zero Machine*], later renamed *Elea 9001*. As soon as Christmas Eve 1955, Adriano Olivetti, while reassuring his employees about the future of their traditional products, coincidentally announced a new project: ,

> We have been methodically working for four years in the field of electronics, where only the greatest American factories have been dominant. Because the rapidly growing science willy-nilly influences the desire for progress of today's society, we have focused on developing its scientific aspects. We cannot be absent from this sector, which is in many ways a relevant and decisive one. Our products, however, are not in danger: as the aeronautics industry did not stop the development of the motor industry, so electronic computing machines will not replace, at least not for a long time, neither the summing machines nor the mechanical computing machines. Instead, they afford new job opportunities for Italian technicians and workers.[3]

As for the third center of the so-called "year zero" of Italian information technology (Bonfanti, 2005, p. 3), in January 1955 the central location of CNR received twelve tons of different materials. Once assembled, they composed the Ferranti-Inac (Finac) machine, an example of Mark I that Ferranti Ltd. had built in Manchester. *Il Messaggero* captured the news with this title: *È a Roma la macchina elettronica che calcola a tempo di record e gioca a scacchi* [*In Rome, an electronic machine computes in record time and plays chess*]. As for the INAC of Picone that I previously mentioned, in 1951 de Finetti spent a year studying the possibility of providing the institute with an electronic computer. In the end, it was purchased, with CNR funding, for three hundred million old liras. Complex vicissitudes characterized the Roman and the Pisan computing acquisitions. Their histories have recently been rewritten in detail.[4]

2. Civilization of the "New" Machines: Prophecies and Worries

During the Fifties, a peculiar, albeit rare, figure of poet-engineer was behind the spark of interest for "intelligent" and "non-intelligent" machines that some sectors of Italian culture displayed. Under the direction of Leonardo Sinisgalli, Finmeccanica published a bimonthly house organ for five years. Finmeccanica controlled about thirty companies in different sectors, including: aeronautics and astronautics, calculus and application, chemistry, shipyard and railways, and metallurgy. The first issue of *Civiltà delle macchine* was published in January 1953. Its covers carried a few reproductions of Leonardo's studies on flying. "A strong food," Sinisgalli will describe it later, "of an unusual taste, a new and rich dish, an object made with artisanal care and extraordinary graphic precision: a few pictures and lots of drawings" (1978). Sinisgalli asked the authors—many of whom had contributed to Vittorini's *Politecnico*—to write long, exhaustive investigations that would rapidly spread knowledge of the unnoticed transformations occurring in Italy and throughout the world.

Although prone to the myth of his archaic and agricultural Lucania, Sinisgalli was thoughtful enough to pay attention (and to attract the attention of others) to the outburst of machines on the scene of young post-war Italy (Pogliano, 1993; Antonello, 2005, pp. 124–168). In the first issue of the journal, Giuseppe Ungaretti wrote to Sinisgalli that computers enshrined a "metric prodigy" capable of easily solving equations that would normally take mathematicians years of work. Such "articulated form" had stirred in creators and manufacturers alike "an emotion almost identical to aesthetic pleasure." Yet all human beings, including the most rigorous scientists, were none the less creatures capable of emotion; it did not appear right to the poet that the metrics of mechanization upset nature, eliminating qualities such as love, pain, tolerance, mercy, irony, and fantasy that machines do not know. People like Carlo Emilio Gadda, who had lived "among the manufacturers and conductors of machines," and had spent time designing, drawing, and testing them, were well acquainted with their constant fascination. Their defects, mistakes, and horrors were undeniable, but machines had also extended mankind's biological possibilities, and introduced an all new energy into history. Alberto Moravia believed that machines brought "infinite benefits"; the fact that they were becoming more and more perfect, complicated, and maybe even "human," did not appear to be a reason to change his opinion on the matter. On the contrary, he thought that the problematic aspects of mechanization stemmed from a negotiable imperfection: "like Achilles' spear—he declared— we are sure that machines will be able to heal the wounds they caused in the world," though he lacked the training to explain how they could do that. Replacing human beings with automata could finally end the servitude of workers (Scheiwiller, 1989, pp. 3–9).

From the pages of Sinisgalli's journal, with its unique blend of science, art, and literature, stemmed a steady flow of enthusiasm for machines. In 1956, Dino Buzzati praised him precisely for this reason:

> You have realized a truly brand new thing, with an unprecedented formula.

> Your being a poet, your engineering culture, and your incurable passion for mathematical adventures have blended together with amazing results. Right where the connection between the artistic and technical world, between ghosts and digits, could appear more difficult or even absurd, you have set a bridge that links them, disclosing a range of unpredictable correspondences and affinities, a whole range of secret communicating vases. (*ibid.*, p. 17)

At the beginning of its second year, this exceedingly original undertaking garnered also the approval of Giuseppe Luraghi, then manager of Finmeccanica. Writing to his lifelong friend, Luraghi remarked that intelligence had the miraculous ability of generating more miracles: "automata, these bunches of electric wires, nerves, and bowels of exact gears that are not prone to dyspepsia, will become our masters and will be more beautiful than we are." *Civiltà delle macchine* turned to the just appeared automata with the uppermost respect.

In 1953 Enzo Paci pointed out that these machines were remarkably different from those previously built or known. His research used Existentialism's and phenomenology's categories to question the place of human beings in a science-dominated world. For instance, William Grey Walter's electronic turtles (invented a few years earlier) proved capable, like real animals, to explore their environments, move around obstacles, and return to their cages to recharge their batteries. The so-called "thinking" machines were high-speed computing devices capable of self-correction on the basis of information about the outcome of their previous actions. Paci had drawn on the work of Wiener—whose *Introduzione alla cibernetica* Einaudi had just released—in order to explain the notion of "feedback" and its remarkable consequences. Remarkably, the principle of "feedback" led to increasingly more amazing analogies between the inorganic and organic worlds, until the resemblance of the human brain to the electronic computer became obvious. Like the human brain, it had memory, was capable of learning, and had the same logical functioning. Did such qualities have us fear a complete mechanization of thought? Paci's answer sounded reassuring, since the new feedback machines had more in common with organisms than with the old machines; the relation between the two types of machines was mindful of the relationship between classic (deterministic) physics and quantum (probabilistic) physics. In *Statistical Machinery,* W. Ross Ashby wrote about an "evolutionary" behavior of the new machines, while Wiener suggested calling them "Darwinian." Some pushed even further and defined them as "means of philosophical knowledge":

> The feedback machine is a self-ruled organism: therefore, if we consider the new machine as a model, nature as a whole reveals itself as an organic, self-ruled system in which probability makes possible self-regulation. At its highest point, the political and ethical history of mankind manifests itself as a complex organic process where mankind must continuously struggle

> against disorder. In order to do this, human beings must be free to choose, find new paths, and *invent* new means.[5]

In its first four years, *Civiltà delle macchine* devoted over twenty articles to the different aspects of the "interscience" that Wiener had named cybernetics. Each epoch has its automaton—as the physicist Carlo L. Calosi noted in the first issue—and in time such automaton perfects itself, getting not just progressively closer to, but often surpassing the living model. With the advent of electronics, the automaton of the second half of the twentieth century "had muscles, senses, a nervous system, and—within limits—intelligence." Mankind assigned it a set of instructions to carry out certain operations and produce the desired results; as far as the automaton is concerned, it executes the task. Calosi wanted to stress the importance and irreplaceable contribution of mankind: on the one hand, he reassured those who feared "a crazy attempt to deprive humankind of its distinctive qualities," and on the other he disagreed with those who thought an automaton could replicate all human capacities. His outline of this new companion, the automaton, followed the profile of the perfect worker:

> The automaton has its own personality traits: quite accurate and constant in its work, it is disciplined in carrying out orders, and has a solid memory. The precision of its work surpasses that of any human being. Its senses are far more developed than those of any living creature; it receives more information than them from the surrounding external environment. It sees and hears things that human beings, because of the distance from the source or because of the nature (frequency) of light and sound, cannot perceive. Its nervous system and its information processing centers, from which the decisions to act stem, are far readier than those of any other living creature. (Calosi, 1989, p. 160)

If a technician such as Calosi was careful to set boundaries to the overflowing interest for cybernetics, others instead dreamt of spreading it even more widely. In 1954, Vaccarino put forth the construction principle of a machine "capable of using our words in (and for) those situations in which we use them." He applied the analysis of language that the so-called "Italian Operative School," whose members Sinisgalli hosted in his bimonthly periodical because he was curious about their original style of thought, had been carrying out. Instead of being phonetic, the longed-after "talking" machine would express meanings by means of a semantic process on symbols. Vaccarino was aware that to most people the prospect would appear "absurd and crazy" and that detractors would argue that these machines, sort of novel version of *homunculus*, should, before *speaking*, learn how to *think*. Indeed, such was the intended attempt, though no one would now expect it to happen by way of the distillations and calcinations of the alchemist's vase. In order to demarcate the field of research and to make this new "synthetic man in miniature" believable, he surmised that the idea of "thought" as something mysterious or even supernatural, as a manifestation of an immortal soul as in the Greek philosophical tradition, had to be dismissed. Instead, the *Centro italiano di metodologia ed analisi del linguaggio* [*Italian Center for Methodology and Language Analysis*], which had been founded in Milan a few years earlier, was developing an alternative and "operative" conception of thought:

> When we say we think, we do not mean, as philosophers argue, that some knowledge process is taking place in us; instead, this verb's ambiguous meaning [to think] covers the special operations that we carry out .... To imagine a machine that correlates each thing it produces (for example, the observed things) with a differentiating symbol, which the machine chooses so that it coincides with the symbol that a current language already uses for the same purpose, is not so difficult. That will be the talking machine, the machine that semantizes what it produces. (Vaccarino, 1989, pp. 161–164)

Since 1949, the Milan Center published *Methodos*, a quarterly periodical divided into two sections, one devoted to methodology and one to symbolic logic. The first issue opened with a piece entitled *Betrachtungen zur Axiomatik* by Hugo Dingler, a German mathematician, physicist, and philosopher whom Ceccato—the Center's animator and the journal's director—designated as the journal's favorite commentator and with whom he entertained a thorough correspondence (Ceccato, 1952; Dingler, 1952; Oliva, 2001). Ceccato pursued the mission of converting thought into an "operative technique" with extraordinary perseverance. His perspective had nothing in common with the "theoretical" attitude of philosophers: "the issue is to consider whether the things we talk about contain a process and, if so, what kind of process this is" (Ceccato, 1953, p. 250). *Methodos* had many original features, including translations into English of articles in Italian; perhaps more importantly, it showcased authors who were mostly unknown in Italy.

As Somenzi thoughtfully observed in 1955, seven years had passed since Wiener's announcement of cybernetics and Von Braun's work on the application of a multistage rocket to astronautics. Considering the rhythm with which scientific and technological research usually develops, the public opinion expressed some impatience toward these *homunculi* and spacecrafts that did not appear to have gained much traction. Somenzi tried to address the apparent delay by way of an "economic" explanation: if the manufacturing of artificial individuals and satellites had guaranteed a military victory over one of the powers warring over the control of the world, such dreams would have been realized in a short time. In fact, some of the world's super-powers had quite possibly already undertaken secret research programs that were more advanced than the currently known ones. One factor could not be neglected: "the possibility of broadcasting TV, commercials, or propaganda programs that could be sent from a single satellite over entire continents constituted an economic and political mainspring"(Somenzi, 1955, p. 40). As for cybernetics, the developments in the mathematical theory of information and its applications in the fields of communication and cryptography, as well as in the fields of theoretical physics, biology, economics, linguistics and psychology, had not failed. In any event, the design of mechanical automata imitating human and animal functions that so far had been limited to living matter undoubtedly prompted great interest. On this point, Somenzi was careful:

> Without taking away from people the hope that cybernetics will sooner or later reach the desirable tools for the "intellectual prosthesis" of the mentally deficient, we think crucial to study the different types of electronic brains not in order to explain and reconstruct the cerebral phenomena that made such results possible, but to reproduce its functions through the composite tool/man that uses it, thus sparing to man the part of the work done by the tool .... Computers have not been manufactured to study the mental behavior of the accountant or mathematician, but to help them (as the abacus did) with their work concerning commercial, technical, and scientific purposes .... But which company would deem worth of generous funding the project of a talking machine able to discuss philosophy? In cybernetics, instead, such a robot would be the "greatest" achievement. (Somenzi, 1955, p. 41)

In a long lecture that Ceccato read at the Royal Society during the third "Symposium on Information Theory" (September 1955), the dream of a "talking" machine that Vaccarino had mentioned the year before was back with plenty of details. Once freed from the proclivity to "philosophize," a kind of original sin that everyone appears to display, "the whole mental sector of human activity" opens up, he said—first to the anatomist and physiologist, and then to the manufacturer of mechanisms. One of those mechanisms was the automatic translator, whose potential workings Ceccato illustrated: the device would save time in reading the word to be translated, in looking it up in the dictionary and, finally in writing it (Ceccato, 1989). Enrico Maretti, a young engineer whom Ceccato included in an ambitious enterprise that Sinisgalli will finance the following year, had collaborated on the study of mechanical translation. The final product was exhibited at the Finmeccanica stand of a Milan exhibition on automation. The poet-engineer christened the "mechanic model of mental operations" *Adamo II*. It purportedly showed the efficiency of even a fragment of the brain conceived as an organ of thought. In *Civiltà delle macchine*, Maretti explained the long series of attempts, of "doing and undoing" that the artifact had required. Language being the main access to brain's activity and its "isomorphic" product, the researchers had chosen twenty-three mental operations that they had deemed to be the most frequent and crucial. *Adamo II*'s purpose was to showcase the formation of three notorious philosophical sentences: *Cogito ergo sum*, *Verum et factum convertuntur*, *Das Für-sich-sein im Anderssein ist der Prozess* (Maretti, 1958; Maretti, 1989; Ceccato, 1987).

Adamo II had already been announced during a philosophical meeting at the Università Statale of Milan in April of 1955. The Catholic philosopher Gustavo Bontadini sarcastically recounted the episode in a very polemical report aimed at his "anti-metaphysical" colleagues (such as Banfi, Paci, Cantoni, and Carbonara):

> In the last hour of the meeting, after Ceccato had spoken, a great deal of interest stirred up, and everyone gathered around the speaker, bombarding him with questions. Ceccato spoke about the cybernetic machine that could translate a secondary school Latin lesson or the radio-broadcast of a soccer game. It concerned the manufacture of the *homunculus*. Even though almost no one among the philosophers had understood Ceccato's technical content, no other presentation aroused as much interest as his. The manufacture of the *homunculus* would be a prelude to the manufacture of God. If we predict that mankind will go to the moon in thirty years' time, then the manufacturing of God might well take a few centuries. The question is just of time, however—of measure. (Bontadini, 1955, p. 301)

As attested in an anonymous note (*Le prime accoglienze di Adamo II* [*First Receptions of Adam II*]) that follows Maretti's article I mentioned above, at the exhibition inaugurated on 8 April 1956 at the *Museo della scienza e della tecnica* [*Science and Technical Museum*]—an international CNR meeting on the problems of automation—that "brain fragment" began "to exert the appeal of its red lights" on the audience. Many newspaper and magazine articles announced the cybernetic *homunculus*—"a cranial box without eyes, ears, and nose," as Sinisgalli wrote. Responses went from amazement and distrust to a concern over its implications. Dino Buzzati wrote about these reactions in the *Corriere della Sera* as follows: "If manufacturing a machine that reproduces human thought is at least theoretically possible, should we conclude that mechanical laws rule human beings? And would their freedom be lost?"

A few months later, in a surprising coincidence ten young mathematicians and logicians met at Dartmouth College to examine programs that would allow machines to simulate each and every aspect of learning, or each and every other feature of human intelligence. At that same meeting John McCarthy notoriously suggested the term *Artificial Intelligence* for the new field of research that he would later ask the Rockfeller Foundation to fund (McCarthy et al., 1956). Over one thousand participants took part in the Milan one-week-long session, a convention that produced three volumes for a total of 2,500 pages. A discussion of the most recent technological innovations took place in this crucial occasion, which showed also Italy's weakness and unpreparedness towards these challenging developments. Gino Martinoli sounded a quite dramatic alarm:

> It appears to me that we live unaware of the abyss that is open in front of us: a society that, compared to the civilization and economic structure toward which the rest of the world goes in the wake of great technologi-

> cal developments, has an education system that is inadequate in size and courses of study. Our country may end up simply watching such evolution and transformation without taking part in it, and this because, even more than means, we may lack educated people. New techniques, new plants, prodigious machines will be eventually imported to our country, sure, but foreign technicians will have planned, studied, and built them. We could envisage a new form of "colonialism," not of foreign armies treading on our land, but of foreign engineers and technicians controlling and dominating those machines .... In spite of the appearance of formal freedom, machines whose control and management has escaped the hands of Italian children will slavishly dominate and enslave our country. (1958, p. 2319)

Gustavo Colonnetti, appointed president of CNR after World War II, with the difficult task of rebuilding the whole national system of scientific research, had immediately called for facing head-on the deep turmoil that automation would bring to the economic and social structure (Colonnetti, 1958). During the meeting, others explained that, because after all the human brain devised and controlled all servomechanisms, the current technological revolution—summarized as "less labor and more brains"—would triumph over physical labor without humiliating and discouraging mankind. Thanks to a broadening of markets, a beneficent revolution would increase, not diminish employment, while the smaller number of workers that the productive process would need in the short run would shift to the tertiary sector.

A curious group picture shows teachers and students from the 7th Theory of Information course seminar that the *Scuola internazionale della Società italiana di fisica* [*International School of the Physics Italian Society*] organized in Varenna. It lasted from 7 to 19 July 1958. Wiener stands out at the center of the picture, along with Eduardo Caianiello to his right: after earning his Ph.D. at Rochester, Caianiello had been professor of theoretical physics at the University of Naples for two years. Setting off that initiative, he defined the theory of information as the "mathematical core" of cybernetics, which he described as an attempt to synthesize and bridge over different branches of knowledge, a wide field comprising an almost infinite variety of themes that would be subject to a unitary method. The great development, which had already happened in other countries, brought him to foresee a second industrial revolution not less unsettling than the first one. As far as the country in which the course was taking place was concerned:

> So far, Italy has been totally absent in the active fields of research and their application. This is indeed a painful situation, since many of these investigations require a quite small amount of money. The seminar that I have the honor to lead has been conceived and organized with the hope that it will jump-start some serious research in the field and some effective collaboration in the national and international areas.[6]

In 1961, after three years of discussions with colleagues with whom he had been sharing the same neuro-anatomical, mathematical, and theoretical physical premises, Caianiello published the *Outline of a Theory of Thought-Processes and Thinking Machines*. The implicit conviction was that the human brain, though terribly complex, obeyed dynamic laws that, if we observed the operations of single neurons, were not necessarily complicated. Researchers believed that such laws produce collective behavioral attitudes—correlated to thought processes—in great collections of cells (Caianiello, 1961). Following up on the international collaboration he had called for during his 1958 course, Caianiello kept on his relationship with Wiener, who was at times a guest at the Laboratory on Cybernetics of the Institute of Theoretical Physics at the University of Naples in the following years. The *Rendiconto* [*Report*] of the Academic Year 1961–1962 shows a group of researchers already involved with different theoretical and experimental projects that included: a physical model of neuronal nets (*Educanda*), a machine that recognized forms (*Procuste*), and studies on memories and works of neuro-anatomy and physiology (Braitenberg et al., 1964). The International School of Physics, which the Institute of Theoretical Physics created and NATO financed, offered courses that kept researchers in touch with the international scientific community. Traces of this are found in the lectures Caianiello published in 1962, which were dedicated to the *Cybernetics of neural processes*, and in the later *Automata Theory* (1965; 1966). The Laboratory became part of CNR in 1968, with an independent main office in Arco Felice. In 2003, ten years after Caianiello's death, the *Istituto Italiano per gli Studi Filosofici di Napoli* [*Italian Institute for Philosophical Studies in Naples*] held a meeting on Caianiello's "scientific legacy": the proceedings include a brief testimony by Carla Persico Caianiello, who recalled her husband's affinity for oriental cultures and his collaboration with Soichiro Honda in arranging the well-known international symposiums *Discoveries* since 1976 (Termini, 2006).

3. Not only Philosophers

On the political side, 1960 represented a watershed in the history of the Italian Republic. The country shifted toward new political coalitions that would mark, warts and all, the incoming decade and that would influence the culture of the country. Its economy entered a phase of great expansion—a "miracle." The same turning point applies to other processes concerning scientific culture in Italy. The public enthusiasm for the sciences, which had after all been instrumental to the Allies' military victory, gradually started to fade. In 1960, Mauro Picone chaired a committee that proposed to associate all those involved in the different areas of automatic calculus. On 4 February 1961 they registered the statute of AICA [*Associazione Italiana per il Calcolo Automatico*, (http://www.aicanet.it/)]. AICA was bound to be quite active. Again in 1960, during a meeting in Padua, the philosopher Augusto Guzzo vented a discomfort that had been repressed for too long. Mankind can enjoy tools, he said, "but, please, to tie mankind, this great thing that mankind is ... to the car—and why then not to the reactor or to the rocket? The childish epinicion [a celebratory ode] of science does so" (Guzzo, 1962, p. 180). If science is a great adventure, and if we are the ones who make it, we cannot let it rule and become its slaves. Guzzo acknowledged, with satisfaction, the wearing out of the anti-metaphysical "chatter" that an "irresponsible" Neopositivism had nourished; Geymonat's speech at that same convention showed the extent of the turnabout. It is difficult to understand why Geymonat chose to announce his conversion to Realism in front of an audience composed of Catholic philosophers who could barely wait to make use of it.

The *embrassons-nous* that took place in Padua and appeared to unify members of schools usually in conflict turned out to be ephemeral and equivocal. The nineteenth National Congress of Philosophy opened in Bari in 1962 with many Catholic philosophers participating. Since the topic was *Philosophy Confronting the Sciences*, Guzzo could easily reaffirm the unshakable reasons of a philosophy that was at least "a reflection on that same science that wants its death." Philosophy did not have to be forgiven for still existing; instead, it should show pride in asking "what science is and how it can rise." Guzzo opened the convention, which took place partly in Lecce, in a region (the Salento) famous for its marvelous olive trees and grapevines—sacred plants of a still lively and prosperous civilization that had so far been removed from "science as most perceive it and from flashy technology." Had Northern Italy been better acquainted with the South and had it loved it more, it would not have hurried so much to "northernize" it (*ibid.*).

A year after Wiener's death in 1964, the newly-founded *Istituto internazionale delle comunicazioni* [*International Institute of Communications*] in Genoa organized a memorial meeting in his honor. The four-day meeting involved over one hundred participants, and took place in the Aula Magna of the College of Engineering. Wiener had been a guest of the local Institute of Physics in 1961. He loved the Riviera Ligure so much that he had almost decided to retire there from MIT, though unfortunately this remained only a wish. The physicist Giovanni Polvani, president emeritus of CNR and of the convention, recalled how we had come to see information as the phenomenon permeating the whole universe, common to all living and non living organisms. I want emphasize Polvani's crucial role, and especially his promotion and support of cybernetic investigations. In the 1967 Pisa meeting that he opened—and which gave birth to an institute affiliated with CNR two years later—he spoke about the "very wide and widely different fields, an immense range of facts and phenomena" that cybernetics would soon cover (Baldocchi and Lenci, 1971, pp. 1–4). For the most part, this was due to the late Wiener's personality and work, which his friend Giorgio de Santillana eulogized in Genoa. The most remarkable feature of the convention that we can glean from the daily press covering it, such as the *Secolo XIX*, is the quantity and quality of the contributions. Warren McCulloch, Dennis Gabor, Jerome Lettvin, Marvin Minsky, Seymour Papert, and Joseph Weizenbaum all presented papers. The articles extolled the vitality of a discipline that reached the living core of all sciences and created a sort of free zone where the universal currency of cybernetics would make all kinds of intellectual exchanges possible. On that occasion, a remote consultation between a human female patient who was at the meeting and the psychiatrist that Weizenbaum's group had incorporated into the program of an MIT computer, took place. Ceccato too gave a lecture; the philosopher Evandro Agazzi warned against the application of terms we normally use for higher human activities to machines.[7]

The voices of the philosophers were markedly loud in that decade of increasing upheaval. After the national convention of 1962, the Perugia gathering of 1965 faced issues that the emerging scientific knowledge had similarly brought forward: the general theme was *L'unificazione del sapere* [*The unification of knowledge*].[8] As Franco Lombardi said in his foreword, not only philosophy had not exhausted its task but, aware of the huge progress in the sciences and of its critical role, philosophy was now ready to intensify its dialogue with specific sciences. The physicist Giorgio Salvini, who was present at the meeting, shared the same opinion and highlighted the "dramatic necessity" of a debate among specialists working in increasingly fragmented disciplines. Someone had to feed aims and wisdom to science: "wisdom was once associated to philosophers. Maybe, it could still be that way" (*Atti Perugia* 1965, p. 459).

But the community of Italian philosophers did not present a unified front. Gathered in Pisa for its 21*st* meeting in April 1967, it discussed and split on the theme they had chosen: *Man and Machine*. The event greatly surpassed the previous ones in dimensions and consequences: there were more than four hundred participants, more than seventy reports, and dozens of speeches and replies. According to Guzzo, the 73 years old president of the *Società Filosofica Italiana* [*Italian Philosophical Society*] who authored one of the two main reports, a clear difference of opinions emerged: "some have resolutely denied the possibility of automata acquiring a future independent life, while others have passionately prophesied that result." Guzzo's witty or tendentious synthesis summarized a far more complex and blurred event, whose real protagonist had been Somenzi, the 49 years old secretary of the Society. Somenzi was the main responsible for summoning his reluctant colleagues to be part of the endeavor. Two years earlier, he had been in charge of the editorial operation that introduced authors such as Wiener, von Neumann, Turing, Sherrington, Ryle, and Shannon to the Italian audience. A brand new universe of ideas had managed to stir a lot of interest in spite of being at odds with the national *genius loci*. The anthology on neurocybernetics that soon followed it determined, along with other factors, the choice of the theme for the Pisa meeting.[9]

In his presentation, Somenzi first highlighted the necessity to define machines so that we can place them among all the tools that mankind uses; to this end, he briefly reviewed the changes that occurred in the concept of machine since the Industrial Revolution. Even if some philosophers had realized that we could use them also for purposes as different as the mental work of calculating or keeping time, until the Nineteenth century, he said, machines were mainly to replace animals and human beings in the supply of physical labor. Although Babbage had already worked on the project of a universal computing machine, discerning "energy" from "information" became possible only in the mid-Twentieth century. Eventually, Turing generalized the project in its most abstract form. Since then, according to Somenzi, a general notion of machine as a "system capable of assuming a single state among different ones, a list that pairs up a particular action to each one of such states, and the set of rules to switch from one state to another" had taken root (1967, p. 55). Somenzi was trying to make the significant point that philosophers could no longer deny that dealing with such machines meant to go far beyond clocks, locomotives, looms, and player pianos. The advent of an "ideal automaton" that we could use also as a mathematical model of human mental activity or as a more efficient artificial substitute of the human brain, entailed "that a machine could think and that men are machines of the same kind," radically changing the terms of the debate. Somenzi quickly listed the achievements and failures of cybernetics; in the end, he hoped that scientists and philosophers would collaborate in examining the limits and potential of intelligent machines, as well as meditate on thought and language "as machines," keeping an eye to a future where the distinction between nature and technology would probably get blurry.

President Guzzo, a Catholic and Idealist philosopher who must have tried his best to rein in his irritation toward the invasion of machines, gave the plenary speech. More than ten years earlier, he had completed the third part of his philosophical system with a thick volume titled *La scienza* [*Science*], where he argued that we had to devote a long standing and just attention to technical and scientific experience, without which mankind would truly not be such (Guzzo, 1955). He claimed that the essence of human beings, however, would never be dissolved in the field of science. A strong limitation to the appreciation of technology as one aspect of artistic creativity, the issue returned in 1967. By assimilating the automatic machines and the electronic processors to the thousand year old *poiesis*, Guzzo could deny their brand-new originality. He was especially concerned about the risks connected to the worship of technology: human beings must oversee themselves and the machines they keep on building, in order not to forget and alienate themselves in them. They should always be free to change their techniques and to invent other machines, "they the lords of the Sabbath, not the Sabbath the lord of their minds" (Guzzo, 1967a). In response to the thesis of the Secretary, Guzzo said:

Such is reality nowadays: there—and not in a vague reference to "generic

> machines"—I place my considerations on the power that modern mankind sometimes perceives as "omnipotence," hoping to rule the universe through the automata that are overwhelming the world. Somenzi gives voice to the great expectations of this awe-inspiring movement and his wish to build artificial brains; at the same time, however, besides its achievements, he points out also its failures. (1967, p. 104)

Space limitations forbid me to give an exhaustive report of the Pisa meeting (documented in three volumes). Instead, I limit myself to a brief summary of the positions there expressed, following the example of those who commented upon them during and after the meeting. The panorama paints two opposite sides and different in-between nuances. On the one hand, the "conservatives" were intent on rejecting the reducibility of the mental to the cerebral and the generic plausibility of intelligent machines; on the other hand, a group of "progressives" trusted the potential of thinking automata. In between the two ends of the spectrum lay a series of different arguments about intelligence, conceived as a function of the nervous system that is nonetheless not easily transferable to artificial subjects (Statera, 1967; Urbani, 1997; Di Giandomenico, 2000; Pogliano, 2007).

Civiltà delle macchine offered a harsh review of the Pisa meeting: given its theme, the journal argued that a more dialogical and less declarative style should have prevailed. Instead,

> we review the convention with a sense of uncertainty, of a missed opportunity, with the pronounced persuasion of a two-faced, Janus-like discourse that cohabits and grafts itself upon itself, without ever looking at itself. In this way, though unintentionally, philosophers and scientists did silence one another.... We witness the discomforting and sterile paradox of an incommunicability that is more emotional than theoretical and conceptual.... We have the impression of witnessing a conflict between two different ways of thinking that amounts to a clash between the past and the future. (g. j. 1967, pp. 2–3)

This evaluation was perhaps too harsh, since philosophers, willing or not, had dealt with machines for at least a few days. But I can recall another event that shows the intensity with which Italian culture was experiencing the metamorphosis in the sciences. From the 16 to 18 October 1967, the *Accademia dei Lincei* in Rome called together thirty-four speakers from diverse fields of knowledge to present their views on the scientific, technical, and social aspects of electronic automation. The mathematician Beniamino Segre served as president and opened the sessions acknowledging how different the opinions were on the "spiritual" importance of the new electronic computing machines. Such a diversity had surfaced a few months earlier in lively debates at the 21st Philosophers' National Convention:

> There are those who right-on consider the present time as the Cybernetic Age and who predict or fear the domination of the machine on mankind;

> others believe in the unlimited possibilities of the new tools, while some see insuperable limitations to such possibilities. It should be noted, though, that the most cautious and reserved in their appraisals, in spite of the brilliant practical implementations they have already achieved or that they will achieve soon, are those technologists from whom we might have expected enthusiastic judgments and forecasts. (Segre, 1968, p. 10)

After Segre, the Minister for Scientific and Technological Research, Leopoldo Rubinacci, pointed out that all technical innovations unavoidably provoked anxiety; rivers of ink had been printed about the possibility that the new automata could one day prevail over their manufacturers. After all, mankind's recent history was a dyad of "human beings and machines, machines and human beings—a pendulum of mutual influence in perpetual oscillation" (Rubinacci, 1968, p. 15).

On that occasion, the mathematician Alessandro Faedo, Chancellor of the University of Pisa, presented himself as the historian of the "automatic calculators" (from the abacus to the circuits with transistor); he lamented the "disablement" of Italian universities that, lacking autonomy, were unable to organize courses "for those young people who will have to use the new machines or give technical assistance to the customers."[10] Luigi Dadda presented *Situazione e prospettiva del calcolo elettronico in Italia* [*Situation and perspective of electronic computation in Italy*], and was among those who had introduced that kind of computation at the Milanese Politecnico a few years before. At the end of 1965, almost 1,200 machines were set up, namely 18.2% of those operating in the EEC (European Economic Community) countries, while the worker density per million was 82 against 112 of the EEC average and 660 of the USA. The researchers estimated a "discrepancy" of around 8.5 years relative to the USA (Dadda, 1968). A list of reports regarding the current and future changes that computers brought about in the scientific fields of mathematics, particles physics, space research, astrophysics, and meteorology, followed. Among those, three talks regarding linguistic applications stood out. After all, since 1949, the Jesuit Roberto Busa had been using punch card machines, later replaced by electronic computers, for lexicographic analysis in the Gallarate Center.

On the second day of the convention, Giuseppe Montalenti opened a session on the applications to biology saying that the use of computers had allowed research in genetics that would have otherwise been impossible; there was also the simulation of continuous and rhythmic processes, like the movement of gene frequencies in populations or their growth; but the most relevant application remained the modeling of the elementary functional phenomena of the neural system. Montalenti had no doubt that mental functions, even the "superior" thinking functions, were accessible to a "mechanist investigation" (Montalenti, 1968). Valentino Braitenberg and Antonio Borsellino, representives of two local areas quite active in that line of research—respectively, Naples and Genoa—took up

Montalenti's words. During the discussion, Braitenberg and Montalenti declared themselves sure that biologists were those who had gained the most from the fusion between their science and computers. Braitenberg first evoked a new enthusiastic "cybernetic atmosphere" and gave some examples of studies stemming from it (1968, p. 216); Borsellino, hoping that the "pessimists" would not reply with a prejudiced rejection, entrusted the "optimists" with the task to prove that the brain carries out the activities of the mind without residuals. To the pessimists who had been bringing attention to the substantial technical failures of the existing intelligent machines, Borsellino replied that responsible for those failures were overly ambitious goals pursued with overly limited resources:

> too few input "sensors," with only one channel, too few computing levels, too limited memories, etc., to compete (speed apart) with the huge wealth in sensor multi-channel inputs, with the complexity of the computing levels of the natural intelligences.[11]

Again Somenzi dealt with the philosophical issues that artificial intelligence raised by avoiding the *a priori* assertion that only human beings were intelligent. Yet, it appeared that the efforts scientists and technicians made to explain the comparison between the brain and the electronic computer did not defuse the mounting fears about the "materialist" consequences of the potential successes achieved in the simulation of mental activities. Somenzi, ever since he had chosen to be artificial intelligence's defender, tried to address the objections that anxious critics had been putting forth (Somenzi, 1968). In the discussion that followed his presentation, he simplified the problem in the following terms:

> As a matter of fact, the most direct purpose of current research in artificial intelligence appears to be the production of instruments superior to the average human brain in the execution of well-circumscribed operations, such as the prediction of the process of determined natural phenomena according to their past, leading to the eventual formulation of hypothesis, theories, and laws on their genesis and evolution or the discovery of new facts or demonstrations in the mathematical sciences. In order to coordinate the huge quantity of data that scientists gathered from electrophysiology and brain biochemistry in a unique interpretive scheme, we should put the data from psychology, linguistics, and other sciences of human behavior in the electronic computer only when such techniques will apply to the study of the human brain. (*Atti Roma* 1968, p. 242)

Somenzi had also the help of de Finetti, who was convinced that we should correct many mistaken ideas about different aspects of electronic automation. The task was urgent, because the range of possible applications of such automation was becoming more and more flexible and widespread. It therefore became necessary to mix sufficient imagination with adequate competence and realism to realize them in the best possible way. Some "metaphysical biases" against the tasks supposedly reserved to human beings, which machines had presumably usurped, were also present. Without dwelling much on this point, de Finetti remembered how machines would relieve human beings' intelligence from wasteful routine tasks and from the insufficiently rapid elaboration of the vertiginously increasing masses of data. In any case, de Finetti claimed that it would be preferable to base each opinion on the dispassionate examination of advantages and disadvantages, instead of holding "on an *a priori* thesis boasting the supremacy of the human being, alone or helped by the machines he has created" (1968, p. 324). During the conference, others spoke at length of automation as a way to free human beings from their boring, difficult, repetitive, and banal tasks and to enable them to make the most intelligent and responsible decisions—for instance observing the crucial effects of automation in the industrial arena (Quazza, 1968). The investigations dealt with the economic premises, the consequences, and the social aspects of automation. Federico Caffè sounded a warning note about the risk of enhancing the already jarring social disparities or of replicating the sufferings that had characterized the first industrial revolution and, later, mass production. We should not neglect, he and others said, the psychological effects on workers (Travaglini, 1968; Caffè, 1968; Meschieri, 1968). From the peculiar and privileged observatory of IBM in New York where he worked, Eugenio Fubini stressed that "today we can only start to see how electronic computers will modify the course of our existence." He foresaw a not so distant future and an evolved society when each individual would be familiar with the language of computers in order to partake in the advantages of an environment that would be built upon them. A technological "discrepancy" between Europe and the United States was already a reality, and he wondered which relations we should encourage between the two Atlantic shores to bridge the gap. According to Fubini, to intensify and strengthen the continents' interdependence was the better strategy:

> European technological skills will develop far more quickly if the "Old Continent" will continue to put to use the information and knowledge flowing from America, instead of placing its efforts along criteria of national or local self-sufficiency. I do not mean to become the supporter of the *Americanization* of Europe, but instead of the *Europeanization* of technological procedures and commercial methods that have proved to be extremely profitable in America. (Fubini, 1968, p. 403)

The last session of the Roman conference, which Guido Calogero concluded highlighting how 1967 had been a propitious year for the "spirit of cooperation" among philosophers and scientists coming from different disciplines, was entirely "philosophical." The philosopher Calogero declared himself satisfied that the free spirit of the recent Pisan meeting had carried on also to the last four days of this convention. To remember—as the mathematician Faedo had done at the beginning of the conference, and Marino Gentile (professor of philosophy in Padua) was doing at the moment—that the names of two great philosophers, Pascal and Leibniz, stood out at the beginning of the history of automation, was perhaps appropriate, too. As "unskilled" as he was, Gentile expressed gratitude to his colleagues in the sciences who had offered such a "marvelous" outlook—principles so simple to begin with, but generating such a great complexity. He separated two fundamental approaches regarding automation: the *misoneistic*, which defended a real or fictitious tradition, and the *mythicizing*, which inclined to exalt the new for its own sake. These attitudes were opposed but also complementary, and perhaps altogether avoidable if we rejected all suggestive language and brought back all implementations to the underlying scientific principle (Gentile, 1968). Less "unskilled" than Gentile, Barone also suggested ruling out "enthusiasms or preconceptions" in order to separate the current and future products of science from the prophecies of science fiction. Barone too recalled the meeting of a few months earlier, listing the issues that automation had put in front of philosophers who over the course of the century had become increasingly receptive to the linguistic and analytic dimensions. The interest in the creation of a new Adam was deep and understandable, and it brought back to the fore the old mind-body problem and, more generally, an extremely innovative drive to overcome older schemes and structures. Up to this point, at least a few philosophers displayed enough good will to learn from scientists. But, in reverse, did philosophy contribute to scientific and technological research? Barone was obviously convinced that it did and he first spotted it in the "mental agility" and "critical skill" stemming from an understanding of how the conceptual and linguistic realms are never definitive or absolute: two qualities that he said were essential to people who invent and project. We could only avoid the danger that society would transform itself into a "mechanism of automata, following the rigid tracks of defined programs," if analysts and programmers dared to critically evaluate their artifacts and kept in mind what was outside their field (Barone, 1968, p. 509).

With his usual verbosity, Guzzo resumed the thread of the lecture he had held in Pisa in April. In front of the new machines, he said, the disagreement between "old" and "young" people was acute—these terms, however, did not necessarily pertain to real life age differences: "the more the eldest give vent, or worse, brood, gloomily and with hostility, the more jaunty and mocking the polemic young people become toward them." In Pisa, Guzzo had already spoken of "slave" philosophies (that technological progress had hypnotized and ruled) and "savage" ones ("with eyes closed to experience and with ears intentionally stuffed to avoid hearing the sound of the contemporary civilization they reject") (Guzzo, 1968, p. 519). He sang again the praises of a third path, proper to "civil" philosophies that are instead ready to confront progress without idolizing it. Worshippers and haters of machines were likewise fomenting the mythology surrounding them. They only pretended to know them and ignored their nature of "things" that worked when and how we decided to make them work. The copious literature on the miracles or threats of mechanicism, the dream of machines that were beneficent and healing or nightmarish, were equally signs of the contemporary neurosis. These opposites were the hallmarks of adolescent behavior. Guzzo preferred to rely on lucid and adult minds and on the awareness that the lectures at the Lincei conference exemplified, which he appreciated because they conveyed a new conceptual framework:

> it is no longer time—neither we regret it—for scientists to incur as a duty in the anti-philosophical dispute; likewise, it is no longer time for philosophers to show indifference, or even contempt or ironic detachment toward the work of scientists. The past generation of anti-philosopher scientists and anti-scientist philosophers is no longer the present generation. Today, physicists are capable of doing good natural philosophy and they say so. The mathematicians have reached such a critical awareness of their work to know well that sometimes they are doing philosophy. On their side, the philosophers of the present generation are aware of the relevance that the work of scientists has for philosophical thought. (Guzzo, 1968, p. 524)

Scientists welcomed that opening, with a few symptomatic comments. For instance, Conversi claimed that he had been the first to propose and invite a "philosophical delegation"; he recommended "agnosticism without prejudices" as the only rational stance toward disputes. They had to discuss issues with serenity; the fear that machines were "stealing the art of thinking from us" had to be exorcized. If a great deal still belonged to the realm of science fiction, rising up against the idea that perhaps some of it could, in the near future, become reality, was futile (*Atti Roma* 1968, p. 532–533). Borsellino, for his part, was glad to see that a "peace treaty between science and philosophy," first elaborated in Pisa, was finally signed in Rome; he hoped good consequences would follow. But he did not refrain from commenting on how that truce (provided it became effective) was especially significant for Italian culture, since the devaluation of sciences that had been "carried out during a certain historical epoch" (*ibid.*, p. 534) had caused so much damage. The last one to take part in the discussion was Guzzo. He objected that if a peaceful spirit does not support them, treaties are simple pieces of paper. Better to talk about an "attitude," a "conversation" between scientists and philosophers, which should take place after those philosophers have studied science ("to the best of their abilities, it goes without saying") and before having a say on such a serious matter. As for a recent past that had allegedly witnessed the devaluation of science, "such has not been the case," added annoyedly Guzzo, "and at any rate talking of such matters is not tasteful" (*ibid.*, pp. 544–545). Unsurprisingly, a few weeks after the meeting at the Lincei, Italo Calvino started a lecture tour choosing to talk about *Cibernetica e fantasmi* [*Cybernetics and Ghosts*]. Calvino's approach was at once natural and original (Antonello, 2005, pp. 169–230): we see the world in its different aspects—he told audiences at the time—more and more as discrete instead of continuous, "a series of discontinuous states, of combinations of impulses on a finite number (a huge but finite number) of sensors and control organs." Electronic brains could not produce all human functions, but they supplied a convincing theoretical model for the processes of memory, mental associations, imagination, and consciousness. Cybernetics had radically changed the image of those processes:

> In place of that prismatic cloud that we carried in our heads until yesterday, whose movement of condensation or dispersion we related to impalpable psychological moods, dim landscapes of our souls—in place of all that, today we feel the rapid flowing of signals on the complicated circuits that connect power relays, diodes, and transistors that stuff our cranium.[12]

He involved in the discussion literature too, despite the fact that such perspective raised "lamentations interspersed with shouts of blame" in most literary people. Calvino claimed to have always known, more or less instinctively, that the author was a machine and that writing was a combinatorial game. He said he had always known that a machine could take his place, and did so not just with a serene and regretless disposition, but with a sense of relief and self-confidence. Calvino was also aware, however, that at one point something unexpected, the "hidden ghosts of individuals and society," entered the game. The strength of modern literature laid precisely in its willingness to give voice to the unconscious, to the vast sea of content that remained unarticulated and repressed.

NOTES

1. Somenzi graduated in physics in Milan in 1940, with Giovanni Gentile Jr., with a thesis on superconductivity; the following year he met Vaccarino while attending the school for officers of the Air Force Engineers in Orvieto (Vaccarino, 1947–1948; Somenzi, 1947a; Somenzi, 1947b; Somenzi, 1948).

2. de Finetti, 1952a; de Finetti, 1952b. The title of the article took up the title of Edmund Callis Berkeley's book (1949).

3. The *Messaggio dell'ing. Adriano Olivetti ai dipendenti* [*Message by Adriano Olivetti to employees*] was published in "*Notizie Olivetti*", January 1956, pp. 2–6, and Caizzi, 1962, p. 248 reported only an excerpt of it. The laboratory of Barbaricina soon moved to an industrial building outside Milan, in Borgolombardo (Rao, 2005). Gallino, 2003, pp. 15–27 discusses the farsightedness of Adriano Olivetti—who died in 1960—as well as the alternate vicissitudes of his company, which culminate in its liquidation of the 1990s.

4. See Böhm and Dainelli, 1958; Caracciolo di Forino et al., 1958; Nastasi, 2005.

5. Paci, 1989. Paci's lecture opened the academic year at the University of Pisa, on 6 November 1953.

6. Caianiello, 1959, p. 344. The lectures and the seminars published in the *Rendiconti* [*Reports*] of the course do not include works by Wiener, who was the guest of honor. Different aspects of information technology were presented in the following order: B. MacMillan, R.M. Fano, D. Slepian, D.H. Huffman, P.E. Green, H.E. Hans, Y.W. Lee, D. Gabor, H.L. Gelernter and N. Rochester, M. Halle, B. Mandelbrot, V. Breitenberg, G. Moruzzi, E. Newman, A. Feinstein, S. Watanabe, L. Löfgren, W. Reichardt, and B. Hassenstein.

7. The four articles appeared in *Secolo XIX* from 26 to 29 October 1965 (issues 254–257). They are anonymous, except for the third one, which Massimo Zamorani signed. Only in 1964 Ceccato started a long and durable collaboration with *Il Giorno*, with a series of articles that would be later collected in a volume called *Cibernetica per tutti* [*Cybernetics for everyone*] (1968).

8. The introductions to the meeting (by Paolo Filiasi Carcano, Marino Gentile, Vittorio Mathieu, Enzo Paci and Ugo Spirito) appeared in *De Homine* in 1965 and in a separated volume by Sansoni. The other presentations are in *Atti Perugia* 1965.

9. Società Filosofica Italiana, 1967. The quotation is from Guzzo (vol. III, p. 31). In 1965, Somenzi edited a collection bearing an evocative title—*La filosofia degli automi* [*The philosophy of automata*]— which saw two more modified and enhanced editions: see Somenzi and Cordeschi, 1986 [1994].

10. Faedo, 1968, p. 33. In 1969 Faedo succeeded in starting a course in Information Science at the Science Faculty of the University in Pisa, followed by those of Turin, Bari, and Salerno.

11. Borsellino, 1968, p. 226. Unfortunately, at the moment I do not have much documentation on the activity that the Borsellino's group carried out in Genoa.

12. Calvino, 1980, pp. 167–168. *Cibernetica e fantasmi* carried the subtitle *Appunti sulla narrativa come processo combinatorio* [*Notes on fiction as a combinatorial process*] and was the text of a conference that Calvino held in different cities for the *Associazione Culturale Italiana* [*Italian Cultural Association*], from 24 to 30 November 1967. He then presented it again abroad.

Five

PROCESSING CULTURES: "STRUCTURALISM" IN THE HISTORY OF ARTIFICIAL INTELLIGENCE

Patrice Maniglier

The Western intellectual tradition has long been trying to turn "spirit"—the collective term for mental or spiritual phenomena, as the case may be—into the object of a rigorous science.[1] Yet, for a long time this project did not merge with psychology's own, as some restrictive historical interpretations have claimed (Gardner, 1985). Instead, "spirit" became the object of a series of disciplines which we may call, according to the country and the epoch, "the moral and political sciences," "the human sciences," "cultural studies," or "the social sciences." As Stefano Franchi and Güven Güzeldere have shown (2005), a historical study of Artificial Intelligence (AI) must integrate dimensions that historians of the discipline have too often disregarded. The problems that Artificial Intelligence confronts have ancient roots and go well beyond the scope of the engineering project that we now know as AI. However, it is useless to go all the way back to Plato and failing to notice that Montesquieu wrote a book titled *L'esprit des lois*; that the concept of spirit (*Geist*) is at the heart of German idealism with Hegel, Schelling, and then Humboldt all the way to Cassirer; that Comte suggested a politics of the mind; that Durkheim, picking up the Kantian challenge to science, turned to sociology to explain the human capacities to speak truthfully and act morally that Kant had ascribed to transcendental philosophy, thereby claiming that the scientist, not the philosopher, should study the *a priori* conditions of knowledge.[2] To see what we need to see, a long-range view in space and time is not sufficient; we also need a broad scope.

Even a long-ranging and broad view at the classics, though, will fail to discover any attempts to apply *mechanical* models to the mind. No specific theory defines the space that the contemporary disciplines of the mind occupy: instead, a new link between *mind* and *machine,* which to the majority of classic authors would have appeared unnatural, defines it. Even Descartes—whose heritage many protagonists of the "cognitive revolution" are keen to claim—flatly rejected the idea that the human mind could be mechanical and that a machine could then simulate it. Descartes held a historically old-fashioned view of machines that he based on devices, such as clocks, that react only to purely mechanical factors.

According to a conception that will later become the theory of the reflex arc, imagining that machines could model the behavior of animals as they react to external stimuli became possible. But the same approach does not work with human beings who, before acting, decompose the external stimuli and recompose their behaviors from smaller components. Human beings possess what animals do not: freedom.[3] This combinatorial model allowed the symbolic paradigm to prevail in cognitive science, on the basis of the new machine that Alan Turing (Chomsky, 1969[1966], pp. 18–24) developed. Many have concluded that the theoretical objectification of mental phenomena through mechanical models is inseparable from a specific machine—Turing's universal calculator—and from a specific theory portraying the mind as a computer that processes internal representations. Fodor called this model the "computational-representational theory of the mind" (2000).

This theoretical model would not have reached the level of popularity it enjoys without the benefit of a virtuous circularity with the human sciences. Generative linguistics, in particular, has provided a decisive contribution to the credibility of the model of the mind as a processor operating on internal representations. At the same time, the interpretation of its models in terms of computer programs provided a handy justification for linguistics itself. A linguistic theory enabling computer programs to simulate, more or less satisfactorily, aspects of human verbal behavior, enjoyed a decisive advantage over its immediate rivals. The epistemology of cognitive science made possible a double exchange that, since the invention of the "Turing test," turned Artificial Intelligence from an engineering discipline into an experimental field where we could test and evaluate allegedly scientific *theories* of the mind.

I will start my analysis assuming that the polarity that produced the double experimental field I just described defines the space of "cognitive science." On the one hand, it must be possible to *simulate* the mind through the resources that some form of Artificial Intelligence provides. On the other hand, the "mental" phenomena cannot be understood unless the *cultural* phenomena are understood as well. In short: the contemporary approach to the problem of the mind is how to simulate cultural phenomena.

The issue is whether the link between cognitive science and the human sciences favors the dominant computational-representational model. During the last few years, we have seen the emergence of a double series of challenges against this model. They came from the technical side of Artificial Intelligence and from the theoretical side of the linguistic theories to which AI is more or less connected. The challenge had to be double because the originality of the contemporary understanding of the "problem of the mind" lies precisely in the complex circle tying together theoretical models and technological simulations. From the technical side, the critique of the symbolic paradigm came from a variety of dynamicist approaches: those formulated in terms of connectionist networks (Rumelhart and

McClelland, 1986a); the morphodynamic and neo-cybernetic models like those that catastrophe theory put forth (Petitot, 1985a; 1985); enaction theory (Varela, 1989), and, finally, the roboticist models that artificial life offered (Brooks and Steels, 1995).[4] On the theoretical side, the challenges came from the internal revisions to standard generative theory (and in particular from the so-called "Principle and Parameters" approach), but also from the radical challenges to linguistics' framework that cognitive linguists such as George Lakoff (1987) and Ronald Langacker (1987) have advanced (not to speak of the contemporary revival of corpus linguistics and of the hermeneutic approaches to language).[5]

Several theoretical and technical considerations motivated this twofold critique. In this chapter, I would like to discuss one of them that we seldom stress. The symbolic paradigm appears unable to provide a satisfactory account of the *cultural* and therefore *variable* character of mental phenomena, which is nonetheless one of their crucial features. First, I will clarify in which sense the symbolic model is unable to deal with cultural variation, and I will discuss the objections coming from within the field of generative linguistics (William Labov's sociolinguistics), and from Artificial Intelligence's models that descend from the connectionist paradigm or from robotics (Luc Steels). Second, I will show that this twofold critique amounts to an imperfectly aware rediscovery of Structuralism, the scientific paradigm that held the human sciences back from the adoption of generative grammar. Contrary to an interpretation that we commonly hear, Structuralism did not amount to a search for the linguistic invariants that a community share (*la langue*, the structure of a given language underlying the stylistic or dialectal variations occurring in its spoken instances). On the contrary, Structuralism was the effort to understand the principles according to which the mechanisms that allow us to learn a language (or a different dimension of cultural life) lead us also to transform it. Better yet, without us being aware of the transformation process, they prompt our everyday use of language to transform it. The New York encounter between Claude Lévi-Strauss and Roman Jakobson, in 1943, marks the birth of Structuralism as a transversal movement that cuts across the human sciences. The series of meetings that would officially christen Artificial Intelligence were about a decade away from it.[6] In my view, we can interpret Structuralism as a precursor of the contemporary research programs I mentioned above: while Structuralism privileges an approach to the mind in terms of *mechanisms* (although it does not offer a specific technological model for it), it also holds that mental phenomena can occur only within cultures that are constantly undergoing social and historical variations.

In short, I would like to show, through an archaeological excursion into the early history of AI, that the paradox summing up Artificial Intelligence's challenge comes down to the following question: how can we build a historical machine? A machine lying beyond the opposition between the cultural and the mechanical fields would allow us to conceive the mind as a device devoted to the generation, hence the uncontrollable variance, of cultures. The paradox lying at the root of Artificial Intelligence, I will argue, is how to construct machines that, far from being controlled and controllable, may instead open up new spaces of unpredictability, contingency, and history.

1. The Symbolic Paradigm and its Variationist Critiques

A. Generativism Faced with Historical and Dialectal Variation

Creativity plays such an essential role in the definition of language that Chomskian linguistics puts forth, that we may find surprising to hear it accused of "fixism." Noam Chomsky himself exploited this aspect of the theory in his attack against the linguistic Structuralism fashionable at the time. But Chomsky distinguishes two kinds of creativity: a *rule-governed creativity* that does not allow the speakers to leave the language they speak; and a rule-changing creativity that *changes the rules*. The whole point of the debate is whether the distinction is legitimate. No element of the generative model makes cultural variation strictly unthinkable. Its representation of the changing mechanisms and the way the theory conceives change, however, remove the possibility of empirical observation. Within the generative theoretical framework, modifying the language means changing its rules: the application of the rules cannot provoke their own transformation. Since we conceive language's operations as the application of imperative and univocal rules to mental representations (similarly to the symbolic model that Allen Newell and Herbert Simon were developing the same time), dealing with the continuous and involuntary process of language transformation becomes impossible. Change can only be discontinuous: there must be a sudden jump from an unchanging language to another equally unchanging language. The two languages must be totally different, because they do not have the same rules, and the set of rules that govern languages also define them. Therefore, it is as if a language were never *in time*. In one of the founding texts of sociolinguistics, Uriel Weinreich, William Labov, and Marvin Herzog (1968) advanced the paradoxical conclusion that languages, in generative theory, cannot have any history: "after all, if a language has to be structured in order to function efficiently, how do people continue to talk while the language changes, that is, while it passes through periods of lessened systematicity?"[7] We could push the paradox a bit further and say: languages do not ever truly change, they only succeed each other.

The problem concerns not only diachronic change, but also dialectal synchronic variations. The well-known thesis about the so-called *homogeneity* of linguistic competence is one of the corollaries of the definition of language as the set of basic representations plus the transformational rules that operate over them thereby producing the infinite set of all possible performances. The homogeneity thesis states that all linguistic subjects speak the same language and never speak more than one language at the same time. We interpret a dialectal variation as the coexistence of a plurality of languages that are similar and yet numerically distinct. English rules found in grammar books are only valid for the standard and normalized language television newspeople and linguistics professors speak. But other languages exist, such as the language that the New York's African-American population that Labov studied spoke. We would be then forced to say that they are two *different* languages, both worthy of study but that different rules govern. For instance, in African American Vernacular English (AAVE) a rule establishes that the third person singular of the present tense carries no syntactic marker (*she write*): instead of a mistake, we face just a different rule linked to other rules that contributes to the overall consistency of the linguistic system.

Labov remarked that no exclusive usage of this or that dialect comes to pass, because an individual can preferentially, but not systematically, use AAVE in a determinate context with a frequency that we can statistically evaluate. More than a change of code, we observe a variation in the linguistic rules. Sometimes those rules are applied, and sometimes they are suspended while different ones replace them according to relatively stable measurable ratios. In order to deal with such phenomena, Labov suggested the paradoxical concept of *variable rule*. He showed that we can suspend them with a measurable frequency, rather than changing the rules as we apply them. Contrary to the prescriptions of generative grammar, this result proved that we do not apply rules imperatively and mechanically. They are closer to sociological *regularities* than to symbolic *rules*. Labov noted also that a twelve-year old boy explaining the game of "skelly" could switch from code to code about eighteen times over the course of the following, short example: "*An' den like* if you miss onesies, *de othuh person shoot to skelly; if he miss, den you go again. An'* if you get in, you shoot two twosies. *An'* If you get in twosies, you go to *threesies. An'* If you miss *threesies* then the person tha' *miss skelley shoot* the skellies *an' shoot* the onesies: *an'* if he miss, you go *f'om tthreesies to foursies.*"[8] In this context, does it make any sense to speak of *two* different languages? Or isn't the concept of dialect ill-construed? We should perhaps say, with Weinreich: "native-like command of heterogeneous structures is not a matter of multidialectism or 'simple' performance, but it is part of unilingual linguistic competence" (*ibid.*, p. 101). As Weinreich stated, we must break down "the identification of structuredness with homogeneity" (*ibid.*), and stop construing linguistic structures as the sum of individual practices and as an object that stands separated from those practices. Instead, we should think of linguistic structures as intrinsically heterogeneous, variable, and multiple entities.

As a consequence, linguistics had to renounce the concept of rule and abandon the ideological support that the symbolic paradigm coming from Artificial Intelligence had provided. Cognitive linguistics followed this strategy, especially with the 1980s synthesis of Lakoff (1987) and Langacker (1987). They tried to articulate a different kind of linguistic theory, which would share the project of inscribing language within the larger realm of cognitive phenomena while abandoning the concept of cognition as formal manipulation of mental representations. Cognitive linguists do not identify knowing with computing. They equate cognition with *re*cognition. Cognition is a particular form of perception, categorization, and re-identification. For instance, Langacker shows that we should not rush to declare ungrammatical expressions such as "A bicycle is wanted by me" and "That bicycle is wanted by my brother," against the correct expression, "That bicycle must be wanted by somebody." Similarly, the acceptability of the following sentences becomes increasingly more controversial:

Sheila was kicked by her mother
? Sheila was wanted by her mother
?? Sheila was resembled by her mother[9]

Nonetheless, Langacker remarks, reaching a decision on all these cases is essential, otherwise we would not know how the passive voice functions. More than an either/or decision, the inclusion of a linguistic expression within a language is a matter of degree. On the basis of these examples, cognitive linguists rejects the definition of language as a set of rules applicable to uninterpreted symbols. Instead, they propose to define language as the set of symbolic forms emerging from general (non language-specific) cognitive routines that are neither "formal" nor "meaningless." These forms are part and parcel of the cognitive relationship between the speaker and her or his perceived reality. In the last example, for instance, the verb "to resemble" does not bear the passive voice because, even if syntactically transitive, it is not an action verb. Grammar, Langacker claims, is a form of schematization of reality. Language becomes a particular form of perception, a modality of seeing and showing, instead of a module of human computational abilities. The communication between linguistics and cognitive science does not go through the formal analogy that we can establish between the processes at work in the syntactical or formal processing of linguistic representations. Instead, the two disciplines communicate because they both deal, although differently, with *the recognition of forms*. Cognitive linguistics—and Langacker's in particular—has successfully shown that even the highest grammatical categories are not innate. On the contrary, they emerge from the ordering of a great variety of linguistic patterns. As I will show below, Saussure's linguistics shares a similar intuition.

The deep reworking of Chomskian linguistics that we know as the "Principles and Parameters" approach has abandoned the concept of rule, holding fast to a view of language as uninterpreted symbol processing.[10] Is Chomsky's new theory perhaps similar to the view that underlies cognitive linguistics, we may wonder? After all, this new version of generative grammar sacrifices the concept of linguistic structure [*langue*]. Instead of positing the existence of a set of rules, it assumes universal innate constraints defining a Universal Grammar that each and every individual shares. These constraints are multiple and not necessarily consistent, hence the different ways in which individual languages parametrize them and the different ways they solve the incompatibilities that the constraints produce.[11] In principle, there is no "language" to speak of: people communicate directly, from universal grammar to universal grammar. "Languages" are just unsubstantial epiphenomena, like clouds of dust raising from the battlefield. Variations are completely on the side of linguistic performance, and significant variations rest upon a new hierarchy of constraints.

This new generative model entails a problem that is symmetric to the old one: instead of freezing languages into overly restrictive rules, it makes languages disappear altogether. Nothing is left between the linguistic faculty and speech acts. I will not discuss the details of this new model, but I would like to emphasize a philosophical paradox that applies to all variants of generative linguistics. Chomsky understands language [*langue*] as the inner theory that a subject builds about the language that the community to which s/he belongs to uses. Languages as such—in the sense of the theoretical objects the linguists construct—are never social facts: they are purely psychological internal phenomena. The object of linguistics has been prone to the extreme psychologization and individualization of linguistic facts, and to the use of terms such as "Inner Language" (IL). Yet, this position is intimately paradoxical. Knowing a language would amount to knowing the language that we use within a community, but this language we use is just the inner language of this or that individual. When I know a language, therefore, I only know the knowledge of something, which, in turn, is the knowledge of the knowledge of something ... and so on, in an infinite regress. In short, defining language as a system of *inner* knowledge means defining it as the knowledge of *nothing*. This is the paradox that a system of specular knowledge, lacking constraining criteria, encounters. In the *Philosophical Investigations*, Wittgenstein used similar arguments to deny the pertinence of the notion of "mental processing." To follow a rule *all by yourself*, he argued, is impossible, because to know whether the rule has been correctly applied or not would become impossible. Rules exist only where communities and communal usages are in place.[12] Speaking subjects must know an *external* language that must have a degree of autonomy from the individuals who speak it. Such external language, not the language views that specific speaking subjects may have, is the linguists' proper object of study.

Even leaving aside these philosophical considerations, however, the linguistic models that the "Principles and Parameters" approach inspired are no longer an argument in favor of an approach to cognition based on the model of the symbolic processor. We know that a von Neumann architecture requires sequential processing: the application of an algorithm that erases or replaces determinate characters on the basis of syntactical (mechanical) constraints transforms the internal representations. Jerry Fodor has generalized this sequential approach to information processing with the concept of modularity. He represents the whole of cognition as a sequential process (1983). In the "Principles and Parameters" approach, however, multiple constraints acting in parallel toward an optimal solution produce good linguistic expressions. No true "transformation" in the classic sense takes place, since this very concept caused all kind of protests. In the 1960s, the virtuous circle that bound together the different cognitive sciences vanished. Fodor himself recognized this point against those who believed that the cognitive disciplines had finally reached their Holy Grail (2000). In the new model, we no longer see a phonological module applying its rules to transform a phonetic representation into a phonological one, followed by a morphological module, a syntactic module, a semantic module, and finally a pragmatic module. On the contrary, all these dimensions cooperate simultaneously to determine a *well-formed expression* that we cannot accept as such because of specific rules but because of, better yet, *in spite of* the pressure coming from simultaneous and heterogeneous constraints. This model requires a parallel instead of a sequential form of processing. A break between generative linguistics and the cognitivist paradigm to which it had so much contributed had then occurred. Unsurprisingly, Paul Smolensky's theory of Harmonic Phonology (with Géraldine Legendre), soon to be generalized into Optimality Theory (with Alan Prince), has provided one of the most convincing approaches to linguistic change in terms of constraints and optimization of "Principles."[13] Smolensky is one of the leading figures behind the ascent of connectionism, the main challenger to the predominance of the symbolic paradigm in Artificial Intelligence. Yet, Smolensky's theory provides an argument about the best approach to the modelization of cognitive phenomena. He was trying to determine whether parallel or sequential processing fits best the modelization of linguistic data. I will now switch to the recent history of the problems that Artificial Intelligence encountered when it faced the issue of cultural variation.

B. Cultural Machines

The same objections I have raised against the "computational representational" model in linguistics we can also find in the critiques of Artificial Intelligence's classic symbolic paradigm. AI's inability to address the intrinsic variability of some behavioral patterns turns into an argument in favor of a radical paradigm shift that eliminates the concepts of rule and of internal representation.

Scholars rarely mention connectionism's better treatment of variation among the reasons responsible for its rapid success in the 1980s. Although the argument, which affects the modelization of learning, does not appear explicitly in the summa of 1988 (Rumelhart and McClelland), it discretely operates at the heart of the theory. While claiming to be the heir of Jean Piaget's neo-Structuralism, Jeffrey Elman grasped the importance of the modelization of change.[14] The idea was simple: Elman wanted to show that learning is a genuine *self-transformation*, not the acquisition of new rules or of new objects for those rules. Learning transforms the very structures according to which the mind works. He argued that we must overcome the antinomy between, on the one hand, an empiricism that presents a series of contents that we just need to harvest, and on the other an innatism that relies on nature (or perhaps on lived experience) and provides a series of forms. Instead, forms themselves change and, most importantly, *they allow themselves to be changed*. Therefore, we need a system that, instead of passively suffering change like a butcher's block that splits after many repeated hits, can actively organize and prepare its own transformation. We can never rest content with a theory of the mind unable to provide an account of how the mind's architecture opens itself up to its transformation. The *same* mechanisms allow the machine to work at a given time the way it does and also to transform the very architecture of its mechanisms. The application of the same procedures causes their transformation. These minimum requirements apply to any theory of learning. To learn does not mean to acquire contents—it means to *change oneself*. When we learn, we do not file new items in ready-made boxes—we change the overall architecture. "Learning is structural change," write Bates and Elman (1993). Therefore, to find a definition of "structure" that is intrinsically open to its own transformation becomes imperative.

Elman thinks he can find the desired definition in connectionist models. Learning is not a secondary property of connectionist networks—it constitutes their essence. Structural change *is* their reality. Between their normal performance and a change of competence there is no essential difference. Significantly, Elman often stresses how the same architectures—ultimately, the same mechanisms—may produce different behaviors: "Connectionist models illustrate how the same learning mechanism can give rise to behaviors which differ dramatically over time" (Elman et al., p. 43). We should not juxtapose a theory of inner and hence invariant structures with a theory of learning through the experience that we gain in variable contexts. On the contrary, we need to build a theory of open structures that are capable of change. What can connectionism offer to this project? First, it dispenses with the concepts of rule and of representation. The models follow rules that are not built-in—the model itself discovers them. Likewise, no special area exists where we store the items we have learned and whence later, if recognition problems arise, we may retrieve them. The model's connections contain the system's memory as the record of their successive corrections. Memory consists of subjective identities instead of objective contents (like the doves in the ironic aviary of Plato's *Theaetetus*). A learning-capable system will not, for this reason, lack an original, well-specified architecture, nor will it resemble the empiricist's blank slate. Quite the opposite: the system does have a specific architecture that includes a multi-layered neuronal network plus a so-called back-propagation rule that allows it to change the synaptic weight according to the difference between the actual and expected results that the model produces. Instead of acquiring new rules or new representations, the networks learn by changing the weight of their connections—in fact, by changing themselves. Pattern recognition, for instance, does not consist in breaking down the stimuli into a small number of elementary traits first, and then trying to match the decomposition with one of the previously memorized categories. It consists in letting the network match a large variety of input activation patterns with a small number of connections-dependent output patterns. The network's behavior is not correct or incorrect: instead, it is more or less correct, more or less similar to the desired output. As a result, the system can deal with unforeseen situations that may even transgress its categorial system. David Rumelhart, Paul Smolensky, James McClelland, and Geoffrey Hinton (1986, pp. 22–23) offer a simple example of this phenomenon: a system designed to recognize the different rooms of an apartment (bedroom, kitchen, living room, bathroom) on the basis of the furniture will find a stable configuration for a "studio"—a more complex room that may have the bed in the kitchen. A classic cognitivist model would be unable to classify the object, while the connectionist model finds no difficulties in generating what is in fact a new category. This may occur because instead of a memory that store them, we have categories that emerge locally from the system's specific and concrete operations.

The philosopher Paul Churchland has been the strongest defender of connectionism against classic symbolic cognitivism. He has stressed that connectionism can model the intrinsically cultural and consequently variable character of human cognition. For instance, in "Reductionism, Connectionism, and the Plasticity of Human Consciousness" (1992), he argues that there is no contradiction between the naturalist and even materialist approach to the mind he defends, which reduces the mind to the brain and rejects the validity of folk-psychology, and the conception that mental phenomena are impossible without culture. Churchland even argues that the true challenge facing a naturalization of the mind consists in the description of everyday actions. Physically observable features, each one isolated so that a simple analog device can independently measure it, do not make up the cultural phenomena we are sensitive to—a gesture, a look, a word. We may measure this way only the features of the *physical* world, as for example does Dretske's magneto-tactile bacterium (1986), the most complex *non*-cognitive organism.[15] *Phonemes* provide an excellent example of "complex, subtle, context-sensitive or stimulus transcendent features" (1992, p. 133). We are unable to recognize any trait of the pure phonetic signal as a vocalization feature (a vibration of the vocal cords that distinguishes a 'b' from a 'p,' a 't' from a 'd,' and so on). At the phonetic level, vocalization features are so variable depending on context, so distributed along the signal, so altered, that we cannot imagine a specialized detector able of identify their presence or absence along the lines of a thermostat recognizing temperatures. In fact, no matter how subtle its sensors' operation, such a machine would have troubles *perceiving* the phonetic data, let alone *interpreting* them. On the contrary, Churchland stresses that we can train a multi-layered neural network to recognize variations such as these because, instead of breaking it down into individual features, the network analyzes the input signal as a whole and in parallel. Scholars have often said that we may prefer the connectionist networks to classic symbolic system because of their neurological compatibility. Alternatively, I think the main argument in their favor may be their ability to model human *cultural ability*. Human beings may learn *any culture*, exactly in the same sense in which connectionist networks can learn any arbitrary difference in their inputs.

I do not wish to provide yet another apology of connectionism but rather, in order to open up a different historical perspective on cognitive science, I wish to emphasize that we often misunderstand the *problem* these models address. Luc Steels has explored a similar perspective that tries to join a renewed theory of Artificial Intelligence to a variationist linguistics firmly grounded in the study of cognition. Steels and Pierre-Yves Oudeyer (2000) have applied the model to diachronic phonology. Since standard connectionist models do not address the role of the *teacher* who teaches the network to discriminate between "good" and "bad" output, the *meaning* of the differences they "memorize" is completely external to the network. Steels, on the contrary, is interested in having networked computers interact with each other with no human intervention. We cannot tell the overall system what to learn in advance. We do not consider the phonological features intangible external data, but results emerging from the interaction among multiple feature detectors (which in turn other networks may have extracted) and from the reciprocal validation that computerized agents give each other. Steels's model, as a phonological example will illustrate, is simple. Imagine an imitation game in which an agent A randomly produces an articulatory coded sound. An agent B tries to randomly reconvert what it has perceived into an articulated sound, with A providing positive feedback if the sound received back from B is equal (or at least similar) to the sound A originally produced, and negative feedback otherwise. The situation constitutes an inverse mapping problem between the articulatory and acoustic spaces. A correction rule allows the system to modify the agents' performances and to obtain a progressive convergence between sound productions that originally were purely random. Darwin's evolutionary theory explicitly inspires Steels and Oudeyer's use of individual variations and a selection principle. The model is interesting because we do not provide phonological categories in advance: they emerge from the interactions among the agents. As discontinuities appear within the articulatory and acoustic spaces, the agents reciprocally produce, so to speak, each other's cartography. The model approaches the modelization of *cultural* emergence, at least if we admit that we can consider a culture a sensible *milieu* that subtle parameters, which do not belong to any linear function of the five senses, constitute. Cognitive agents who can decompose variable and continuous signals into invariant and discrete entities reciprocally validate those parameters.[16]

These emerging categories cannot be stable, because they are not based upon extrinsic parameters nor upon a natural necessity. They rely upon the agreement—or upon the *feeling* of an agreement (since no external parameter measuring the agents' consensus exists)—among agents that face reciprocal categorizations generated from an originally random linguistic corpus.[17] In the two-agent model I illustrated above, the inverse mapping problem is simple enough to quickly bring the system to converge onto a stable state. In a large multi-agent system including a great number of interactions, multiple concurrent solutions to the same problem come to coexist. Often, the agents are forced to innovate, because they do not have a global view of the network. The innovations spread, according to predefined conditions, so that the system's own make-up contains the seed of its evolution (Steels, 2000). We need to add nothing in order to conceptualize the variation of a repertoire of forms: the mechanism presiding over the constitution of these systems is the same mechanism leading to their transformation. We need no additional rule or specific transforming action: the system models the normal functioning of a cognitive population that continuously and involuntarily changes its life-world. These categorial systems are intrinsically open to variation, because they are themselves local results emerging from a great number of interactions. The lesson is clear: the modelization of cognitive phenomena may have to *start* from the very concept of culture.

Intelligence is a collective phenomenon. The mind exists only in the masses. It does not reside in my, or for that matter in anyone else's, head. It is *between* us. Intelligence depends upon our ability to understand each other, which means that it depends upon our ability to create sensible *milieux* whose parameters our exchanges establish. We can consider the artificial models I discussed above *cultural* simulations, because they produce effective regularities binding together only a specific collectivity whose boundaries are not neatly marked.[18] The intelligence arising from this simulation would brilliantly pass even Turing's test. Turing argued that intelligence is the quality we attribute to agents who *say what is to be said* within specific contexts. We could generalize the approach and claim that intelligent agents *do what needs to be done:* spirit (or mind) is the ability to share conventions and uses that are not necessarily explicit—in the case of Turing's test, uses that concern the differences between sexes as they seep into a language that an agent must be familiar with in order to "imitate" (Turing, 1950). The current horizon of Artificial Intelligence seems to open up on the simulation of emerging cultural phenomena within populations.

Nowadays, we widely share this diagnosis. "Memetics" is a project of this kind, though perhaps a bit rough around the edges. Several social theories of the mind have been recently proposed in philosophy (Dennett, 1991; Brandom, 1994; Searle, 1995; Descombes, 1996), together with a re-evaluation of philosophies of "objective spirit" such as Hegel's (by Robert Brandom, for instance) or Humboldt's (by Vincent Descombes). There are great differences among these philosophies. Their arguments are incompatible and they do not even share the same position regarding the status of modelization and its relevance for philosophy. Nonetheless, they all support, for different reasons and with different means, a view of the mind as something inseparable from social and cultural phenomena. My intention is not to argue in favor of a specific philosophical thesis on this problem. I just wish to provide yet another historical perspective on the problems arising from the encounter between a hypothetical science of the mind (which comes together with the very concept of "artificial intelligence"), and the Humanities and social sciences. I want to stress that the modelization of cultural phenomena, which we often misunderstand or altogether forget, may turn out to be a positive terrain for this encounter. The movement that we have came to characterize and also "mis-characterize" as *Structuralism* opened up this space a long time ago. The new reading I propose below will clarify what the simulation of cultural phenomena, the veritable *experimentum crucis* of Artificial Intelligence, stakes out.

2. Symbolic Function and Cultural Variation: a New Reading of Structuralism

A. The Sign as a Mental and Historical Object

Structuralism, no doubts about it, attempted to articulate a theory of the human spirit, a goal that Jakobson and Lévi-Strauss explicitly posited. Coincidentally, Structuralism apparently joined the human sciences in a decisive rejection of history and in the common hope of explaining cultural phenomena as the result of purely systemic formal constraints. As for Structuralism's relationships with Artificial Intelligence, they were purely anecdotal.[19] We have learned that Structuralism's opening gesture was the rupture with the decidedly historical approach to language that nineteenth century linguistics had practiced. Was anyone more responsible than Saussure for the ban of all research on the controversial problem of the origin of language from twentieth century linguistics? Was not Saussure the author of the synchrony/diachrony distinction that led to a representation of the history of languages as a succession of intrinsically immobile states, as if a language would never change, and change itself could never be the object of scientific study? Saussure argued that whereas speech is an individual phenomenon, language is a social one. But he said so because if speech were to be *identical* in *all* individuals, then in order to gain access to a language the study of one individual would suffice. William Labov labeled "Saussurian paradox" this recourse to a "sociological" argument that leaves aside all sociological work. Chomsky used the same argument—within a different framework—in support of his conception of "I-language" that I discussed above. To hear that Structuralism may have something to contribute to the modelization of historical and cultural phenomena may then be surprising.

Yet, the *Cours de linguistique générale* leaves the reader with the impression that the thought of the master from Genève has a few contradictions, uncertainties, and hesitations. In a chapter unfortunately seldom read, and titled "Mutabilité et immutabilité du signe," Saussure writes:

> Linguistic structure [*langue*] we take to be language [*langage*] minus speech [*parole*]. It is the whole set of linguistic habits which enables the speaker to understand and to make himself understood. But this definition fails to relate linguistic structure to social reality. It is a definition which misrepresents what a language is, because it takes into account only how the individual is affected. But in order to have a language [*langue*], there must be a *community of speakers*. Contrary to what might appear to be the case, a language never exists even for a moment except as a social fact, for it is a semiological phenomenon. Its social nature is one of its internal characteristics. (Saussure, 1983, p. 77/112)

The text could not be clearer: the "social product" exists in the community, not inside the brain of individuals. In his manuscript notes, Saussure even speaks of a "collective soul." By separating language from speech he does not mean to privilege invariance over variation. On the contrary, he bets that language itself is a variation. Following the passage I quoted above, Saussure goes on to show that *langue* is not only a social, but also a historical phenomenon. Far from being an exception to historical and dialectal variation, language constitutes itself as language only through the two factors that Saussure called Time and Mass. A linguistic theory unable to portray language as a social and simultaneously historical phenomenon would be inadequate, says Saussure, because language is a system so complex that no one can master its becoming (*ibid.*, p. 76/110–11).

How can we reconcile these texts with those that appear to extricate language from historical and dialectal variation? We know that Saussure did not write the *Cours de linguistique générale*, which two of his students, Charles Bally and Albert Séchehaye, notoriously put together. Since the 1950s, the increasing rediscovery of Saussure's lectures notes and manuscripts has allowed us to gain a much better insight into the ideas of the master of modern linguistics (Godel, 1957). Saussurian exegesis has established that Saussure's problem was not to turn his back to nineteenth century historical linguistics, but to advocate an "epistemology of comparative grammar."[20] Saussure was not trying to find language's identity behind the diversity of speeches. On the contrary, he was trying to understand why, *as we speak a language, we end up speaking a different one*. Comparative grammar—the discipline he practiced—discovers this basic fact when it shows that languages as different as Latin, Gothic, and Sanskrit have all originally been the "same" language. Saussure stresses the point forcefully in his inaugural lecture at the University of Genève: "French does not descend from Latin, it *is* Latin" (Saussure, 2002, p. 152). They are not two different languages—they are the *same language*. The discovery that the repetition of linguistic signs leads to their transformation led him, retroactively, to show that we can never guarantee synchronic identity. As a result of this essential and structural reason, languages are naturally "abandoned to variation." Saussure's intellectual goal was to uncover the structural features of language that open it up to an orderly variation whose traces, like fragments of an object that has never been whole, the comparatist collects. The editors of the *Cours* later reversed the order in which Saussure delivered his lectures. He *had begun* with an exposition of historical and geographic linguistics and *concluded* with synchronic linguistics. The discussion of synchrony had given him the means to address the problem of "diachronic identity," as he called it. Namely, what must have been the identity between *calidus* and *calidus*, for [*calidus*] to become [*chaud*] (warm) when repeated over and over?[21] Again, language is *in* time. But language has not been thrown into time as if into an element that wears it from the outside. Language belongs to time intrinsically, for reasons that pertain to the very nature of its identity.

Far from reproaching contemporary historical linguistics for indulging in an overly historical approach to language, Saussure claims that it has considered language's historicity as deriving from its "material" and "physiological" aspects, like pronunciation differences. As a consequence, the only feasible account of the historical character of language had entailed a rejection of language as a symbolic activity. Variation had become the unconscious component that prevented the reduction of language to a purely communicative function and brought into language itself resistance and opacity. These last characteristics showed that languages, far from being subjective functions, possessed an objective reality that linguistics could study by following the strategy that the natural sciences had applied to other natural phenomena.[22] Only an alteration of the articulatory gesture, as a difference in someone's imitation of someone else's speech, could explain linguistic change. Such was the research program of the "Neogrammarian school," the movement that dominated linguistics at the end of the nineteenth century. The Neogrammarians attempted to support the reconstruction of the history of language with an experimentally controlled physiological theory that would explain linguistic change through "phonetic laws." In turn, the set of physiological and psychological constraints applicable to all individuals grounded these laws.

Saussure was undoubtedly the first one to have seen the impasse of the Neogrammarian program. Linguistic phenomena can never be reduced to specific spatio-temporal events such as individual articulatory gestures. Although the whole theoretical and philosophical tradition has typically ignored it, the reason is simple: linguistic events are *incorporeal*. Signifiers themselves "*are not physical in any way* [*incorporel*]. They are constituted solely by differences which distinguish one such sound pattern from another" (Saussure, 1983, p. 117, emphasis added).

The problem of diachronic identity picks up where the problem of synchronic identity had left off:

> Every time I utter the word *Messieurs* ('Gentlemen'), I renew its material being: it is a new act of phonation and a new psychological act. The link between the two uses of the same word is not based upon material identity, nor upon exact similarity of meaning, but upon factors the linguist must discover, if he is to come anywhere near to revealing the true nature of linguistic units. (Saussure, 1983, p. 107/152)

We must understand that no simple "variation" exists, because *everything* changes. Keeping a minimal "kernel" and distinguishing essential from accidental change is impossible. Linguistic entities are realities as *un-observable* as they are un-measurable. Rigorously physical criteria would make the description of a linguistic sign's identity or its unity, its individuation in the phonic chain, simply impossible. Linguistic history after Saussure has shown that any measuring apparatus that we deploy to provide empirical demarcations of the sign is fruitless. Jakobson mentions it in the first of the *Six Lectures on Sound and Meaning* he gave at the New York School of Social Research, to an audience that included a very attentive Claude Lévi-Strauss (1962, v. 8, p. 327). The relationship between phonic substrate and phonological value remains an open issue. The morphodynamic, connectionist, and catastrophist approaches tried to solve precisely this problem.[23]

In order to address the problem of intrinsic and inherent variability of linguistic phenomena, Saussure introduces the apparently old-fashioned concept of sign. Linguistics is not part of the new discipline Saussure calls "semiology" because language shares the same signifying *function* with fashion, culinary practices, mythological tales, and ritual prescription. Instead, linguistics is part of semiology because language shares with them the same *ontology*. This peculiar object that displays a wealth of *variable identities* characterizes the domain of semiology,[24] and provides its the very definition: "a science which studies the *life* of signs as part of social life" (1983, p. 15/33, translation modified, emphasis added). The science of semiology studies how, as long as we use signs and they "circulate," they never cease to change. To all those who tried to reconstruct a general linguistics on the basis of the results of comparative grammar, Saussure countered, not without a touch of irony: "The 17th and 18th century philosophers dealing with language have given us the answer already: language is made up of *signs*." This statement presupposes the reconstruction of the very concept of sign. In particular, it presupposes a clean break with the idea that the sign is the *material means* we use to express or communicate a mental idea. Instead, the signifier is as mental as the signified. By stating that *the same factors give language a meaning and a history*, Saussure attempts to reconcile the two great traditions that today are still dividing the linguistic domain: the tradition stemming from the study of rational grammars, which study language as the more or less perfect expression of a subjective mind or reason; and the tradition stemming from nineteenth century historical grammar, which claims that since linguistic expressions have a material dimension that abandons them to history and separate them from immaterial thought, we cannot address languages from the standpoint of rationality or, more generally, of the thinking faculty. Saussure juxtaposes the two traditions in his unified interpretation (Saussure, 1983, pp. 811–83/117–118). The integration of linguistics with semiology achieves a double result: on the one hand, it reaffirms that language and thought are intimately bound (since everything in language is mental); on the other hand, it stresses that we will never properly perceive such intimate link as long as we ignore "TIME, the unforeseen factor absolutely ignored by the philosophical tradition" (2002, p. 231). Time is intrinsic to the psychological function of language. Signs make up language, but signs *are* themselves parts of *thought*—not just material carriers that we use to convey thought. More surprisingly, thought exists only in time and it never stops changing. It is as if what allows us to think also led us to think something else without us even noticing it...

B. Thinking Machines and Historical Machines

How does semiology, Saussure's new theory of signs, account for the fact that a sign is a thought that can always become a different one? The task is a difficult one. One of the axioms of Structuralism is that the positive particular features of a sign never define it, only the differences from other signs do. This approach may end up removing all dialectical or stylistic variations. It does not matter if French speakers roll or not their *r*'s, as long as they share enough phonemes with those they wish to communicate with. As the speaking subjects ignore or not even perceive variations at the speech level, we can hardly see how they could have any consequence on the linguistic structure.

But Saussure's theory of the sign is arguably more complex than this description, and relies on a thesis that we might call the *double determination of the sign*. According to Saussure, two different levels and two different logics determine any sign. First, a sign gets determined as a "*term*," and second as a "*value*." Even though the editors of the *Cours* failed to emphasize these technical terms at all, Saussure did define them. Two objective modalities common to all signs, we must consider them as *emerging forms* bound to two differentiating mechanisms that Saussure explicitly spells out. On the one hand, we have a playing out of *differences*; on the other, a playing out of *oppositions*.[25] The conflict between these two modalities explains the essentially and intrinsically dynamic character of any language and any semiological system. At the outset, a sign is a concrete set of phonic, articulatory, and brightness *qualitative variations* (not to be confused with the distinctive traits) that generically characterize a sonic experience as a complex *nuance*. Then, according to a process Saussure invariably qualifies as mechanic, automatic, or instinctive, the "mind" [*"l'esprit"*] compares the terms made up of purely sensible nuances. The mental comparison *opposes* the terms to each other and produces the "values," which we may think of as pure positions within a closed relational system. Only at this point, by abstracting from the nature of those differences, we may study the system as a *"form"* and not as *"substance."*

Let us look closely at this two-step process. Let the sign "clear," a precise, determinate, qualitative, and singular *nuance,* be produced in front of me. More precisely, this sign ("clear") is an association between different sets of qualitative nuances: acoustic ones (a tendency of sound to diffuse within the acoustic space), but also articulatory ones (the inner sensation of the gestures of my articulatory apparatus), as well as an optical one (a singular impression of "clearing up" that merges synaesthetically with the previous ones), and so on. We can justifiably call "signs" these synaesthesiae, because the singular acoustic nuance that, at any given moment, characterizes the sound "clear" pronounced before me would not exist if another *heterogeneous* qualitative variation did not accompany and remark it—the sensation of "clearing up" that goes always with it, such as when, for instance, my parents used to draw the curtains and say: "Today is a clear day." The opposite is true as well: the qualitatively heterogeneous variations that synaesthetically merge into "clear" (linguistic sounds, for instance) determine my experience of light variations. Qualitative forms *emerge* from the joint determination of a qualitative variation by a different one. These forms are truly synaesthetic. They are packages of mixed qualitative variations that are mutually "compared," thereby acquiring an oppositional value: we classify the "terms" thus obtained on the basis of their constitutive qualitative differences. Only at this stage—only after the classification—they start functioning as "distinctive features." For instance, we can organize coach/bus, train/metro, and road/street in a series under the variation URBAN/EXTRA-URBAN. Because language is indeed responsible for the ordering, via comparison, of the two series of terms, it follows that the series would be meaningless outside the English language. Saussure builds a serial and qualitative grammar, where the grammatical categories are not *a priori* logical functions (as in Port-Royal's and Chomsky's rationalist grammars), but they are instead the ratio behind series of concretes variations (as in the connectionist models).[26] Saussure redefines each single term: no longer just the sum of distinctive features it represents, it also occupies a position within a system. Such position defines the term by *opposing* it to all others: "enseignement" [*teaching*] is no longer just the sum of the phonic and semic distinctive traits it combines, but it is also not "enseigner," [*to teach*] or "éducation" [*education*]. More precisely, the relative distance of the term from the other terms is what defines it.[27] Saussure calls this the "value" of a term, namely the product of a "post-processing," whereby we determine the *same* entities twice, according to two different logics and, additionally, through a parallel process that operates on multiple qualitative dimensions (2002, p. 87–88).

Saussure's theory lets us understand how the simple repetition of these complex entities—their *use*—entails their very modification, even though *we can determine each linguistic utterance only as the realization of a linguistic possibility that, in a certain sense, logically precedes it.* We carry out the serial organization of the "terms" (the post-processing) on the basis of the concrete qualitative differences that constitute them. We can alter these differences—for stylistic, physiological, and sociological reasons—without changing the structure of the oppositions. This is why the overall equilibrium of the system is *susceptible* to change, but *only inasmuch as a new interpretation intervenes*, inasmuch as someone will create—on the basis of new differential traits—a previously nonexistent series of oppositions that will modify the possible analysis of linguistic performances. Saussure provides numerous examples of this phenomenon. Take, for instance, the creation of plurals by vocalic alternation in Germanic languages: *foot/feet*, *Gast/Gäste*, and so on. First, plurals are expressed similarly to Latin: *fot/*foti*, *gast/gasti*. Then, phonetic reasons prompt an *Umlaut* to modify the central vocal: *fot/*feti*, *gast/gesti*. Finally, the final *i* is dropped (Saussure, 1983, p. 83/120). The process creates a form expressing the plural—not through declension but through the inflection of the root—that the original language did not have. It deals with the *value* that an *interpretation* conjoining two qualitative forms produces. The interpretation acts in the second step. We might even say the process relies upon a "mistake"—a consideration that only attests to the blindness of the qualitative and serial logic that is typical of linguistic usage. The theory of value wants to be a *grammar of usages*. Being an *a posteriori* logic, it depends of the *quality* of the terms. Its ordering does not originate from the projection of pre-given structural schemas upon a relatively indifferent substantial reality. On the contrary, it arises from the later establishment of series of similarities and dissimilarities among the terms. *Every* grammatical phenomenon is based upon this logic, because the grammatical functions are mere super-series among the criteria underlying the oppositional series. More importantly, Saussure builds his system *a posteriori* and the formal dimension of language emerges from the ordering of qualitative data. Even the highest grammatical functions (like "noun," for instance), depend upon the qualitative reality of spoken languages. Far from being a prerequisite of linguistic activity, grammar is a product that becomes manifest when the sensible world orders itself.

According to Saussure, these ordering processes are rigorously *mechanical*: they are based upon an innate human faculty. The great originality of the founder of semiology—and the reason why we should include him into the history of cognitive science—lies in Saussure's suggestion of the existence of a *non-specific* faculty—namely, the semiological faculty. This faculty deals with the processing of sensible data and its functioning mode gives rise to the emergence of historical and social products such as languages (and fashion, religion, and so forth). Saussure always distinguishes between *language*—a universal ability that would eventually fuse with the semiological faculty ("*facultas signatrix*")—and the *languages* in whose absence we could not exercise that faculty. *Languages* are *involuntary by-products* of the exercise of our linguistic faculty. We cannot consider them ensemble of rules or operations. They are systems of particularly complex emerging forms. Undoubtedly, one of the most important points of Saussure's cognitive theory is the distinction between the level of universal and mechanical *operations* and the level of variable and emerging *forms*. For Saussure, language itself is an innate, physiologically identifiable faculty located in the Broca's area of the brain: he sees language as a set of "*principles* ruling" languages (2002, pp. 145–146), "of possible operations of the human instinct applied to language" (*ibid.*), "of psychological forces" (p. 158), a "function," a "power," an "organization," an "apparatus" (p. 178). In short, language is a set of previously assembled mechanisms that specific stimuli trigger. Saussure imagines that a sort of *mechanical procedure* characterizes the semiological faculty. He strives to rethink history within the framework of a mechanical conception of thought. He is not after *historical laws*. On the contrary, he is keen to distinguish the level of laws (or of processes) from the level of forms (or of events). The mechanisms *are* blind, and even more so as they never act directly upon the forms they change. Saussure does not accept a naturalist conception of the mind that we may understand as a collection of blind mechanisms setting off in coincidence with precise conditions and always acting locally. Nor does he welcome a culturalist approach that would endorse the "objective spirit" and frame human cultures as living collective spiritual organisms that evolutionary tendencies inhabit, projecting on them their teleologies. He is after a third position that straddles the schism between individualizing cognitivism and the holism of the objective spirit.

Structuralism's posterity has ascribed Saussure's obstinate upholding of the mechanical and blind character of the processes that preside over change to his archaism, a testament to the difficulty of breaking free from the constraints of the age.[28] In alternative to Saussure's, Jakobson will propose a much more teleological concept of system. He introduced the concept of "Structuralism" on the basis of the "traditional teleological position" and with the explicit goal of turning finality into a respectable scientific category, thus making available to Western science a characteristic tendency of Russian science.[29] In his "Principles of Historical Phonology," Jakobson defended the structural approach from the standpoint of historical considerations: the system shows a life of its own, an internal tendency. A correct account, he claimed, requires the introduction of the notion of system: "When we consider a linguistic mutation in the context of linguistic synchrony, we insert it into the sphere of teleological problems" (1929, p. 218).

On the contrary, I view Saussure as a precursor of cognitive science precisely on account of his rejection of all teleologies. He developed an extremely original conception of what we might as well call *objective spirit*. According to Saussure, the objective spirit is not a living thought, a collective subjectivity that palpitates through institutions, language, and social habits, as Jakobson, echoing Hegel and Humboldt, stated. The objective spirit is a collection of thoughts, of thought-things, or, in Husserlian terms, of "noemata" without "noesis." It is a set of ideas that create themselves and "therefore force the mind to embark upon the special road" that the randomness of phonetic transformations has provided: "The 'genius of language' counts for *nothing* when facing a single fact like the suppression of a final *o*" (2002, p. 216). A linguistic system is the *outcome* of the serial organization of qualitative elements that emerge from the mutual determination among several dimensions of qualitative variation. Signs may have a "life," but not because sign systems are organisms. Cassirer (1945), like Jakobson, inscribes Structuralism within the tradition of objective spirit that sees cultural systems as organic forms (although in a spiritual and not in a biological sense). Saussure tried to warn us in advance against this tendency. The determination of signs as *thought-effects* occurs to the extent that individuals *use them* throughout their interactions; as a result, such usages may have an effect on the signs' actualizations. Semiology "studies the life of signs at the heart of social life" precisely in this sense. A general semiology will be a theory of the universal conditions that allow signs to become thoughts that necessarily change within the contingencies of history. Specific "semiologies" will study the modality of construction characteristic of particular sign systems on the basis of the "substances" they work upon and of the particular histories of thought they give rise to.

Saussure attempted to apply his semiological program in his studies on legends that have remained unpublished for a long time and on which we are now slowly gaining a slightly more synthetic perspective. Legends and languages obey the same set of strange conditions: no matter how much we strive to repeat what we have heard, we end up changing them. This is why *structural* conditions that throw legends, like languages, into history, similarly rules them:

> Every time the poet, due to a lack of memory of his predecessors or otherwise, gathers a legend, he collects for this or that scene only the *accessories*, in the proper, theatrical sense of the word; after the actors have left the stage, this or that *object* remains—a flower on the stage that is retained in memory and more or less tells what has happened.... The effect of the imagination on memory lapses is the main factor of change together with the effort to otherwise remain within the tradition. ("Unpublished manuscripts," p. 3)

This is how we must understand the concept of "life." The system has its virtue, a kind of intrinsic vitality resembling, more than an organism's, the activity of the anthill. What gives the anthill a life is not ants' ability to reproduce, but their ability to regenerate themselves from their refuse. No force can resist change, yet a force, once change has occurred, can get things starting again. Saussure compares language to an "anthill where we plant a stick and that would be instantly repaired from its breaks" (2002, p. 267). Language is a "dress patched with pieces of its own material" (1983, p. 235), "a machine that will always run no matter the damages that may be inflicted upon it" (1967, p. 192, n° 1144). This is Saussure's formidable intuition about the nature of language: language is a machine, indeed, but a machine whose debris fuel it. Claude Lévi-Strauss will rediscover the same formula in *The Savage Mind*. Saussure's position has nothing in common, obviously, with a system of innate forms applied to passive representational contents—as Chomsky would have it—nor with an organism endowed with its tendencies—as the functionalists (including Jakobson) would.

C. Culture at the Boundaries of Simulation

Which conclusions can we draw for an archeology of Artificial Intelligence from a retrospective consideration and a new understanding of Structuralism?

To reinstate Structuralism into the history of Artificial Intelligence without stressing its proximity to cybernetics and the general theory of systems may be surprising. The hypothesis that we should study cultures (or cultural forms such as languages, mythological systems, political institutions, social organizations, and so on) as systems that one rationality governs independently of the effects on individual lives—in short, as holistic forms—would appear to favor the second option. On the contrary, I would like to emphasize that Structuralism holds an anti-holist position: it stages an always divided, fragmented, and always *non-totalizable* totality, where the local usages never cease transforming the global structure that still conditions them (and whose absence wold even prevent us from perceiving them). All the same, Structuralism had excellent motives fancying itself as a close neighbor of cybernetics. The first definition of the term "Structuralism"—due to Jakobson, as I said above—stressed that this transdisciplinary scientific movement wanted to reacquaint sciences with final causes. Norbert Wiener's cybernetics found in the concept of *feedback* the means to articulate a mechanical theory of teleology that was bound to be crucial to a scientific theory of intelligence and to a technology capable of producing intelligent beings. What could an intelligent behavior be, if not a calculating behavior anticipating results, correcting itself, and therefore giving proof of some rationality?

Here is what a re-reading of Structuralism can teach us: more than in the achievement of control, the secret of human intelligence lies perhaps in its ability to abandon itself to entities that an uncontrollable becoming characterizes. We assume that human intelligence is symbolic and that in order to determine the entities to which it responds, it always operates in parallel upon several dimensions. But we should not seek a *positive* approach to intelligence in the modelization of entities capable of anticipating the consequences of their behavior. Instead, we should seek it in the modelization of entities capable of becoming sensitive and of producing unsubstantial entities that a life of their own inhabits. These entities are "unsubstantial" in the sense that we can extrapolate them from the physical variations of the environment only by the mutual validation that the agents of a community give each other. They transcend any one's full understanding, because they transcend a global rationality in order to enter their unpredictable history. Researchers often claim that the major challenge facing Artificial Intelligence is the simulation of *consciousness*, understood as an agent's ability to know what it does. On the contrary, I think the simulation of *culture*, in its double meaning of consensus and mixed variation, is its true challenge. I am not suggesting we should reject cybernetics' thesis about the profound continuity between life and intelligence, but that we should not pursue such continuity through the creation of artificial *super-organisms* that replicate at the collective level the same type of rationality typical of individual agents (or of auto-telic systems).[30] We should pursue the continuity between life and intelligence through the creation of artificial *milieux* and artificial organisms of a specific sort: those able to constitute themselves as they find in the environment the conditions of their existence and the means to participate in the purely contingent evolution that characterized them and their milieu.

At stake is not simply a matter of bringing the concept of intelligence back to its social and cultural dimensions. A careful reading of Turing's test would suffice to prove it: being intelligent means being capable of mastering at least a few social conventions. We call "intelligent" everyone who behaves *like us*. To claim that the mind is just a subjective faculty that processes and mechanisms operating at the individual level constitute, is hardly sufficient. Nor is it sufficient to claim that thinking means being able to participate into varied social and cultural systems like language, religion, and the political system—all forms of the "objective spirit" that some thoughts lying beyond the reach of cognitive operations typically make up. We need to change our conception of "objective spirit," and the distinction between the two legacies of Structuralism I have discussed above may help us in this task. We need to do away with the conception of the objective spirit as a set of autonomous systems whose evolutionary laws we can reconstruct at the *systemic level*. We need to conceive it as a system of emerging coherences that mechanical processes operating at a lower level, themselves blind to the effects they engender, produce. As in Smolensky's sub-symbolic paradigm, the processes operate at the local level, but their effects are global.

The concept of "sub-symbolic" operations is crucial to the connectionist paradigm, whose artificial neurons are neither a realistic representation of biological units, nor logical representations of symbols or parts of symbols. They are instead the traits determining—at a higher logical level—the emergence of symbols that the computational process does not busy itself with but that a global, perceptual, and passive level still produces. Computation operates at the local micro-logical level, so to speak, while patterns appear at the global and yet ineffective macro-logical level. I think the distinction between micro- and macro-levels is crucial to the epistemology of the sciences of the mind. This epistemology tends to pit hermeneutic approaches to cultural phenomena against theoretical or "nomological" approaches. The first approach characterizes the social sciences as so many efforts to understand as closely as possible the singular, historical, unrepeatable phenomena that all human phenomena *essentially* are (as if the most proper task of the social sciences were to tell us who and what we are). The second seeks causalities, mechanisms, processes, and laws that we pattern on the paradigm of classical physics. We usually classify Artificial Intelligence in the second category. We might even put it this way: within the project of the social sciences (leaving aside those who are after the individual genesis of social facts) two seemingly divergent orientations are in order. One project tries to uncover the systematic and super-organized nature of social phenomena, which constitutes a supplemental level of organization; another project insists on the random, unpredictable, unmasterable, and untotalizable character of anything related to history and culture. The first tries to find *more rationality* in the social systems, while the second looks for less of it, telling itself: "contrary to what happens in the natural sciences, where the objects are predictable, masterable, and controllable, the social sciences deal with objects that are intrinsically labile, slippery, and out of control...."[31]

My reinterpretation of "Structuralism" points toward a third option that lies beyond the hermeneutic and the nomological perspectives. It accepts the presence of *mechanisms* underlying cultural phenomena, but it denies that the search for these mechanisms implies that mechanical laws can describe the phenomena that appear at the cultural level. We should keep the level of laws and processes separate from the level of events and entities. Underneath history are processes, but processes do not rule history. *Some laws allow us to have a history, but history itself is lawless*. We are machines capable to enter, make up, and modify cultures, but cultures are not machines. The sooner we understand that a great current in the social science upholds this thesis, the sooner will we be ready to open up a new phase for the long overdue and all too necessary encounter between Artificial Intelligence and the social sciences.

A new reading of Structuralism allows also a new interpretation of the epistemology of the human sciences and a better definition of the possible meeting point between them and Artificial Intelligence. Yes, the human sciences present us with a different form of scientific rationality. And no, they do not doom us to reject the search for the laws of the mind/spirit compound. They lead us toward a scientific rationality that *instead of seeking to model specific phenomena, strives to model the ability to let unforeseen phenomena emerge*. Let us call it, not a *competence*, but the ability to let this competence vary in an unpredictable, random, uncontrollable way. The challenge we bring to the simulation paradigm that has long supported Artificial Intelligence is clear. What are we simulating? This or that behavior, or the ability to simulate/assimilate/dissimulate behaviors in general? This or that cultural behavior, or the ability to participate in cultural life and to transform it from the inside, without ever turning participation into *control*? But if we no longer simulate a specific behavior, *we will no longer know what it means to be intelligent*. We will stop asking intelligent machines to replace us in doing what we already know how to do. Instead, we will ask them to invent—among themselves and, more importantly, between them and *us*—unceasingly new cultural forms. We will ask them to take part in the collective events through which intelligent forms continuously reinvent themselves and to redefine what to have a mental life means, because we can always find a new and yet variable cultural consensus. A proper grasp of Artificial Intelligence entails that its machines are not models but partners in social and cultural life.[32]

Intelligence is within culture, because an intelligent being is a being whose culture we share and is sensitive to the same environmental and synaesthetic variations as we are. Since the cultures we share are unpredictable, we cannot model the emergence of any specific one. Instead, we should model the ability to enter into a history that accidents, contingencies, and unpredictable adventures make up, where the adjustment of even the most minute detail may affect and change the whole order of a social system. The simulation of intelligence will have to leave aside specific behaviors and concentrate on the creation of machines able to entertain cultural relationships with us (or among themselves). A political constraint limiting the very possibility of Artificial Intelligence may well be appropriate, since our attitude toward machines may represent our biggest epistemological hurdle. We need to stop asking machines to do better than us what we already know how to do. We will have to learn how to ask them to become social partners that can invent with us *new things to do* and new cultural habits whose ultimate meaning is no one's possession. If Saussure's theory is correct, then we must reevaluate the whole paradigm of simulation. Saussure saw culture as an ensemble of systems of patterns emerging from mechanical processes where usage variations may provoke a rigorously *incalculable* transformation of its structure. In his perspective, teaching a machine how to play chess means allowing the machine to participate with us in the long-term processes that have witnessed the game of chess acquire a history, its rules change, its practices evolve, and its practitioners, before they dropped it to turn to Sodokus, continuously develop new strategies.

The renewed paradigm that, I claim, provides the precondition for Artificial Intelligence's new age has been at work in the human sciences for a long time already. The theories we build about the kind of beings we are transform their objects—namely, *us*—because we discover ourselves through them. Similarly, Artificial Intelligence cannot be a simple description of culture, but should be also and coincidentally a recreation of it.

NOTES

1. The author uses the French term "esprit," which covers the English "mind" and "spirit." Even though the translation alternates between the two English terms depending on context, they refer to a single unified concept in the original. Similarly, the French term "les sciences humaines" is broader than "social sciences" or "the Humanities," roughly covering all the disciplines included in the two more traditional American designations. The translation has (mostly) opted to use the less usual "human sciences" [Translator's note].

2. See the "Introduction" in Durkheim, 1912, pp. 12–28, where Durkheim argues that the universal and necessary character of *a priori* forms of thought is part of every individual's social obligation. See also Karsenti, 2006, on Durkheim's confrontation with Kant and on Comte's "politique de l'esprit."

3. See the end of the *Discourse de la méthode*, part V, where Descartes states that "whereas reason is a universal instrument which can be used in all kinds of situations, these [mechanical or animal] organs need some particular disposition for each particular action; hence it is for all practical purposes impossible for a machine to have enough different organs to make it act in all the contingencies of life in the way in which our reason makes us act" (1988, v. I, p. 140 = AT VI, 57). In short: humans enjoy an infinite number of possible actions in spite of a finite number of organs.

4. For a philosophical synthesis of the dynamicist approaches, see van Gelder, 1998; Pobojewska, 2001; van Gelder and Port, 1995.

5. In the French context, where this approach is particularly developed, see François Rastier's works (especially 1987) and the theory of semantics forms that Pierre Cadiot and Yves-Marie Visetti (2001; 2006) developed. For a philosophical discussion of their approach see Salanskis, 2003.

6. Lévi-Strauss often mentioned cybernetics and game theory as an influence upon his structural methodology. I will come back later to the meaning of this influence. For a more detailed analysis of the relationship between Artificial Intelligence and Lévi-Strauss see Franchi, 1997.

7. Weinreich, Labov, and Herzog, 1968, p. 100. This is also the starting point for Labov's 1968 theory, one of the most far-reaching attempts to reintegrate the study of variation into linguistics.

8. See Labov, 1968, p. 189. The passages allegedly in AAVE are in italics. Labov himself notices that the attributions are sometimes arbitrary.

9. See Langacker, 1987, p. 50 and p. 37, respectively, for the first and the second example.

10. In 1981, Chomsky proposed this new model as a radical paradigm shift, even though scholars still debate whether the new theory represents a clean rupture from classic generative grammar.

11. Chomsky, 1981 and Chomsky's introduction to J.-Y. Pollock, 1997.

12. See Wittgenstein, 1973, especially secs. 198, 199, 202. For an original and yet synthetic discussion of these issues see Laugier, 2006.

13. See Botlanski, 1995, and Prince and Smolensky, 2004.

14. For the link between Elman's work and variant linguistics see Laks, 1996. I owe more to this text than this brief footnote may let transpire. Laks is among the few authors who stressed the connection between Structuralism and the new paradigms emerging from the critiques to the classic symbolic paradigm.

15. See Dretske, 1986, sec. 3. The magneto-tactile bacterium is a marine organism endowed with an internal magnetic needle which—when living in the Northern hemisphere—points him toward the North Pole and therefore away from the oxygen-rich shallow water that would be fatal to it.

16. I have tried to defend this interpretation of a science of culture—not the most popular position on the subject, nowadays—in Maniglier, 2006a.

17. Wittgenstein's enigmatic sentence—"'So you are saying that human agreement decides what is true and what is false?'—It is what human beings *say* that is true and false; and they agree in the *language* they use" (1973, sec. 241/p. 88e)—may be relevant to this point. See Laugier, 2006 for a suggestive reading that associated Wittgenstein's position to Kant's aesthetic judgment, which is likewise postulating the existence of intersubjective agreement.

18. See Lestel (2001, pp. 107ff.) for a discussion of culture within an ethological context. For instance, Lestel states: "Behavioral variations recorded by primatologists can only be considered as belonging to 'culture' if they cannot be explained either by genetic determinism or by environmental determinations."

19. Even though Lévi-Strauss cites Wiener, von Neumann, and Shannon, he never uses their work as for the construction of his models. He shares with them the ambition to build a general scientific theory of the "spirit." Lévi-Strauss reported that when he learned that one of his neighbors in New York (whom, he later found out, was Shannon) was working on an "artificial brain," he found it "a bizarre idea" and didn't pay any attention to it (1983, p. 347).

20. See Bouquet, 1997; Fehr, 2000 (orig. Fehr, 1997); Maniglier, 2006b; Utaker, 2002.

21. "It is a mystery on which basis two words may change completely (calidus: chaud; <Germ> aiwa: <all.> je), while we can still affirm their identity. What does it come down to? Precisely! We immediately enter into the main problem of language. Even though the answer is not that easy" (Saussure, 1967, p. 413).

22. Franz Bopp, the founder of comparative grammar, originally presented this argument, which the Neogrammarians in Saussure's time still used. Michel Foucault turns it into one of the interpretive keys of his history of the human sciences (1966, esp. pp. 306–307).

23. See Laks, 1996, for a connectionist approach to the problem, and Petitot, 1985a—the author who turned it into a central tenet of his theory—for an approach from the standpoint of catastrophe theory.

24. In a manuscript notes on legends that Lévi-Strauss (1968, p. 259; 1995, chap. X), often cites, Saussure writes: "It is true that as soon as we delve into this problem we see that in this domain—as well as in all domains close to linguistics'—all the incongruities of thought come from an insufficient reflection on what identity and its features are when we deal with an non existent being like a word, or a mythical person, or a letter of the alphabet. They are all different forms of SIGN, in a philosophical sense ill-understood by philosophy itself" (2003, p. 387).

25. For a detailed demonstration of the philological validity of this interpretation and a precise explanation, I can only refer to Maniglier, 2006b.

26. See P. M. Churchland, 1992a, p. 174ff. on how the NETtalk voice recognition software can reconstruct the categorial hierarchy of the English phonological system on the basis of the distance between activation patterns that singular letters trigger in its input, without any explicit phonological representation.

27. See Saussure's schema (1983, p. 125/175), which suggests a sort of a sketch of a network.

28. As Jakobson put it: "For Saussure, changes in language are blind and fortuitous, and 'alien to the system of language.' But observation has shown on the contrary that changes cannot be understood except in relation to the phonological system which undergoes them" (1978, p. 49); see also Lévi-Strauss, 1973, p. 26, for a similar point.

29. See Jakobson, 1929b, now in Jakobson, 1988, pp. 55–56.

30. See Kroeber, 1917 for the thesis about cultures as super-organisms.

31. I am exploiting and reframing Rickert's (1910) celebrated distinction. Nomothetic sciences attempt to discover general laws, while ideographic sciences try to characterize singular events.

32. I owe the idea of this reversal in the traditional approach to Artificial Intelligence—from the theoretical as well as from the engineering standpoints—to Warren Sack, who practices it as a software developer. See, in particular, his interview with Joseph Dumit (2000).

Six

ARTIFICIAL INTELLIGENCE WITH A NATIONAL FACE: AMERICAN AND SOVIET CULTURAL METAPHORS FOR THOUGHT

Slava Gerovitch

In their proposal to the Rockefeller Foundation for the pioneering conference at Dartmouth College that in 1956 founded Artificial Intelligence (AI), John McCarthy, Marvin Minsky, Nathaniel Rochester, and Claude Shannon, wrote that "every aspect of learning or any other feature of intelligence can in principle be so precisely described that a machine can be made to simulate it" (Norman, 2005, p. 710). Decades later, the aspirations of AI remained the same—to grasp the universal principles of thought in order to implement them in a computer. In 1984, Patrick Winston articulated the goals of AI research as follows:

> One central goal of Artificial Intelligence is to make computers more useful. Another central goal is to understand the principles that make intelligence possible .... Artificial Intelligence excites people who want to uncover principles that all intelligent information processors must exploit. (1984, p. 3)

At the same time in the Soviet Union, a budding AI community formulated goals that sounded remarkably similar, "to understand how the human being thinks, what are the mechanisms of thought" (Smirnov, 1975, p. 3). Scientists on both sides of the Iron Curtain understood AI research as a search for fundamental principles of human thinking.

American and Soviet scientists believed that a general, universal, ahistorical mechanism of human thought existed. Yet, as these scientists themselves belonged to different cultures, they had distinct and culturally specific intuitions of human thinking. The "human beings" whom they took as universal categories were, in fact, people who belonged to specific cultures. Their AI models reflected the specificity of their cultures. According to the recent feminist critique of Allen Newell and Herbert Simon's general theory of human problem solving, their "view from nowhere" was in fact "a view from somewhere." A limited set of problem-solving protocols that served as a foundation for the theory represented "the behavior of a few technically educated, young, male, probably middle-class, probably white college students working on a set of rather specialized tasks in a U.S. university in the late 1960s and early 1970s" (Adam, 2005, p. 339).

Commonly accepted patterns of behavior—actions that we perceive as typical and normal—and also different strategies of handling daily situations that we call "common sense" are the foundation of everyday practice in any society. John McCarthy has famously called AI systems "programs with common sense," implying that a fundamental universality of common sense knowledge underlies human thinking (1968). As the anthropologist Clifford Geertz suggested, however, common sense is "historically constructed and ... subjected to historically defined standards of judgment. It can ... vary dramatically from one people to the next. It is, in short, a cultural system" (1983, p. 76). Geertz warned against "sketching out some logical structure [common sense] always takes, for there is none" (*ibid.*, p. 92), thus unfortunately undermining McCarthy's basic premise.

Everyday practice serves as a mediator for constant exchange of cultural symbols and shapes the cultural vocabulary for any given group. As the anthropologist Hervé Varenne has put it, "American culture is whatever one cannot escape in the United States" (1986, p. 6). Soviet Union had its share of things people could not escape, too. For American urban professionals, cultural influence "spreads through a diffuse network of everyday experiences that range from reading *The New York Times* to using bank cards on automatic teller machines to watching MTV" (Hayles, 1990, p. 4). Soviet intelligentsia's everyday experience looked quite different. They never read *The New York Times*, used automatic teller machines, or watched MTV. They read *Pravda* and underground literature, sat at Party meetings, and stood in lines in food stores. What was typical and normal to them, looked peculiar and exotic to Americans, and vice versa. Yet, even if common sense is not universal, AI models do tell us something—if not about the fundamentals of human thinking in general, then perhaps about specific cultural constructions of common knowledge.

Cultural influence manifests itself not only through typical patterns of behavior and strategies of everyday life, but also through language—via the metaphors by which we live and think, including thinking about thought itself (Lakoff and Johnson, 1980). In this chapter, I will discuss the different cultural metaphors for thought prevalent among American and Soviet intellectuals and explore their connections with specific AI systems. I will maintain that deep cultural factors lay beneath considerable differences in the approaches to AI that American and Soviet scholars developed. While looking for general principles of thinking and behavior, AI specialists implemented in their models their own cultural stereotypes.

1. Different Cultural Metaphors of Freedom: Choice vs. Creativity

The American anthropologist William Beeman has made the following keen observation:

> There are few things all Americans can agree upon. Near the top of the list, along with mother and apple pie, lies "freedom of choice." The power of

> this concept is shown by its ability to be adopted without public comment by groups as diverse as the Advertising Council, pro-abortion forces, and the anti-busing movement. This concept comes close to being sacred for Americans. (Beeman, 1986, p. 59)

Americans face the necessity of choice every day. If we consider such an everyday situation as shopping, the main problem for American customers is making the right (we may say, "healthy") *choice* among an appealing variety of foods and goods. The ability to make the right *choice* is also a crucial part of academic training in the United States. College students choose most of their courses from among a great variety of courses that are offered; routine multiple-choice tests require selecting one right answer among several possibilities. The computer scientist Edward Fredkin in his youth developed "an exact science" of passing such tests. He recalls: "We had a multiple-choice exam, and my bet was that I could pass if I were only shown the answers and none of the questions. Some of the answers went, like, 347, 492, 513, 629—and I did pass the test" (Wright, 1988, p. 15). The AI pioneer Marvin Minsky was also quite good at passing tests, and on this basis he was sent to an experimental school for gifted children (J. Bernstein, 1982, p. 23).

By contrast, most everyday situations in the Soviet Union left the citizen no choice at all. Higher education curriculum prescribed a fixed, pre-determined sequence of courses for every major. The only choice students had was in selecting a preferred athletic activity. Multiple-choice tests were rare. Instead, the student had to spell out all the intermediate steps, and if the algorithm was inefficient (or just different from the textbook's), the grade was lowered, even if the answer was correct. And finally, the Soviet way of shopping posed a different sort of problem for the customer. The problem was not what to choose, but how to find anything at all. With the shortage of many foods and household items, people could obtain sought-after products only via back channels. An ordinary Soviet citizen had to create a unique, long chain of informal social interactions through a network of friends, relatives, friends of relatives, and relatives of friends, so that s/he could find a desired washing machine or a television set at the other end (see Ledeneva, 1998).

The Russian cultural tradition had long treated the notion of choice among preset alternatives with suspicion. While the Americans valued the "freedom of choice," the Russians often spoke of the "burden of choice." The Russian philosopher Nicholas Berdyaev viewed choice as restrictive, instead of liberating:

> Philosophical textbooks generally speak of freedom as identical with "free will," that is to say, as the possibility of choice, of turning right or left.... For me freedom has always meant something quite different. Freedom is first and foremost my independence, determination from within and creative initiative .... The very condition of choice may result in a sense of repression,

> indecisiveness, or even in the complete disappearance of freedom on the part of man. Liberation comes when the choice is made and when I have begun to create. (1965, p. 173).

The Soviet government, even though it forced Berdyaev into exile in 1922, shared his disdain for the freedom of choice and did everything it could to free Soviet citizens from the burden of choice, political or otherwise. Remarkably, the dissidents, for all their opposition to the Soviet regime, felt equally uneasy about the notion of choice. In his 1959 *Free Philosophical Treatise*, published abroad, Aleksandr Esenin-Vol'pin wrote that, "we value freedom conceived as the possibility of choice; but certainly not because we like to choose (the necessity of a choice is sometimes simply horrible and almost always unpleasant!), but because we desire to choose without compulsion." Trying to find a "scientific expression for the concept of freedom," he suggested that "'freedom' cannot always be understood as a possibility of choice" (1961, pp. 135, 137). Facing a narrow spectrum of alternatives, the Russians often perceived it as inherently incomplete. They felt that to widen the spectrum and to create a new, yet unknown solution, was necessary.

Cognitive psychological theories that American and Soviet scholars developed reflected the different cultural values of choice and creativity. The Soviet psychologist Andrei Brushlinskii, for example, rejected the idea that thinking involved a choice among preexisting alternatives. He argued that true thinking must produce a new alternative:

> Actual live thinking, for example, solving a task or a problem, always takes the form of prediction of an initially unknown, future solution. This prediction ... makes the act of choosing among alternative solutions unnecessary. The situation of choice among alternatives is characteristic of the formal-logical correlation of the *products* of the mental process of thinking, but it is not characteristic of the *process* itself. (1979, p. 62, emphasis added)

The American cognitive psychologist Jerome Bruner, by contrast, described concept attainment as a process whose every step "can be usually regarded as a choice or decision between alternative steps" (1973, p. 151). Bruner's work showcased the "cognitive revolution" in psychology that we closely associate with the work of the American AI pioneers Herbert Simon and Allen Newell, who placed choice at the heart of their "heuristic search" model of intellectual activity.

AI specialists in the Soviet Union and in the United States sometimes drew on psychological theories, and sometimes psychologists drew on AI models. More habitually, however, AI specialists ignored psychologists' findings, believing that knowledge should flow from AI to psychology, not the other way around.[1] When AI and psychology agreed, this often happened because they both relied on the same cultural stereotypes.

2. Bureaucratic Man: Striving for Control of Social Environment

Herbert Simon, trained at the University of Chicago as a political scientist, early on strove to find the "atomic phenomena of human behavior in a social environment" (1951, quoted in Crowther-Heyck, 2005, p. 5). Explicitly referring to everyday experience, he found this atom of intellectual activity in the act of choice:

> [N]one of us is completely innocent of acquaintance with the gross characteristics of human choice, or of the broad features of the environment in which this choice takes place. I shall feel free to call on this common experience as a source of the hypotheses needed for the theory about the nature of man and his world. (Simon, 1955, p. 100)

Simon drew on a wide array of mathematical theories that offered different formalizations of choice in well-structured environments—such as econometrics, game theory, operations research, utility theory, and the statistical decision theory—disciplines that his biographer Hunter Crowther-Heyck has termed "the sciences of choice." All these theories assumed the act of choice to be free and rational: an individual acted upon his environment, but the environment did not affect the individual's goals or preferences.

Simon also borrowed from another set of disciplines—sociology, social psychology, anthropology, and political science. These "sciences of control," by contrast, emphasized the malleability and docility of an individual that the social environment mold and over which group and societal pressures hold sway. The "administrative man" of the sciences of control appeared utterly incompatible with the "economic man" of the sciences of choice.

Simon's experiences with the International City Managers Association and the Bureau of Public Administration shaped his faith in liberal reform through expert management and led him to believe in the active role of the intellectual in shaping the social environment. Drawing on the sciences of choice and the sciences of control, Simon developed a theory of "bounded rationality." We could solve complex problems by reducing them to a limited set of alternatives and choosing rationally among them. Belonging to an organization limited an individual's choices and thus made rational decision-making possible: "The rational individual is, and must be, an organized and institutionalized individual" (1997, p. 111). Eventually, Simon arrived at a comprehensive "bureaucratic worldview":

> Simon came to define mind and machine, organism and organization, individual and institution, all as highly specialized yet tightly integrated hierarchical systems, each locked in a continual struggle to adapt to its environment as best it could, given its limited powers. To him the mind and the computer were model bureaucracies, and a bureaucracy was a model mind. (Crowther-Heyck, 2005, p. 7)

In his 1956 paper, "Rational Choice and the Structure of the Environment," Simon used the metaphor of a maze to introduce a mathematical model that described how an organism, although based on incomplete information, could, making a sequence of rational choices at branch points, meet a multiplicity of needs (1956, pp. 129–138). This was not merely a convenient description. Extrapolating from his personal experience to the entire humanity, Simon regarded a sequence of rational choices as a "universal" model, a philosophy of life:

> A philosophy of life surely involves a set of principles .... Principles can provide a book of heuristics to guide choice at life's branch points, a thread to keep one on the right path in the maze .... In this chapter, I have been describing my life, and also my personal life philosophy, but I have also been describing the life of Everyperson. (1991, pp. 360, 363)

In the 1950s-1960s, Newell and Simon developed the "heuristic search" approach that quickly became the dominant paradigm for American AI research. According to their model, problem solving activity consisted in finding a path from the initial to the goal state within the *problem space*. This space looked like a branching tree or a labyrinth; at every step of the process, the problem-solver had to choose one of the alternatives—one of the branches that diverged at the point of choice. In the absence of complete information about the labyrinth, or if the labyrinth was too large to make a feasible calculation, Newell and Simon suggested using *heuristics*—rules of thumb—that would help us make the right choice. They believed labyrinth search to be a universal model of intelligence and considered their computer program, the "General Problem Solver," to be a general "theory of human problem-solving" (1963, p. 279).

As Simon and Newell's conceptualization of human behavior grew increasingly formal, the model situations they were drawing on became increasingly circumscribed and regulated: from semi-independent decisions by workers in big organizations to semi-automatic actions of machine-bound operators in air defense control centers to chess players' limited repertoire of permissible moves. In different computer implementations of the heuristic search model—the theorem-proving Logic Theorist, a chess-playing program, and the "universal" General Problem Solver—Newell and Simon tended to focus on situations with complete, unambiguous, computer-friendly descriptions.

Newell and Simon redefined the problem of choice: they no longer spoke of "making decisions," but of "solving problems." If the decision-maker could consider different goals, the problem-solver had to focus on the assigned problem. Decisions turned into "a less contentious, less political, process of allocating 'processor time' to different tasks. Choices were now less decisions about which set of values to accept and more decisions about what set of data to process" (Crowther-Heyck, 2005, p. 214). Politics was reduced to technology: the liberal aspiration to control and purposefully transform the environment turned into a purely technical task of simplifying search in a labyrinth.

When elaborating her cultural "grammar" of American storytelling, the anthropologist Livia Polanyi has emphasized "control" as one of the most important categories of American life. "Proper people" as everyday conversations portray them, are those who "can *control* the world sufficiently to be happy and have power" (1985, p. 140, emphasis in the original). By contrast, in the Soviet case, instead of something you could control, social environment was something that could potentially control you. A Soviet cultural grammar would probably say that "proper people are those who can sufficiently escape control by the world to be happy." In a formalized and abstract form, Soviet AI models translated the everyday struggle for intellectual autonomy of independent-minded intelligentsia.

3. Soviet Controversy over "Thinking Machines"

The idea that computers could perform intellectual tasks stirred a serious controversy in the Soviet Union in the early 1950s. In the paranoid Cold War context, they often viewed scientific and technological innovations coming from the West with great suspicion. In reaction to the popular discussions of "thinking machines" in the West, the Soviet press condemned this idea as a potential technological threat and an ideological subversion. Soviet journalists berated the capitalists for their hidden agenda to substitute a robot for a striking worker and to replace a human pilot who refused to bomb civilians with an "indifferent metallic monster." Soviet philosophers, for their part, attacked the idea of "thinking machines" as "idealistic" (detaching thought from its material basis in the brain) and "mechanistic" (reducing thought to computer operations). Soviet critics lumped all controversial uses of computers under the rubric of "cybernetics" and labeled this field a "reactionary, idealistic pseudo-science." Despite its glaring logical contradictions—they portrayed cybernetics as idealistic and mechanistic, utopian and dystopian, technocratic and pessimistic, a pseudo science and a dangerous weapon of military aggression—the campaign had a serious impact on Soviet research. As a result of media frenzy, work on "thinking machines" became ideologically unacceptable, and scientific calculations became the limit that early Soviet computer applications had to observe.[2]

But the anti-cybernetics campaign did not dampen the interest of Soviet scientists in computer systems that could perform intellectual tasks. All the first Soviet large electronic digital computers had been installed at defense research institutions, which were relatively immune from ideological pressure and which also gave their employees access to most recent Western publications. Early Soviet champions of cybernetics and AI came largely from these institutions. The mathematician Aleksei Liapunov led the computer programming department at the Division of Applied Mathematics of the Mathematical Institute of the Soviet

Academy of Sciences in Moscow. This Division (after 1966, the Institute of Applied Mathematics) performed calculations for the Soviet nuclear weapons and rocketry programs. These calculations were double-checked against the results that the Computer Center No. 1 of the Ministry of Defense, which was under the supervision of the computer specialist Anatolii Kitov in charge of research and development, obtained. In 1955, taking advantage of the thawing political climate after Stalin's death, Kitov and Liapunov teamed up with the leading mathematician for the nuclear weapons program Sergei Sobolev and published an article in the journal *Problems of Philosophy*, in which they publicly dismissed ideological charges against cybernetics and effectively legitimized research in this field. Trying to calm down the controversy over cybernetics, they avoided such odious terms as "thinking machines" and preferred to speak vaguely of "the theory of automatic high-speed electronic calculating machines as a theory of self-organizing logical processes similar to the processes of human thought" (Sobolev, Kitov, and Liapunov, 1955, p. 136).

As the cybernetics movement grew in strength, bringing under its umbrella all sorts of mathematical models and computer applications in "cybernetic biology," "cybernetic physiology," "cybernetic linguistics," "cybernetic economics," and many other fields, the pendulum of Soviet public attitudes toward "thinking machines" swung in the other direction.[3] The Soviet press began extolling the intellectual abilities of the computer, portraying it as an all-powerful magic tool for solving any problem. Articles entitled "Thinking Machines" and "Bordering on Science Fiction" mushroomed on the pages of newspapers and popular magazines. Journalists quickly dismissed the previous ideological critique by claiming that it applied only to capitalist society:

> If in the capitalist world the introduction of "thinking" machines means the growth of unemployment, exploitation of workers, and fear of the future, in a socialist society, by freeing people from hard, uninteresting work, machines would provide an opportunity to focus on something lofty and joyful—to think, to create, and, in particular, to create new "thinking" machines. (Petrovskii, 1956, pp. 23–24)

Projects exploring intellectual capabilities of computers proliferated. Liapunov set up a machine translation group at the Division of Applied Mathematics, Kitov organized a similar effort at the Computer Center No. 1, Mikhail Bongard of the Institute of Biophysics in Pushchino near Moscow proposed an original approach to pattern recognition, and Nikolai Shanin at the Leningrad (now St. Petersburg) branch of the Mathematical Institute developed a theorem proving program.[4]

Early Soviet research on AI was done largely out of pure enthusiasm, without sustained institutional support. Despite the media hype, the Soviet government showed little interest in "thinking machines." The leaders of the cybernetics movement distanced themselves from AI aspirations, trying to cultivate an image of the computer as an efficient tool and not as an autonomous agent. The chairman of the Cybernetics Council of the Soviet Academy of Sciences, Engineer Admiral Aksel' Berg publicly proclaimed that electronic computers "will be increasingly providing help to man, but will never replace him and *will never think*" (1964, p. 87, emphasis in the original). Computer time remained in short supply, and supervisors did not look favorably on computer programmers' attempts to divert valuable computational resources to investigate problems that aroused their intellectual interests. For example, Aleksandr Kronrod, head of the computational mathematics laboratory at the Institute of Theoretical and Experimental Physics in Moscow, used his control over the Institute's sole M-20 computer to launch work on a chess-playing program. The Institute's administrators viewed spending computer time on games as "overindulgence" (Landis and Yaglom, 2001, p. 1000).

Language reflected the tenuous position of Soviet AI. To stress the metaphorical meaning of the expression "thinking machines," quotation marks always accompanied it. The very term "artificial intelligence" remained controversial, and researchers avoided it. They preferred more neutral-sounding vocabulary, such as "cybernetic psychology," "the study of information processes," or "heuristic programming" (Boiko et al., 1967). In 1964, when the mathematician Dmitrii Pospelov and the psychologist Veniamin Pushkin brought together computer specialists and psychologists interested in AI for a regular colloquium at the Moscow Power Engineering Institute, they named their field "psychonics." The psychonics group directly challenged the Simon-Newell model of thinking and put forward an alternative approach.

4. The Freedom Not To Choose

The term "psychonics" was formed by analogy with bionics. While specialists in bionics hoped to imitate the "design" of living organisms in engineering systems, Pospelov and Pushkin aspired to use psychological knowledge to construct intelligent computers. Pushkin conducted several eye-movement tracking studies of chess players and concluded that each player, rather than searching for the solution in a preset problem space, constructed a different mental model of the position on the board. He asserted that initially we do not structure the human problem space as a tree, and that the process of finding a solution involved creating a new problem space and not, as Newell-Simon's labyrinth model did, "pruning useless branches" (Pushkin, 1971, p. 204).

Soviet AI specialists disliked the labyrinth model not for its inefficiency, but for its departure from their cultural expectations. Even without knowing the conceptual origins of the "General Problem Solver," they associated it with the "bureaucratic apparatus" of labyrinth search. While some followed Newell and Simon's logic and asserted that "the human being thinks by exhaustive search," many others suggested alternative models, for example, thinking as a chain of associations (Kronrod, 2001, pp. 168, 139).

Pushkin and Pospelov conceptualized thinking not as a search, but as a reflection of/on the problem. They argued that we often formulate the descriptions of the current situation in different terms than the description of the goal. In the case of chess, for example, the location of specific pieces on the board describes the initial position while the goal state—a checkmate—requires a higher-level description that involves the inability to move the checked king. The human chess player must be able to go back and forth between low-level and high-level descriptions and build and manipulate different mediating models of the situation. Pushkin and Pospelov argued that situation modeling, not labyrinth search, was the basic intellectual procedure: "Among all the existing words and notions used to describe productive thinking, the most adequate, the most suitable is the Russian word *soobrazhenie* (reflection/imagination) .... The solution *reflects* the situation, based on the *images* or models of its elements" (1972, pp. 140–141, emphasis added).

For Pospelov and Pushkin, humans manifested their creativity in abandoning the old labyrinth, re-conceptualizing the problem, and constructing a new problem space. For example, we could not construct four equilateral triangles out of six matches if we sought the solution on the plane. Constructing a new labyrinth of solutions—in the three-dimensional space—would produce the answer (1972, p. 139).

Where Newell and Simon started with a ready-made structure of the problem, Pushkin and Pospelov suggested that structuring the problem was an essential intellectual step in finding a solution. Building an adequate model of the situation was more important than powerful search algorithms. Pushkin and Pospelov proposed a "semantic language" for formal descriptions of the situation at different levels of generality and developed a system for building relational situation models. They called it *giromat*, borrowing the term from the Polish science fiction writer Stanislaw Lem, who had imagined an intelligent machine capable of adapting its structure to the nature of the problem it was trying to solve.[5] Pospelov and his team implemented this approach, which combined technological and human elements, in computer systems controlling the loading operations in a sea port and other industrial operations.[6]

Pospelov and Pushkin's critique of the labyrinth theory echoed the Soviet cultural perception of choice as a restraint on creativity. In one of his books, Pospelov quoted the Polish dramatist Slawomir Mrozek's mocking exposure of choice as a fetish of freedom. In Mrozek's 1961 play, *Strip-tease*, the following dialog occurs between two people sitting in a room in front of an open door:

> MAN A. What is freedom? It's the possibility of choosing. As long as I am sitting here and know I can leave via that door, I am free. On the other hand,

> the moment I get up and go out, I have made a choice. In other words, I have limited my field of action. I've forfeited my freedom. I become a slave of my going out.
>
> MAN B. But sitting here and not going out, you're also making a choice. You're choosing sitting and not going out.
>
> MAN A. You're wrong. While I'm sitting, I can still go out. But by going out, I will have excluded the possibility of sitting.
>
> (Mrozek, 1967, p. 55, quoted in Pospelov, 1986, pp. 13–14)

Freedom consists in not making a choice, for any choice is a false choice. Interestingly, Mrozek made *his* choice: he left Poland two years after publishing *Strip-tease* and settled in France. For Eastern bloc intellectuals, the rigidly structured labyrinth of choices that the government offered seemed overly restrictive. Some chose to emigrate; some, like Pospelov and Pushkin, chose to expose the limitations of choice-driven behavior and to create new problem spaces.

5. Intellectuals under an Oppressive Regime: Striving for Autonomy

The historian Oleg Kharkhordin has argued that in the late 1950s-early 1960s the Soviet government changed its tactics of political control of the population. Moving away from the punitive and haphazard terror of the Stalin era, the Khrushchev leadership created a system of "voluntary" mass organizations and entrusted it with crime prevention and workplace discipline enforcement. As a result, a pervasive rational system of preventive control among peers emerged, in which the entire population was enlisted in mutual surveillance. Everyone was watching and being watched at the same time:

> People's patrols ... constituted a fearful core of mutual surveillance, which one joined from time to time in order to enforce the standards that one otherwise reluctantly suffered when they were imposed by others. The people's patrols were more meticulous and closer to the people: they were the people policing itself, and thus escape was hardly possible from their omniscient gaze and omnipresent power. (Kharkhordin, 1999, p. 286)

Soviet intellectuals were watched (and watched others) especially closely, and they developed sophisticated strategies of living under the surveillance. Recent studies of Soviet intelligentsia undermine the Cold War stereotypes of the Soviet scientist as blindly supporting or passively resisting government policies (see Gordin, Hall, and Kojevnikov, 2008). A more typical figure would be a physicist working on nuclear weapons during the day and reading underground literature at night (Rassadin, 2004, p. 217). Interested in results, the government allowed the scientists some intellectual license, as long as it was limited to their subject of study. The historian David Holloway has called nuclear weapons laboratories "islands of intellectual autonomy" (Holloway, 1999, p. 175). One theoretical physicist later recalled:

> [P]hysicists constituted a privileged caste, an aristocracy. There were fewer controls on our freedom than on those of any other member of Soviet civil[ian] society. The only laws we felt restricted by were those relating to the conventions of scientific work. Relatively speaking, we were free people. (Mark Azbel, quoted in *ibid.*, p. 187)

Mathematicians and computer specialists working on defense projects enjoyed a similar privileged status. As priests in a temple of an all-powerful goddess, the Computing Machine, they carved out areas of intellectual autonomy in the climate-controlled, limited-access rooms that housed the mammoth-size computers of the first generation. The mathematicians Izrail Gelfand and Mikhail Tsetlin of the defense-research-oriented Division of Applied Mathematics used their portion of intellectual freedom to engage in a study of the central nervous system.

In 1958, Gelfand and Tsetlin organized an informal regular seminar on mathematical models in physiology that began its meetings in the privacy of Gelfand's apartment and later continued for many years at Moscow University (Ivanov, 1998, p. 568). Neurophysiologists traditionally assumed that different nodes within the central nervous system coordinated their activities via a complex system of interconnections. This assumption, however, baffled mathematicians: in a large system, the number of connections would grow so rapidly that any mathematical model would become too complex. Tsetlin and Gelfand, by contrast, proposed a model in which every node regarded the activity of all the other nodes as changes in its environment. They showed that individual nodes did not have to interact directly, but could merely observe external changes and follow a simple adaptive algorithm that minimized their interactions with the environment. This strategy resulted in the purposeful behavior of the system as a whole, if we defined purpose as minimization of the system's interaction with its environment. In this model, purposeful behavior of the whole system did not require great complexity from its subsystems. All individual parts acted quite simply: instead of building complex coordination networks, they tried to avoid interaction. Gelfand and Tsetlin called this adaptive mechanism the "principle of least interaction":

> At each moment, the subsystem solves its own "particular," "personal" problem—namely, it minimizes its interaction with the medium; therefore, the complexity of the subsystem does not depend on the complexity of the entire system.... [O]ur mathematical models allow us (to a certain degree) to imagine the interaction of the nerve centers without considering the complex system of links and the coordination of their activity. (Tsetlin, 1973, pp. 150–152)

In Gelfand and Tsetlin's model, all subsystems acted independently, without direction from a single command center. Given a "payoff function," the individual parts (nervous centers) developed optimal strategies on their own. Their actions affected environmental inputs for all the other parts and led to improvements in individual strategies. Once all the parts had minimized their interactions, they achieved an overall coordination, and this resulted in the optimal strategy for the entire (nervous) system. Tsetlin and Gelfand concluded that the principle of least interaction grounded the purposeful behavior of the central nervous system (*ibid.*, p. 152).

Tsetlin and Gelfand noted that their notion of purposeful behavior as striving for minimal interaction was similar to the British cybernetician W. Ross Ashby's concept of intelligent activity (*ibid.*, p. 150). In the 1950s, Ashby had argued that the machine he designed called the *homeostat*, which did "nothing more than run to a state of equilibrium," could model "many intricate and interesting ways of behaving"(1956, p. 84). By randomly reconfiguring itself, the *homeostat* "learned how to minimize pain in its interactions with the environment" (Pickering, 2002, p. 419). Remarkably, while Tsetlin and Gelfand described minimal interaction in positive terms such as "independence" and "autonomy," British cyberneticians often ridiculed Ashby's idea of thinking as a search for equilibrium; Grey Walter "sarcastically dubbed the homeostat *machina sopora*, the sleep machine" (*ibid.*, p. 417).

The peculiar definition of purposeful behavior as minimization of the system's interaction with its environment resonated with the Soviet intelligentsia's drive to preserve maximum intellectual autonomy. Tsetlin argued that his model of the nervous system had the advantage of non-individualized control: the designer did not need to tell every node in the system what it was supposed to do; the system used its freedom of maneuver to self-organize under most general conditions. At a lecture before the Physiological Society in February 1965, Tsetlin explicitly brought up a comparison of free and forced labor to highlight the advantages of self-organization:

> [T]he work of prisoners is more expensive than that of free men, even though the former are much worse fed and clad, and they work no less. The point is not only that the efficiency of prisoners is lower, but that a prisoner must be fed, clad, and watched by someone else. With a free person the matter is different: ... my manager ... doesn't have to think when to change my shoes or linen or what to do with my children. (Tsetlin, 1973, p. 125)

The MIT biophysicist Murray Eden once remarked, "One wonders whether it is a reflection of cultural or social differences that Tsetlin chose to study cooperative phenomena in choosing 'expedient' behavior, while American game theory focuses on competition among the players" (1973, p. xi). Tsetlin's model, strictly speaking, was not a mathematical implementation of socialist ideals. It reflected the intelligentsia's peculiar position within the Soviet system, in which "cooperative phenomena" emerged out of individuals' efforts to escape the control that the environment (the state) or other individuals ("people's patrols") exerted. Eden's suggestion of social and cultural roots of different approaches to game theory, however, is worth exploring in greater detail.

6. Individualistic Games of Capitalism: Minimizing Risks in Pursuit of Payoffs

Different cultures cultivate different patterns of individual and collective behavior. Livia Polanyi has contrasted the different conceptualizations of individual rights in American and Italian societies:

> Some time ago, I had a discussion with an Italian friend. We were talking about *rights*. He insisted that *rights* are given to the *individual* by *society* .... The *individual*, in my American scheme, *has rights* by virtue of being an *individual* person .... The distinction is a crucial one. In one case, the Italian, *society* is the core of the world and must be protected. It is also granted the role of actor much more than in the American scheme in which the *individual* is an actor at the center of the world and *society* is, as an American friend put it, "just a bunch of *individuals* anyway." ... [I]t seemed that everything which could be said about *society* was somehow reducible to the relationship of one *individual* to *other individuals*. (1985, pp. 107–108, emphasis in the original)

The concept of social interaction as a competition between self-interested, rationally calculating, yet cautious opponents was a model for the Hungarian-born American mathematician John von Neumann in his 1926 axiomatic formalization of two-person, zero-sum, games that have a finite number of "strategies" (complete plans of the game). He proved the minimax theorem and asserted the existence of an optimal "mixed," or randomized, strategy that would minimize the maximum loss and guarantee winning the "value of the game." He believed that the minimax strategy captured some fundamental aspect of human rationality: "[A]ny event—given the external conditions and the participants in the situation (provided that latter are acting of their own free will)—may be regarded as a game of strategy if one looks at the effect it has on the participants" (John von Neumann, 1928, quoted in Leonard, 1995, p. 735).

Von Neumann's biographer Steve Heims has traced von Neumann's formalism back to his perception of the world as filled with ruthless competitors who viewed all the other players as cunning enemies:

> The harshness of this Hobbesian picture of human behavior is repugnant to many, but von Neumann would much rather err on the side of mistrust and suspicion than be caught in wishful thinking about the nature of people

> and society. His temperament was conditioned by the harsh political realities of his Hungarian experience. The recommended style of "playing the economic game," the emphasis on caution, on calculation of expected consequences, the whole utilitarian emphasis aptly expresses the characteristic ideals of the middle class in capitalist societies. (1980, p. 296)

Von Neumann himself viewed the self-interested player of his game theory not merely as a useful mathematical abstraction, but as true reflection of human nature: "It is just as foolish to complain that people are selfish and treacherous as it is to complain that the magnetic field does not increase unless the electric field has a curl. Both are laws of nature" (quoted in Wigner, 1967, p. 261).

In 1944, von Neumann and his collaborator, the Austrian-born American economist Oskar Morgenstern expanded the original conceptual framework of game theory to treat problems of economics in their book, *Theory of Games and Economic Behavior.* Like von Neumann, Morgenstern sought universal principles of human behavior. In his 1941 essay, "Quantitative Implications of Maxims of Behavior," Morgenstern outlined a set of formal principles that governed behavior based on the Austrian philosophical notion of "subjective rationality" and took into account an individual's knowledge and interpretation of facts (Leonard, 1995, p. 750). The historian Philip Mirowski has suggested that "Morgenstern's individualistic bias derived from his Austrian background."[7]

Von Neumann and Morgenstern explicitly challenged deterministic decision-making that neoclassical economics enshrined and presented the "solution" of an economic game as a probabilistic "stable set" of possible apportionments of payoff among the players. As Mirowski has argued, they treated mixed strategies as "a representation of the stochastic nature of thought itself" and effectively turned minimax strategizing into "the very epitome of the abstract rationality" (1991, p. 237). Mirowski has further suggested that von Neumann and Morgenstern came to believe that game theory could "simulate the behavior of any opponent and therefore serve as a general theory of rationality," and that in their writings, "game theory and artificial intelligence tended to blur together" (1992, pp. 125, 127).

The historian Robert Leonard has described the transition from neoclassical to game-theoretical economics as "part of a general shift in science which involved, broadly speaking, the abandonment of determinism, continuity, calculus, and the metaphor of the 'machine,' to allow for indeterminism, probability, and discontinuous changes of state" (1995, p. 756). Among the indeterminism that von Neumann and Morgenstern celebrated, one thing remained stable though: the rules of the game. The game itself was defined as "simply the totality of the rules which describe it." Every player had to know the rules and could not violate them: "[T]he rules of game " are absolute commands. If they are ever infringed then the whole transaction by definition ceases to be the game described by those rules" (1944, p. 49).

The pay-off matrix that the players knew and that remained fixed throughout the game effectively embodied the rules that specified to the players which combination of strategies would result in which rewards. "It would not surprise me if the greatest contribution of game theory to social science in the end is simply the pay-off matrix itself that, like a truth table or a conservation principle or an equation, provides a way of organizing and structuring problems so that they can be analyzed fruitfully," wrote one sympathetic economist (Thomas Schelling, quoted in Mirowski, 1991, p. 248). Fixing the rules of the game not only made it possible to derive powerful formal results in game theory; it also provided an anchor for the notion of rationality: the world was too complex for deterministic analysis, but it still followed rules, so a stochastically equipped mind could still calculate an optimal set of strategies.

Game theory initially received a lukewarm reception among economists, but as a general model of decision-making it attracted interest of other social scientists. In the words of the historian David Hounshell, in the late 1940s defense analysts at the USA think tank RAND Corporation came to see game theory as "an ideal approach to solving problems of warfare and decision-making" (1997, p. 253). They embraced the notion that formalization as a zero-sum, non-iterative, two-person game was intrinsically akin to the tactical and strategic problems of the Cold War confrontation with the Soviet Union. RAND analysts rushed to apply game theory to a wide range of problems: strategic bombardment systems/target allocation, missile fire scheduling, and psychological warfare (Edwards, 1996, p. 117). It turned out, however, that "non-zero-sum games consisting of multiple moves in which a changing number of players could learn, mix strategies, and pursue non-rational behavior eluded RAND researchers' formalization abilities," and their superiors found the results "quite disappointing" (Hounshell, 1997, p. 254).

Game theory attracted defense analysts not because it demonstrated particular effectiveness, but largely because it fitted their pattern of thought. They did not perceive the constraints of von Neumann and Morgenstern's formalism; on the contrary, they were perfectly satisfied with the simplest case—two-person zero-sum non-iterative games:

> The strategic community had no qualms about the two-person restriction, since they thought in Manichean terms of Blue versus Red, capitalists versus communists, USA versus USSR. But they also had no difficulty with the zero-sum restriction. In deployment or dogfight models, they believed, winning or losing was a simple offsetting payoff structure .... Moreover, the application of games to the face-off of atomic weapons only reinforced

> the two-person zero-sum mindset, including the all-important stipulation that the game could not be repeated. (Mirowski, 1991, pp. 245–246)

Von Neumann and Morgenstern aspired to create an abstract theory of rationality, but what resulted was a snapshot of the Cold War mindset. The historian Paul Edwards has argued that, "as a metaphor, the zero-sum game represented yet another symbolic enclosure," reflecting the "closed world discourse" of defense analysts (1996, p. 118). While economists debated whether game theory truly reflected actual economic behavior, minimax strategizing proved an excellent match for the strategic community's notion of common sense: their vision of the world as an arena of fully computable and manipulable conflicts with known risks and rewards. Defense analysts stripped game theory of its connotations of indeterminacy and reduced it to an impersonal, formalized, mechanized decision-making tool that could guarantee the winning strategy. Game theory, in its emasculated form, helped them "uphold the appearance of technical, calm, measured rationality in the face of bottomless nuclear horror" (Mirowski, 1991, p. 249).

The strategic community asserted that "the significance of game theory as a decision tool is that it eliminates guessing an opponent's intentions" (quoted in *ibid.*, p. 251). While guessing was the opposite of rational problem-solving for USA defense analysts, it was often the only option available to an intelligent decision-maker in the Soviet Union.

7. Collective Games of Socialism: A Play with Changing Rules

Scholars studying Soviet science in the late Stalinist and Khrushchev periods have remarked the ritualistic patterns of behavior typical of the scientific community. Whether scientists engaged in public discussions of the philosophical and ideological meaning of their discipline or tried to jump on the bandwagon of a fashionable intellectual trend, they had to play a game according to the unspoken rules of public behavior of a Soviet scientist (Kojevnikov, 1998; Krementsov, 1997, esp. pp. 239–248; Gerovitch, 2002). Ritual critique of ideological enemies, skillful manipulation with suitable quotes from Marx or Lenin, and ingenious translation of scientific terminology into an ideology-laden language were among the indispensable strategies of Soviet science. The outcome of debates over the validity of scientific theories often depended on the discussants' abilities to play the game. As the historian Alexei Kojevnikov has remarked, "an important thing about these games was that, in theory and often also in practice, their outcomes were not predetermined, but depended upon the play" (1998, p. 28).

The uncertainty over the rewards and punishments for specific strategies complicated the play. Frequent swings in the direction of Stalinist ideological campaigns often left slow thinkers stuck with old, outdated slogans and made them vulnerable to attacks. On the other hand, ideological campaigns had their own inertia and often became self-perpetuating. Some campaigns were not directed from above; they generated spontaneously but were perceived as intentional, as, for example, the anti-cybernetics campaign. Those scientists who could not properly decipher "signals" from above were often perplexed about the rules and direction of the most recent campaign. Stalin's personal attempts to encourage an open debate over scientific issues faced scientists' reluctance to speak out when the official position was not clearly stated (see E. Pollock, 2006).

Mikhail Tsetlin's theory of collective games of automata exemplifies the fundamental uncertainties of Soviet social games. An automaton is a mathematical model of a finite state machine that changes its state according to its transition diagram and the current input. Tsetlin interpreted an automaton as an agent acting in an environment that randomly penalized or rewarded specific behaviors. Unlike the classic von Neumann-type games, Tsetlin studied games in which the automata faced a world filled with uncertainty:

> It should be noted that the automaton games are discussed here from a viewpoint that differs from the one accepted in game theory. Indeed, it is normally assumed in the latter that the game is defined by a system of pay-off functions previously known to the players.... We thought it interesting to consider games played by finite automata having no *a priori* information about the game, and being forced to shape their strategies for each successive replay in the course of the game itself. (1973, p. 6)

In Tsetlin's games, "the players have practically no information about the game. They are ignorant of the number of other players involved, of the situation at any particular moment and even of what kind of game they are actually playing" (Varshavskii and Pospelov, 1988, p. 97). Tsetlin informally compared his model of an agent operating in an environment with unknown and changing rules to a "little animal in the big world" (*ibid.*, p. 132). His friend, neurophysiologist Nicholas Bernstein used a similar metaphor to describe the fundamental uncertainties of intellectual activity: "To use a metaphor, we might say that the organism is constantly playing a game with its environment, a game where the rules are not defined and the moves planned by the opponent are not known" (1967, p. 173).

Tsetlin discovered that in a changing environment, in which the probabilities of penalties and rewards varied over time, automata that did not have too many states were the most successful. In other words, if the rules of the game constantly changed, it was not a good idea for the automaton to remember too much of its history. The more dynamic the environment, the shorter was the optimal depth of the automaton's "memory."

In his study of collective "distribution games," Tsetlin presented a thinly veiled commentary on the economic strategies of individuals under socialism. First, he considered a game in which a group of automata competed for resources (rewards or payoffs) by choosing different strategies. He designed automata that were completely unaware of the relative strengths of different strategies, but would eventually settle on the optimal strategy by reacting to rewards from their environment. Tsetlin showed, however, that they could increase the average gain if the automata played a game-theory version of socialism—a game with a "common fund" in which all gains and losses of individual automata were added up and then equally shared. The drawback was that the common fund camouflaged the link between individual contribution and reward and placed greater demands on the memory capacity of individual automata. We could "reap the benefits of a common fund procedure starting from a certain level of complexity" of automata memory, he concluded. "If the memory capacity is below this threshold, the introduction of a common fund reduces the average gain" (Varshavskii and Pospelov, 1988, p. 100).

In informal discussions, Tsetlin mockingly translated this rule into the clichéd parlance of the Soviet ideological discourse as "the negative effect of wage-leveling in the conditions of the low consciousness of workers" (*ibid.*, p. 101). The Soviet press often blamed the low quality of consumer products on workers' "low level of consciousness." Soviet propaganda routinely called on the workers to raise their consciousness and to work harder for a common fund. Tsetlin provided a mathematical formalization of this ideological dogma, calculating the precise memory capacity ("consciousness level") that we needed to find an optimal strategy in a game with a common fund.

Tsetlin's analysis implied that different types of games (or economic systems), to achieve the maximum payoff, required different mechanisms of decision-making. In a regular distribution game, the achievement of an overall optimum required each player to pursue individual gains. In a redistribution ("common fund") game, local level decision-making had to be far more complex in order to achieve an overall optimum. One imaginative interpretation of this result was that "capitalism is more profitable when the management system is simple and socialism is more profitable when the management system is elaborate" (*ibid.*, p. 102). Game theory appeared to suggest that, given roughly equal memory capacity of economic agents, socialism required greater social controls, while *laissez-faire* capitalism could work if individual agents were left to their devices. One model explicitly introduced a "manager" into the collective of automata, whose goal was to maximize the total gain. By punishing low-earning automata, the manager could eventually prompt them to occupy states that would be less favorable to them individually but result in greater total gain (Gaaze-Rapoport and Pospelov, 1987, p. 247). Tsetlin's followers tried to mitigate the discouraging implications of their modeling by quoting Lenin's dictum, "socialism equals accounting," but the broader social meaning of restrictions that the demands of a "common fund" placed on individuals were not lost on anyone with a sufficiently high "level of consciousness."

Tsetlin's colleagues turned his result into a fundamental principle of human thinking and behavior. Viktor Varshavskii and Dmitrii Pospelov interpreted memory capacity as a general measure of intellectual ability (1988, chap. 3). They correlated our "intellectual level" with the ability to find an optimal strategy in a game in which our immediate actions, produced at a higher level of organization, had no explicit tie to gains and losses. The writings of Soviet AI specialists paradoxically combined a thinly veiled critique of socialist redistribution and a peculiar definition of intellect as the ability to find an optimal strategy of living under socialism.

The notion of a game with unknown or changing rules was quite familiar to liberal intelligentsia. They played a cat-and-mouse game with the Soviet government, constantly challenging the fuzzy boundaries of permissible discourse. While Soviet laws ostensibly proclaimed many democratic freedoms, the actual practice was to suppress any significant dissent by placing it under the vague rubric of "anti-Soviet activity." The leading dissident, mathematician, and logician Aleksandr Esenin-Vol'pin searched not only for a "scientific expression for the concept of freedom," but also for strict definitions of all legal concepts. He exposed gross logical contradictions in the way the Soviet authorities applied liberal (on paper) Soviet laws to suppress dissent. He gained notoriety in the eyes of Soviet officials by insisting that the government literally obey its laws (see Nathans, 2007). The fellow dissident Vladimir Bukovsky even compared him to a computer that could not tolerate any ambiguity or inconsistency (1999, p. 417). Yet the Soviet government had no intention to part with its privilege to arbitrarily change and reinterpret the legal rules. In March of 1968, Soviet officials forcibly placed Esenin-Vol'pin in a mental institution, giving a whole new meaning to Simon's notion of the rational being as an "institutionalized individual."

Ninety-nine Soviet mathematicians signed an open letter protesting the action against Esenin-Vol'pin and asking for his release, thus abandoning the most expedient strategy of behavior under socialism—to minimize our interactions with the political environment—and entering in a direct confrontation with the authorities. This flagrant violation of the "principle of least interaction" had serious consequences for the signatories (Feinberg, 1996, p. 679). Many mathematicians lost their jobs, including Aleksandr Kronrod, who supported the work on a chess-playing program at the Institute of Theoretical and Experimental Physics (*Khronika tekushchikh sobytii* 1968). Georgii Adel'son-Velskii, a key member of the chess group, resigned in protest (Dobruskin and Fomenko, 2008). The entire chess group moved to another institute, where they eventually developed Kaissa, the first world champion among computer programs (Donskoi, 1999). Ironically, Kaissa's success was due to a quite efficient search algorithm that was based on an original pruning technique.

8. Rats and Butterflies

Two metaphors capture the crucial differences in the cultural stereotypes of thought and behavior reflected in AI systems that the Soviet Union and the United States implemented. Life as a maze—a labyrinth in which we must find the right path—became the central metaphor for American AI. For Soviet AI specialists, the central metaphor for decision-making was not the search in a fixed labyrinth, but the flight of a butterfly, charting its flight trajectory through random streams of air.

The metaphor of a labyrinth evoked the behaviorist pattern of B.F. Skinner's experiments on rats running T-shaped mazes and the popular American cultural image of the "rat race." Marvin Minsky started his research in the early 1950s by building a "basically Skinnerian" electronic machine that simulated rats learning to run mazes (J. Bernstein, 1982, p. 37). Allen Newell and Herbert Simon "internalized the major lessons" of behaviorism (1963, p. 280), and made heuristic search the centerpiece of their theory of human thinking and behavior. Their model of problem-solving drew on Simon's study of administrative behavior, which, in turn, took rats running mazes as a paradigmatic case: "A simplified model of human decision-making is provided by the behavior of a white rat when he is confronted, in the psychological laboratory, with a maze, one path of which leads to food" (1991, pp. 85–86). Simon insisted that the limited knowledge and intellectual capacities of a rat reflected the constraints on human rationality better than the assumption of divine omniscience and perfect rationality: "we need a less God-like and more rat-like chooser" (1954, quoted in Crowther-Heyck, 2005, p. 6)

The Soviet AI researchers Viktor Varshavskii and Dmitrii Pospelov drew on a different image—that of a butterfly. They described a system that simulated the behavior of a moth hunted by a bat. When the bat was too close and the moth could not fly away, the moth started dashing around in a chaotic flight:

> The chaotic flight is a series of passive falls with folded wings, sharp turns, loops and dives. In other words, the moth follows a trajectory which makes it more difficult for the bat to predict its location from one moment to the next. We should mention that in experiments the chaotic flight strategy saved the moth's life 70 percent of the time. (1988, p. 77)

A butterfly fluttering in a chaotic current of life and trying to escape a predator—this image was all too familiar to the Soviet scientists who were trying to preserve their intellectual autonomy.

American and Soviet AI specialists sought general principles—universal, timeless mechanisms of thinking and behavior. Their generalizations, however, relied on culturally conditioned cases. The examples that American and Soviet scientists had at their disposal, were, in fact, culturally specific patterns of social organization and decision-making. When trying to grasp universality, AI models manifested just the opposite—the specificity of cultural patterns.

Without knowing it, science often speaks with a national accent. Cultural symbolic systems can manifest themselves in scientific ideas as clearly as in literature or art. In their simulations of human thinking, AI systems truly reflect both mechanisms of reason and patterns of irrationality, individual creativity and social stereotyping, human nature and human culture.

NOTES

1. Newell and Simon, for example, prophesied in 1958 that "within ten years theories in psychology will take the form of computer programs, or of qualitative statements about the characteristics of computer programs" (1958, pp. 7–8). In 1970, Allen Newell described AI as "theoretical psychology," whose role was to generate problems for experimental psychologists to study (pp.~363–400).

2. On the Soviet anti-cybernetics campaign, see Gerovitch, 2002, chap. 3.

3. On the Soviet cybernetics movement, see *ibid.*, chaps. 4–6.

4. For a brief historical overview of Soviet AI research, see Pospelov, 2003, pp. 407–430.

5. Lem described the *giromat* in his 1955 novel *The Magellanic Cloud* (still untranslated into English).

6. For a historical overview, see Pospelov, 1986, pp. 254–258.

7. Mirowski, 1992, p. 137. On Morgenstern, see Leonard, 2004.

III

MARGINS OF COMPUTATIONS

Seven

THE CARTESIAN-LEIBNIZIAN TURING TEST

Francesco Bianchini

1. Introduction

In this chapter I will consider some aspects of the philosophy and epistemology of Artificial Intelligence (AI). The core of my argumentation will be Alan Turing's methodology to establish whether we may call a mechanical entity intelligent. Two are the issues at stake. First, the simple one: how are we able to determine if machines can think? Second, the more complex one: what do we mean by saying that a human being is intelligent *and* is a machine? Both are epistemological questions. Yet, while the first is at least partially a technological question, the second is clearly a philosophical one.

In what follows, I will attempt to show that modern philosophy started the issue of human intelligence-cum-machine-ness. Contemporary philosophy has continued to wrestle with this issue even after—and because of—the crucial birth of AI in the 1940s and 1950s. Douglas Hofstadter has remarked that "the primary purpose [of the Turing Test] was to convert a philosophical question into an operational question" (1985, p. 525). Nevertheless, after Turing, AI does not get rid of philosophy, nor does philosophy get rid of AI and its achievements. In fact, the two disciplines become more and more theoretically intertwined. In order to pursue this argument, I will introduce the notion of "room argument."

I suggest adopting the expression "room argument" to indicate some thought experiments that have no value of "proof" in the usual scientific meaning of the term, but that are useful for the sense of a theory—in this case, a psychological theory—and the epistemological principles on which we built it. My aim is to provide a likely explanation of mental phenomena and cognitive activity and processes.The "room" in a room argument is a *metaphorical place* that demarcates the outer boundaries of the subject matter that we want to understand and explain. The first (metaphorical) step is getting into the room—the opening move for the formulation of an explanatory hypothesis. The hypothesis concerns the room's contents. In the last five decades, we have, often with similar arguments, made hypotheses about the spiritual, material, organizational, and structural content of the mind. The Turing Test is a "room argument" of sorts, a thought experiment aiming to test some concepts that entail the explanation of what we commonly identify as "mind," and especially the conditions that apply to one of the contemporary identification of the mind—*intelligence.*

The expression "artificial intelligence" dates back to the middle of the twentieth century. It involves disciplines that were coming from different points of views but were dealing with reasoning and thought, all converged on the notion of intelligence. Disciplines as varied as contemporary philosophy, psychology's recent schools, contemporary logic, mathematics as theory of computation, economics as science of social relations, linguistics, anthropology, and biology, they all treat intelligence as a notable concept. The fact that we can consider the notion of intelligence as a point of conceptual intersection among different perspectives of research on mankind and its peculiar characteristics is also why "intelligence" still lacks a solid, formal definition. Although questioned since well before Norbert Wiener's cybernetics of the 1940s, I think a definition of intelligence is necessary to establish when and if intelligence is present in biological or artificial entities.

2. The Room Arguments and the Cartesian "Criterion"

We can consider the "room arguments" as a set of thought experiments that assign a notion, such as the notion of "intelligence," to a specific object. The conditions that such a thought experiment establishes are not definitional traits, but a set of empirical possibilities.

Thought experiments can usually provide some criteria for determining if we can include something in a certain category, such as the concept of "intelligence." The criteria are empirical possibilities, and therefore, broadly speaking, experimental conditions. We can consider Turing's test a *Gedankenexperiment* that establishes a criterion for the application of the notion of "intelligence."

On the basis of Turing's ideas, in the following pages I will reconstruct the "room arguments" that researchers have developed during the last six decades, beginning with Turing's thought-experiment that first inspired me to coin the expression "room argument."

One famous antecedent of these arguments comes from René Descartes. In his *Discourse on the Method* (1637), which is perhaps less well-known than his other works, he states the criteria for discriminating machines from human beings. In particular, he identifies two of them:

> The first is that they could never use words, or put together other signs, as we do in order to declare our thoughts to others. For we can certainly conceive of a machine so constructed that it utters words, and even utters words which correspond to bodily actions causing a change in its organs (e.g., if you touch it in one spot it asks what you want of it, if you touch it in another it cries out that you are hurting it, and so on). But it is not conceivable that such a machine should produce different arrangements of words so as to give an appropriately meaningful answer to whatever is

> said in its presence, as the dullest of men can do. Secondly, even though such machines might do some things as well as we do them, or perhaps even better, they would inevitably fail in others, which would reveal that they were acting not through understanding but only from the disposition of their organs. (1988, v. 1, p. 140=AT VI, pp. 56–57)

Scholars have pointed out that this criterion is a sort of "language argument," a Cartesian proof of the existence of other minds (Sportelli, 2007), a problem that obviously persists in modern and contemporary epistemology. So, *a fortiori*, it takes part in the twentieth century query about what intelligence is and what we have to consider the meaningful outer manifestations of subjecthood to be.

The way Descartes states it and his consideration of possibly pragmatic "other signs" suggest that his criterion is not merely linguistic. Showing an optimistic reliance in the mechanics of his time, Descartes admits the possibility that a machine can produce words. It could not, however, provide "sense" to the words that an intelligent human being utters when s/he engages in linguistic interaction. In this situation, the machine would have to *coordinate words* in such a manner that it can support a dialog and "reply appropriately to everything that may be said in its presence." Descartes (and his contemporary philosophical culture) attributes this trait to any human being equipped with the standard faculty of speech ("even the lowest type of man") and for this reason, at least for Descartes, this feature is as innate as the ideas of the Cartesian gnoseological system are.

The second criterion is even more surprising, because it goes beyond the scope of linguistic practice. Descartes admits, exactly as Turing later, that machines can equal or surpass human beings in some specific tasks. The *manner* in which machines carry out these tasks, however, the "disposition of their organs," (in modern terms, how *we have programmed their functional parts to operate*), would merely show their *mechanical-ness* and their *procedural-ness*.

His conclusion is that machines' behavior would reveal their gnoseological lack: they don't "act *from knowledge*" (emphasis added). Descartes appears to anticipate by several centuries the main objections to the possibility of a genuine "artificial intelligence" that researchers have, since the 1950s, brought to bear upon the epistemological root of cognitive science as a scientific approach to the mental phenomenon of understanding. We can almost view the second criterion that Descartes postulates as a sort of overcoming of the linguistic requirements that Turing's imitation game as a parameter for identifying intelligence exacted.

But, as I will argue later, even though Turing formulated his proposal in the cultural environment that psychological behaviorism influenced, we cannot reduce it to a mere epistemological reflection on the nature of psychology. From the philosophical point of view, we must also consider his proposal as a preliminary reflection on the studies of the mind in search of a "peculiar human essence." This would include studies that confront the question from the perspective of "automation." To put it differently, Turing's "test" does not exclude knowledge but leaves it in the background, foregrounding instead the concrete possibility of building a *machine that can think and act linguistically*.

On the one hand, the Cartesian dualism of substances (*res cogitans* and *res extensa*) appears to engender a profound and intrinsic irreconcilability between the knowledge that humans can afford as opposed to machines'. On the other hand, Descartes's surprising acceptance of a machine that speaks about an "action [...] of a corporeal kind, which brings about a change in its organs" is close to a modern report of the execution of some programs on hardware of some kind. Yet, Descartes claims that even this behavior is not equivalent to the acting by a human being, because the machine's "dispositions of organs"—as also Gottfried Wilhelm Leibniz will clearly grasp—are insufficient to give rise to knowledge.

At the same rate, however, the knowledge of which Descartes speaks does not appear to be a subjective, first-person view, that we endow with the incommunicable flavor of *qualia*. Even within the Cartesian dualistic view, knowledge remains intersub(ob-)jective and communicable, *somehow* susceptible to representation and also, to a degree, if not congruent at least related to symbolical expression. In Descartes's times, such knowledge pertains exclusively to human beings, whose actions manifest it and pass it on to others.

So, the question is: what changes have taken place between Descartes's and Turing's days? From the philosophical point of view and with regard to the possibility and aspirations of a thinking machine and of an intelligent artifact, which researchers have been debating in the past sixty years, what role do we assign to the central point, namely, to the concept of knowledge?

3. The Chinese Mechanical Intricacy of Searle

In September 1980, *Behavioral and Brain Sciences* published John Searle's well-known essay, "Minds, Brains, and Programs," together with the exchanges between Searle and some of his critics from the fields of philosophy, psychology, cognitive science, and AI. Subsequent discussion has largely echoed the original set of essays, focused around the famous Chinese Room *Gedankenexperiment*.

Searle's aim is twofold: to restrict the possibility of intentionality to mechanisms that have exactly the same causal powers as the brain, and to prove that under no condition we can ascribe such a possibility to an artificial, classically computational, system. The attention that Searle's article generated went far beyond a simple evaluation of his conclusions.

Searle's polemical target is, immediately and explicitly, the so-called Strong AI program based on formal symbol manipulation and the identification of programs with thinking processes. The following hypothesis characterizes the theory:

> the appropriately programmed computer really *is* a mind, in the sense that computers given the right programs can be literally said to *understand*

> and have other cognitive states. In strong AI, because the programmed computer has cognitive states, the programs are not mere tools that enable us to test psychological explanations; rather, the programs are themselves the explanations. (Searle, 1980, p. 417)

Searle believes that the evident absurdity of Strong AI justifies its rejection in favor of a weak conception of AI, in which the computer, no matter how powerful, is still a tool, hence an auxiliary and not a fundamental contrivance in order to explain mental phenomena.

From this critical stance towards a precise, though fairly roughly sketched, view of AI, Searle proceeds to the *pars construens* of his argument: since computational artificial systems cannot be intentional, the brain alone, by means of its causal powers, can produce intentionality. To say that thinking machines do not exist, Searle's argues, is too strong; the brain is itself a machine, but a machine whose peculiar powers derive from its biochemical structure. Only a machine that possesses "internal causal powers equivalent to those of the brain" (*ibid.*) can earn the title of a thinking machine that we can consider endowed with intentionality.

By "intentionality," Searle means a property intrinsically linked to the brain and to any machine equipped with a causal structure identical to that of the brain. This move does not rule out the possibility that entities other than human beings may have thinking abilities, but undoubtedly raises the constraints to a quite high and problematical level. At present, only human beings appear to possess these causal powers, though Searle's essay does not specify what such powers consist of. He simply states that "whatever it is that the brain does to produce intentionality, it cannot consist in instantiating a program since no program, by itself, is sufficient for intentionality" (*ibid.*, p. 424). We cannot achieve intentionality, the qualifying feature of "mentalhood," through the mere, formal, and not-human computational power that is typical of computers.

In his attack to computationalism, Searle does not limit his argument strictly to computers. He reinforces his contention and asserts that even a mind that acts computationally by formally manipulating symbols has no intentional power. A position that he reiterates throughout the essay, it reveals him as an enemy of anti-mechanism. Searle's stance does not reject all versions of AI. In spite of the fact that the connectionist approach to cognitive processes was marginal at the time of Searle's essay and debate, he insists that brain's causal power alone has his approval, thus agreeing with connectionism in principle. In those years, old-fashioned symbolic AI, following the reality check that the Sixties and Seventies imposed to its stated goals, appeared a theoretical construction about to collapse. Researchers had to reevaluate the enthusiastic predictions and expectations of the 1950s.

To reduce intentionality in a deictic sense to brain's causal powers and to deny cognitive significance to formal symbol manipulation is equivalent to saying nothing about those powers, and does not warrant the failure of computationalism. Moreover, it does not show the fallacy of all arguments in favor of the thesis that symbol-operating programs can yield mental activities.

To strengthen his argument, Searle resorts to the Chinese Room thought experiment, the core of his essay and the pivot around which he fashions six possible objections to his assertions. The intuitive plausibility of the thought experiment buttresses his argument. In a slightly modified version of it, I will present the structure of the experiment first, and then the exposition of the phenomenon that he wanted to explain—namely, the understanding of a situation expressed in natural language.

In short, the Chinese Room thought experiment consists in imagining a native English speaker with no knowledge of Chinese who is locked up in a room and is busy doing translations in Chinese with the help of an English rulebook. An interface allows the individual in the room, who uses those rules, to communicate to the outside a series of Chinese symbols that are *in some relation* with the Chinese ideograms that someone previously sent to the room via the same interface. These external agents can be indifferently native speakers of Chinese or programmers familiar with the same rules as the person in the room. What happens *inside* the room and how we on the outside interpret it, is what counts. The individual in the room has in front of him/her two batches of papers written in Chinese. By means of the handbook's rules which are in English, s/he can correlate Chinese symbols coming from the outside with the Chinese symbols of his/her batch, and then send back to the external interlocutors a succession of Chinese symbols that are *solely* the output of the handbook's rules.

We can view such a handbook as composed of an instruction set like a program, or better still, like a set of programs. These instructions allow "to correlate one set of formal symbols with another set of formal symbols, and all that 'formal' means here is that I can identify the symbols entirely by their shapes" (Searle, 1980, p. 418).

Rules allow correlations between symbols exclusively on the basis of their form, through a process that, if we were in the room, we would see unfolding as follows: the individual singles out one or more Chinese symbols among those s/he has been assigned; s/he looks it or them up in the handbook, one by one, in succession; s/he checks to see what symbol or succession of symbols it or they correspond to it in the batches; s/he reads the instructions that s/he must carry out once s/he has established the correlation; s/he produces some shapes, by writing new Chinese symbols on the basis of the instructions s/he was been previously fed from the outside; these instruction let him/her "give back some Chinese symbols with certain sort of shapes in response to certain sorts of shapes [given to her/him] in the third batch" (*ibid.*); lastly, s/he sends out these new shapes, the Chinese ideograms s/he has written, through the interface.

In this form, the argument appears faultless. Searle relies on the assumption that we can extend the mechanical description also to an especially high-level cognitive ability: the understanding of natural language narrations. We can think of the two batches of Chinese symbols that the individual in the room has at his/her disposal as follows: the first one, as the description of the prototypical situation of having dinner in a restaurant (with its standard features of arriving, sitting down, looking up the menu, ordering, waiting for the courses, eating, paying, and leaving); the second one, as the description of a particular episode occurring at the restaurant dinner (during which, for instance, a particular person goes to the restaurant, orders a steak, and upon receiving it burnt, s/he stands up and leaves). The Chinese symbols sent into the room from the outside are questions about the story (for example: "Did the person eat the steak?" "Did s/he pay the bill?"); those sent out from inside are answers to those questions. To formulate answers, the individual uses the English handbook that contains the correlations between Chinese symbols. Searle says that his/her behavior corresponds to a computer program's; like the individual, we presume that the computer is able to understand a story and to show this understanding through pertinent answers to questions. Obviously, the person in the room understands neither the story, nor the questions asked or the answers given. The room mechanism makes intentionality disappear; it even *eliminates* it. In Searle's argumentation, this fact signals that we cannot explain intentionality from a computational point of view.

The power of Searle's argument resides in its immediacy. Who could say not to understand his/her native tongue? Conversely, who could fail to acknowledge a lack of understanding of a language that s/he has not only never learned, but that is also so foreign in its syntactic and semantic features and its graphic notation? Of the essence here is our ability to draw the purely formal relation between the symbols of the two languages that let us convey to the people outside the room the idea that they are having a conversation, by way of an interface, with a native Chinese speaker. *But nothing of that sort occurs in reality*, says Searle, thus defeating the claim that the simple formal manipulation of symbols guarantees their understanding, thereby producing intentionality. If an individual, a human being acting like a computer that executes a series of instructions, cannot understand the symbols to which s/he applies the instructions, *a fortiori* a program will not understand them either.

With this argument Searle strongly criticizes Roger Schank's AI research and his cognitive model of short stories understanding. Searle's article discusses especially Schank and Abelson's proposal (1977) of a psychological theory of scripts as structured models of real-world understanding (Schank, 1984). Yet, the polemic purpose is not directly to argue that *scripts* and *frames* are inadequate in catching real-world complexity. As a matter of fact, even if scripts and frames are useful for real-world situations, either through a prototypical informational template or through a standard model of reality (or a part thereof), specific details that we cannot "pre-program" will always be present in the prototype model.

Even a non-monotonic information system can never completely grasp the infinite complexity of real-world situations.

Scripts and *frames* are amongst the most successful systems of knowledge representation because we can periodically inject their standard descriptions with data that relate them to specific situations. By exceeding the limits of first-order predicate calculus, they go beyond intrinsic monotonicity.[1] Even though we could define them as "aware" knowledge tools, they are not the ultimate AI achievement concerning knowledge. What is then the real target of Searle's criticism? To answer this question we have to consider the background context to his *Gedankenexperiment*.

4. Turing and the Intelligence Room

I will take a step back and consider how Turing introduces the famous imitation game with the purpose of evaluating the plausibility of the "intelligent machinery" idea. He is after the answer to a simple and direct question: can machines think? To find out, Turing comes up with a game that consists in imagining three individuals: a man, a woman, and an interrogator whose gender is not crucial for the game. The interrogator is in a room, and the man and the woman are in a separate room. Communication among the two parties occurs only through a teletypewriter. The goal of the game is for the interrogator to determine the gender of the two people in the other room. Interestingly enough, Turing's version of the *Gedankenexperiment* consists in wondering what happens when a machine, in particular a computer implementing a specific program that can simulate human abilities, replaces the man.

The primary criterion for the success of the game, hence of the test, is the impossibility to distinguish between a human being and a machine; for this reason, the vast critical literature on the topic has generally ignored the gender of the human being that the machine would replace. The game can only succeed if the interrogator, who is physically separated from the machine s/he is conversing with, has the sense of talking to a fellow human being, be it a man or a woman. Separation *is*, therefore, the basic condition for the execution of the game; the contact can only be linguistic—and this for two reasons.

First, macro-biological constraints must not be relevant. When compared to the abilities of the machine, the individual's greater or lesser physical abilities should not matter. As a matter of fact, Turing says, "to penalize the machine for its failure to shine in a beauty competition, or to penalize a man for losing in a race against an airplane," would not be right. "The *conditions* of our game render irrelevant lack of ability such as these .... The interrogator cannot demand *practical* demonstrations" (Turing, 1950, p. 435, emphasis added). The *conditions* under which we run the test coincide with the basic *constraints* that the (artificial) *intelligence model* (the thinking machine, according to Turing) must possess in order to be a legitimate participant in the imitation game. Since practicality is outside those means, by definition we must leave out of the game all sensory-motor performances.

By the same token, instruments that realize seemingly intelligent but mechanical tasks, such as for instance the anti-lock braking system (ABS) of cars, are likewise out of the imitation game. We use the ABS system to improve the braking performance of human drivers. It makes use of a feedback system between the driver's pedal pressure and the roadbed's friction during braking; we call this system intelligent because it marks an improvement on human performance. But the presence of a feedback mechanisms should not, for Turing, distract us from the fact that the ABS system is just a useful supplement to the human body. Increased sensory-motor abilities add nothing to intelligence, Turing argues, just as a hang glider that enables flying adds nothing to flying itself.

Second, language unquestionably proves intelligence. Unlike thought—an ostracized chimera of the first half of the twentieth century's psychology and philosophy of psychology—and without the need to affirm or prove its existence, language, which we see as concatenation of phonetic and graphic symbols, exemplifies the productivity and non-repetitiveness of the human mind. Turing began advancing his ideas about a thinking machine at the apex of the behaviorist period; its influence is manifest in the idea of imitation as the standard of identification and in the choice of language as a tangible sign of intelligence. In accordance with the principles of Neopositivism, language is the highest point of a reality we configure as a hierarchy of levels that are ultimately reducible to the physical one.

At the same time, though, Turing, who does not deny the existence of thought, contributes to the dawn of the cognitivist paradigm and in general of the scientific study of mental processes that along with other contemporary scholarship gave birth to AI. The first objection to the imitation game that Turing considers is the following: "may not machines carry out something which ought to be described as thinking but which is very different from what a man does?" (1950, p. 435). What is lacking is not an admission of the existence of thought, but instead an acknowledgment, as far as computers are concerned, of the primacy and preeminence of thought in the simulation of intelligence: "at the very least we can say that if ... a machine can be built so as to play satisfactorily the imitation game, we don't need to take [such an objection] into account" (1950, p. 435).[2]

This is a controversial point. For example, in the early formulation of what would become the cognitive science paradigm Allen Newell and Herbert Simon expressed a different opinion. For them, the replication of human thought processes, and not just the achievement of similar results, was an essential requirement for computer simulation of cognitive activities. Newell and Simon's research methodology initially pursued a heuristics that they could subsequently implement on a general problem solver, like the soon-to-come GPS program (Newell and Simon, 1972). They could obtain this heuristics by accumulating reports ("protocols") of human subjects who are engaged in problem-solving activities. They also rejected the accusation of behaviorism for the methodology that these "protocols" adopt, because programs based on them are individual cognitive models (Cordeschi, 2008).

So, what can we say about the imitation game? Unlike what we would later take the "Turing Test" to be, Turing himself admitted that the game does not represent an ultimate criterion for the presence (or absence) of intelligence (*intelligent thinking*) in a machine. The imitation game is just a *game* useful for testing a human being, not a machine. It tests the interrogator who must successfully distinguish the gender of his/her interlocutor on the other side of the wall. The chances of success of the interrogator should be close to those of a man and a woman, not a machine and a woman, who are playing the game. If it is so, if the game worked and the test passed, this fact in itself—as Gabriele Lolli asserts—would not amount to "a necessary criterion for ascribing intelligence, nor, perhaps, [it would be] sufficient; there is not even a precise way to define the test 'passed,' but only the possibility of establishing better or worse 'game's hands,' in the sense that the interrogator falters to figure out the gender identity more or less like a human interlocutor does" (1994, p. 18).

Over the years, researchers have suggested variants of the Turing Test, sometimes with an ironic intent. For example, Keith Gunderson (1964) asks what happens in the game in which we hit someone who is turned around and cannot see the culprit, who then has to guess who (or, possibly, what) hit him/her. Similarly, Gunderson suggests that if something hits the foot of a person as s/he was stretching it out of a partly-opened door, s/he may find difficult to decide if the blow has come from a person or a falling rock. Does this entail we can seriously ask if a stone can think?

Leaving aside the most trivial renditions of Turing's Test, the most behaviorist aspect of its standard formulation is nonetheless clear. As Bruno Bara (1978) pointed out, we need some revisions to make it work and to define what passing the test means. The most important one is, perhaps, Robert Abelson's extended version of Turing's Test, later called ETTA—Extended Turing Test by Abelson (1968). According to Bara, Abelson takes the essential step of "making formally explicit a basic epistemological principle that is far too easily forgotten: a program must not reproduce a human being as a whole, but a model of him/her" (1978, p. 78). We can view Abelson's operation as a milestone for cognitive modeling—an approach that, we should not forget, researchers will develop as a methodological paradigm only a few years after the advent of AI.

Furthermore, we should keep in mind that Turing does not conceive the imitation game as a test—as *the* Test. It will became such only later: first, as an idea pervading all research into intelligent-acting computers, and then as a *practical test*, a challenge that several "intelligent machines" must successfully clear, as in the *Loebner Prize* (Sundman, 2003). Even though the game has become a test after the passing of its inventor, "Turing has never made use of such a term" (Lolli, 1994, p. 17).

Again, looking at the period when Turing was writing helps us to make sense of the situation. The researchers who pioneered future AI shared a markedly strong interest in games, because games take place in limited domains that are easier to manipulate and because players find easier to describe their reasoning strategies. In the years immediately preceding 1956, when scientists coined the term "artificial intelligence" (at the Dartmouth Conference), the attention they devoted to games—such as tic-tac-toe, checkers, or the more "philosophical" chess—was almost all-absorbing. John von Neumann and Oskar Morgenstern's (von Neumann and Morgenstern, 1944; Franchi, 2004) book on game theory also gave this research a push.

To be sure, Turing's "game" is a peculiar kind of game, a *sui generis* game that he deliberately built to try out the abilities of a program. But this does not mean that he conceived the game as a proper formal test. We should not take the ability of a program to play the imitation game as proof of narrowly conceived cognitive abilities, but as a demonstration, in a more general sense, of an overall ability to manifest some intelligent behavior through language.

On the basis of these remarks on Turing's ideas, what can we say of Searle's Chinese Room argument? Before answering this question we need to consider another argument that does not mention any room, but anticipates Searle's theoretical assumptions.

5. Putnam and the Japanese Telepath

In an essay titled "Language and Philosophy," Hilary Putnam suggests the following: consider a novel written in Japanese with the technique of the stream of consciousness and an individual who, with no knowledge of Japanese, learns a long passage from it by heart. Then, subject this person to a hypnosis session in which s/he has to repeat silently ("mentally") the learned passage "complete with appropriate pauses, intonations, emphasis, etc. If the behavior [of this person] is not violently inappropriate to the content of what is passing through his/her mind, then it might be, in a certain sense, *as if* he/she is 'thinking in Japanese'" (1975, p. 4).

According to Putnam, even a native Japanese telepath capable to perceive the person's stream-of-thought would mistakenly assume that such person is thinking in Japanese. Post-hypnotic suggestion could even lead this individual to believe that s/he is thinking in Japanese, so that his/her beliefs regarding what s/he is doing could not tip off the telepath about his/her ignorance of Japanese. This fact notwithstanding, we can see that s/he "would not be thinking the propositions expressed by the sentences passing through his/her mind, since he/she would not *really* (whatever is his/her 'sense of understanding') *understand* those sentences" (*ibid.*).

This situation recalls what Searle imagined in "Minds, Brains, and Programs." To speculate that he took inspiration from Putnam, who had also defined the Japanese telepath's tale as a *Gedankenexperiment*, is not too far off. Some differences are noteworthy, though. First, Putnam does not worry about the implausibility of his argument, such as the likelihood of learning by heart, let alone repeating with perfect intonation and emphasis, a passage in an unknown language that, in Putnam's experiment, we must read or, better, *mentally repeat*. We can overcome this kind of obstacle by saying that the person learning the passage by heart does so through listening to the sounds that a reader familiar with written Japanese and who reads the story aloud, articulate. It is almost as if the first person were to learn a song, a melody, or a generic succession of sounds.

Even less plausible is the use of telepathy as a tool to grasp the stream of consciousness (here a literary technique whose introspective foundation is hardly warranted) that consists of a series of utterances in the mind of the person who cannot understand Japanese. Putnam has evidently no intention to describe a real situation. The point he wants to make, though in a highly unrealistic setting, is that the pure enunciation of some utterances, or their purely mental thinking, does not guarantee that the person who talks or thinks understands them. We could say the same when we repeat to ourselves nonsensical rhymes or lyrics in a language we do not know. Instead, we should find understanding "in the fact that an *understanding* speaker can *do things* with the words and the sentences he utters (or thinks in his head), *besides* just utter them" (*ibid.*).

Searle's aim differs from Putnam's. Searle's target is a critique of the thesis that by implementing a *generic* symbol-manipulating program we can replicate intelligent thinking, and that such an implementation would endow a computer with "cognitive states." In Searle's view, these cognitive states should correspond *directly* to the real ability of understanding, which he rules out that they could. This said, Searle's debt toward Putnam's *Gedankenexperiment* is unequivocal. I will consider more closely the formulations of the two arguments, so that their differences become clearer through their apparent similarities.

Linguistic mastery. The two philosophers base both cases on the knowledge (or lack thereof) of a language that intuitively is very distant from English, such as Japanese and Chinese. Both cases play with a counter-intuitive situation such as the mastery of a language that is not only foreign, but far from the notation systems of Western languages. Finally, both authors show that this mastery is fictitious and boils down to a merely external, purely mechanical fact that a mechanical method can execute.

But in Putnam's thought experiment the method is mnemonic. Under suitable conditions—such as, for instance, hearing a native Japanese speaker who reads a passage with the appropriate intonation as many times as needed to make an impression in the stream of consciousness of the Japanese-poor listener—nothing prevents him/her from accomplishing the mnemonic feat.

In the situation Searle describes, instead, the point is not to trick an interrogator by means of mnemonic techniques of disputed validity, but through a correct interaction in a linguistic exchange. This leads me to conclude that in this respect Putnam and Searle reference *two different types of linguistic mastery.*

The function of the interrogator. In order to anticipate the elements of the comparison between the two arguments and Turing's imitation game, I will here use the term "interrogator" to designate Putnam's telepath and the Chinese individual or the Chinese-equipped team of programmers standing outside the room where the Searle-executor is locked in. The difference between the situation Putnam describes —a telepath reading someone's thoughts—and Searle's—a native speaker of Chinese who is outside the room and reads the answers that an individual who is locked up in it has given—should not mislead us. Both scenarios frame a *reader of strings of symbols* that are constitutive elements of the two languages. In the Chinese Room experiment an individual receives pages of Chinese symbols. In Putnam's experiment the same occurs, though less explicitly. Putnam's telepath, "hearing" the "inner speech" of the person who has learned the Japanese passage by heart, is identical to the Chinese native speaker interrogator who reads pages of symbols that are Chinese words and statements. If our intent is to prove the validity of the argument with regard to the pure exteriority of the symbolic forms in the understanding of language, what else could s/he do, if not engaging in thought-reading, namely reading the symbols of thought? Since Putnam does not volunteer a further and more in-depth explanation of the telepathic power of the Japanese native speaker, he implicitly assumes that telepathy is nothing but perception of symbolic series inside the minds of others, without any further capability. This is not a (strong) demonstration of the possibility of thought description/explanation as symbol processing. At most, it is the (weak) supposition—which the "telepathic proofs" *do not* support—of the presence somewhere in the mind of the inner correspondent to the external language, by which I mean a biunivocal "symbol to symbol" connection. This inner symbolic concatenation is not the same as processing nor, on the basis of this argument, will processing necessarily determine it. In the end, the interrogator's function, his/her essential role, is to understand the strings of symbols that belong to his/her mother tongue.

The nature of method. A problem for Putnam, it is not one for Searle. Of crucial importance to the Chinese Room *Gedankenexperiment* is the handbook with the symbols of an unknown language that, along with a set of instructions in our known language, allows the correlation between the three batches of Chinese symbols that are in the room. All together, they enable the locked-up Searle/executor to answer the questions that we ask from the outside. The set of instructions that Searle calls "the program" makes possible the linguistic *interaction* with the outside. From the point of view of the individual who acts on the basis of the instructions, this interaction has nothing to do with his/her awareness of the Chinese language. The situation Putnam describes, instead, is different. No *method* formalizes the interaction. At most we can think of a mnemotechnical procedure to learn the passage, as I have suggested with the conjecture of a Japanese reader whose repeated imitation the language-poor person will exploit to memorize the passage. No method allows the interaction for the simple reason that no interaction occurs. The ability to capture the interior monologue falls squarely within the skills of the telepath. No real linguistic exchange occurs between them. If it did, it would prove what Putnam denies through the memorizing method, namely that, besides their mere enunciation through *imitation, knowing means to know how to make something with words.* On the contrary, for Searle the possibility of providing the person in the room with a program made up of a set of instructions that allows the interaction with the outside is not a problem. Differently put, Searle has no interest in the *construction of a method that allows the exchange in the linguistic format of question and answer*, which, instead, is one of the fundamental targets of AI.

In conclusion, if a person locked in a room in a situation like the one Searle describes is unaware of the operations s/he carries out and has no understanding of the linguistic symbols s/he is manipulating, then how we can ever fathom the set of instructions that can make a real interaction with this person look plausible from the outside is not so obvious. The problem is not with the manipulation, but with the formulation of the instructions: the problem lies with how we design, organize, and build the method. Searle's elusiveness about this problem bears out his attempt to make plausible what is not. Putnam, on the contrary, does not hide the unrealistic features of his thought experiment (the use of telepathy), because they are irrelevant to the point he intends to make. The two philosophers give different explanations of understanding. For Putnam, understanding consists in carrying out some operations through language—a view that we can trace back, albeit cautiously, to the definition of "meaning as use." For Searle, the *a priori* assumption that language understanding occurs thanks to each individual's unspecified brain causal powers is tantamount to denying that any program is capable of understanding.

So, on account of a few manifest analogies and a superficial structural similarity, I can trace the Chinese Room experiment back to Putnam's *Gedankenexperiment.* The deep differences I have uncovered, however, should show that Putnam's argument is sounder than Searle's, whose intrinsic weakness the intuitive plausibility of its unfolding just hides from view.

Up to this point I have shown that the structure of Putnam's argument anticipates Searle's, despite the differences existing between them. What has Turing to do with this?

6. Shifting Perspective

In Searle's essay, as his aims, subject choice, and structure of the argumentation bear out, the reference to Turing's "Computing Machinery and Intelligence" is indirect but not completely hidden. I have mentioned the significance of language in Turing's imitation game and its central position in the Chinese Room *Gedanken-experiment*. In order to reveal their discrepancies, though, we should yet again investigate and analyze their similarities.

On the one hand Searle's polemic target, as I said, is the Yale research group that Roger Schank headed, and all programs that simulate natural language understanding, such as, for example, Joseph Weizenbaum's ELIZA (1965; 1978) and Terry Winograd's SHRDLU (1972; 1973). On the other hand, Searle says that we could apply his argument to "any Turing machine simulation of human mental phenomena" (1980, p. 417). Searle's argument targets classical computationalism as an explanation of thought processes, a thesis he calls "strong AI." Since he believes that unspecified "brain causal powers" are the sole root of all "human mental phenomena," no room for weak AI appears to exist either. This is true unless we take this second type of AI as a downsizing of AI's original objectives and not a reformulation of its practices, such as algorithmic structures, computational theories, and simulation models.

Searle's writings, at any rate, bring no explicit attack on Turing. Only by virtue of the central role that language plays in it, we can portray Searle's criticism toward "any Turing machine simulation" as a criticism of the Turing Test. And yet, in Turing's imitation game, language is just the expressive-communicative medium between an interrogator and a machine that pretends to be a human being. The machine exhibits linguistic abilities as an *outer sign* of all cognitive activities. In other words, Turing says nothing about how, in addition to other cognitive abilities such as memory, deductive or inductive reasoning, hypothesis formulation, analogy building, abstraction and creativity in the different fields of mathematics and poetry, a machine produces language as a cognitive ability that it then displays in a conversation with a human interrogator. Turing does not inform us about the role of linguistic performance in a cognitive system; nor does he assert anything about the psychological antecedents or logical roots of language. Searle's objection to computationalism as an explanatory theory of thought processes consists in attributing to the imitation game a higher goal than Turing himself set to it. The imitation game "makes no reference to single abilities, requires no expert interrogator, suggests no test that the machine should pass, but instead a test that that interrogator should pass with reference to the machine" (Lolli, 1994, p. 18). The potential success at such tests would not conclusively prove the presence of any special cognitive activity in the machine, because, unlike the *Test*, which revolves around the machine, the imitation *game* revolves around the human being (the interrogator). More than a test verifying the machine's intelligence, the *Gedankenexperiment* probes the human attitude toward the simulation of intelligent performances. Such an attitude amounts to an evaluation of the theoretical and epistemological principles that lie at the core of the AI enterprise, be it simulative or emulative, symbolic or connectionist, representational, dynamic, situated, or anything else AI could be.

I am not saying that Turing is a theoretician of connectionism. In "Intelligent Machinery" (1948), Turing had shown a clear interest in automatic learning and in the problems of the *continuous* nature of the brain's workings, so dissimilar from the *discrete* nature of computers and in general of any finite state automata. But Turing wrote at the dawn of the computer era and the possibilities that their development and application entailed even for the sciences of the mind could not fail to impress him. His thought was more open than his followers', whose often purely ideological postures marred the whole history of AI, especially in the last thirty years.

In the intentions of its creator, the imitation game is not, and cannot be, an *experimentum crucis* geared to test if a program replicates a *specific* cognitive ability. How have we come to consider it in such a way? What kind of operation does Searle, as he builds his argument, carry out onto it?

Two fundamental steps lead from the imitation game to the Chinese Room. The first one is the transformation of the imitation game in the Turing Test, which amounts to a perspectival shift within AI. Metaphorically, we can interpret this step as the program leaving the room, with the consequent switching of roles between the interrogator and the cognitive model—the program that in the game plays the role of an entity whose nature we have to guess. This switch has momentous consequences because it coincides, *from the point of view of theoretical justification*, with *shifting the burden of proof* from the human being to the program. Therefore, the game loses its game-like nature: we no longer face a human interrogator who must guess if s/he is conversing with a human being or with a machine. Instead, the game morphs into a crucial experiment in which a machine must *prove* to possess one or more cognitive abilities that would turn it a *him* or a *her*.

In his essay, Searle points out that such a paradigm shift exposes AI, for which the Turing Test is now a crucial experiment, to accusations of behaviorism (1980, p. 423). But AI, as far as an analysis of its results is concerned, did dissociate itself from a behaviorist vision of the mental. Instead, it would have been a form of eliminativism, definitely not one of AI 's main trends—be it symbolic AI (whose prime goal is researching *thought mechanisms*) or connectionist AI (whose teleological and mentalist vocabulary is equally functionalist) (Cordeschi, 2002). These accusations, on the contrary, show that the Turing Test is not, and cannot be, a full and exhaustive tool for the evaluation of the achievement of a set goal through the construction of a cognitive model. Similarly, striking a conversation with an interlocutor does not reveal to us the nature of the mechanisms that underlie his/her cognitive abilities. At most, it allows us to attribute to our partner the ability to accomplish some (intelligent) performances. Consider a program that successfully passes the Turing Test, but that we do not evaluate as a model according to other parameters such as its architecture, the function of its components, its ends, and default constraints. Since language is the most powerful means of human interaction and socialization, we might perhaps accept this program into the community of human beings. But the non-private, social nature of language is not sufficient to describe all mental processes, nor it is likely that an exclusively neurophysiological point of view will provide a successful account of them.

Were Searle's criticism limited to putting AI on its guard against the risk that we mistake the duplication or emulation of behavior for the explanation of the phenomenon that we duplicate, it would not be so off the mark—but he pushes beyond this. We can interpret the second step in the transformation of the imitation game as yet another reversal of the observation point. While before we saw a swap of metaphorical roles between the interrogator and machine, now the interrogator-Searle returns to the room where the interrogator and the machine had been previously tested and *judged*. Yet his/her return is simultaneously an entrance into the machine that is still in the room. Dispensing with the interrogator's previous role, Searle can now proceed to judge the machine from the inside. Hence in the Chinese Room argument two kinds of interrogators now emerge: 1) a first-level interrogator that the native Chinese speaker (or speakers who belong to a team of Chinese programmers/interrogators) personifies, who, from the outside, *cannot understand* that the locked-up-in-the-room Searle-executor *does not understand* Chinese; 2) a second-level interrogator, the locked-up-in-the-machine Searle-executor, who observes its workings from the inside, or better still, becomes part of its functioning. The Searle-executor claims he has no understanding of what he is doing, except carrying out "formal operations" on symbols he knows nothing about.

I will now discuss where this whole situation leads us. While the role of the first-level interrogator is equivalent to the Japanese telepath in Putnam's *Gedankenexperiment*, the second level of interrogation is Searle's decisive move against the Turing's general thesis on the possibility of thinking machines (we define the thesis "general," because Searle would accept that only the brain can think, and for him too the brain is a machine). A machine cannot think or exhibit cognitive abilities because a machine cannot manage to understand what is doing. In the specific situation that the experiment describes, the machine cannot understand the *concepts* of the language it produces. If the manipulation of signs follows explicit rules whose application we can implement through a simple comparison between figurative forms (the sign-forms on sheets that compose the narration with the sign-forms that we find in the handbook), no real understanding can occur. What is not explained, however, is the nature of the rules that make up the instruction handbook. As I said earlier, AI strives to solve precisely this emblematic problem, and yet Searle takes it for granted.

Even though the activity inside the room is an essential part of the whole process, the chance of observing the inner set-up of a mechanism, of grasping its workings and yet being cut off from an understanding of the produced phenomenon, all this does not add up to the strongest indictment toward the possibility of AI. Instead, we could consider the entire argument as a *caveat* about cognitive model building. As for its architecture, we have to carefully consider the purpose and the function of every single part of the model, including its likely constraints. Our rationalization must be the effective simulation of a process, not the mere copy of the original. The Chinese Room argument suggests that describing a mechanism so that the description will provide no real explanation of the processes that the mechanism instantiates is always possible.

Searle's argument falls short less because it fails to prove that pure logical-syntactical formalism does not lead to a language understanding that we conceive as a concatenation of syntactically and semantically well-formed statements, than because it lacks a proper account of the instruction handbook (the "program") that the Searle-executor uses in the room. To reduce natural language understanding (or natural language processing at large) to a set of rules, be they expressed in a formal language or not, is not a common given but on the contrary a highly controversial stance. And communicative interaction through questions and answers is a case of linguistic-conceptual understanding. Searle does not say how we may possibly achieve the representation of formally explicit rules that work solely on the basis of the form of the symbols they apply to. If he did, his argument would be disproved. By failing to do so, his argument becomes implausible.

7. The Objections to the Room

Turing and Searle have been the first to raise a series of objections to the room arguments.

I will start with Turing. He imagines the following objections to the idea of a thinking machine that, in his terms, is able to play the imitation game:

1. the claim that "thinking is [exclusively] a function of man's immortal soul"—he calls it the "Theological Objection" (1950, p. 433);
2. the claim that "the consequences of machines thinking would be too dreadful" for mankind; hence we hope that the construction of such machines is impossible—he calls it the "'Heads in the Sand' Objection" (*ibid.*, p. 444);
3. the claim that logical-mathematical proofs such as Gödel's incompleteness theorem, or mathematical hypotheses such as the Church-Turing thesis, show the intrinsic "limitations to the powers of discrete-state machines"—he calls it the "Mathematical Objection" (*ibid.*, p. 444);

4. the claim that we can acquire the knowledge that a machine thinks only by being that very machine and by "feeling oneself thinking," a conviction that exposes us to the risk of solipsism—he calls it the "Argument from Consciousness" (*ibid.*, p. 446);
5. the opinion according to which if a machine is able to do something, then by definition we can mechanize it and therefore it is devoid of interest. Mechanization would amount to capturing only a few of the external manifestations that are typical of thought, not its core. This essentialist opinion causes the research on thought processes to be asymptotic, in some ways repeats (1) and (4). Turing calls generically this manner of dealing with machine thinking "Arguments from Various Disabilities";
6. the claim that machines can do just what we program them for, which is the ancient "Lady Lovelace's Objection" against Babbage's Analytical Engine;
7. the claim of an incommensurable difference between brain continuity and the discrete nature of computers ("discrete-state machines"), which he calls the "Argument from Continuity in the Nervous System";
8. the claim that a set of fixed and pre-established rules govern machine but not human behavior—he calls it the "Argument from Informality of Behavior";
9. the surprising idea that a person with specific mental powers could invalidate the correct execution of the imitation game—the "Extra-sensory Perception Argument" that I previously mentioned.

As scholars pointed out, the primary purpose of Turing's imitation game—unlike what the later literature, dubbing it Turing Test, will take it to be—appears to be more "the discussion of the objections against the possibility of building thinking machines" (Lolli, 1994, p. 19) than the articulation of rigorous criteria for its realization." But not all objections are born equal. (1) (2) and (5) put forth biases against machines in the form of dogmatic convictions that their defenders, as much as anybody else, cannot justify; (4) is an utterly subjectivist philosophical thesis that, if we pushed it to its ultimate consequences, would cause any cognitive science methodology but introspective analysis to be impossible, (9) is a warning concerning the execution of the game that, despite the absurdity of its formulation and Turing's opinion, allows us to draw the effective boundaries of the game field.

The four remaining objections (3, 6, 7, 8) are of a different nature and attack computationalism and AI as a whole *from the inside*. While the previous arguments share an exquisitely *a priori* character, these instead, in spite of missing some specifics, rest on facts. The "Argument from Consciousness" (4), if we understand it in a non-solipsistic way, draws the attention of cognitive modeling to the subjectivity problem and to the qualitative nature of mental phenomena or *qualia*. Provided that they do not become the bulwark of AI's negativistic extremism, *qualia* do spur an epistemological reflection on the principles that underlie the whole field (Dennett, 2005).

(3), (6), and (8) are versions of a single objection against any logical-deductive system that we base on explicit rules: namely, that the system is incomplete when it comes to an understanding of every single aspect of reality. A formal system cannot produce each and every single truth that we can express and it cannot, when applying its rules to statements, axioms, and theorems, go beyond its rigid sequential nature and monotonicity; for this reasons, the system is necessarily "incomplete." As Turing was writing his essay, though, the nascent AI attempted for the first time to overcome such monotonicity. It did so, on the one hand, by exploiting the possibilities that constituents of the structural algorithm such as the procedure call, the conditioned choice function, and the iteration, offer; on the other hand, by adopting heuristic search methods in problem spaces. Finally, (7) shows that Turing and others were farsighted in identifying the mathematical problem at the basis of the opposition between symbolic AI and connectionist AI. This problem ignores who is playing the imitation game, but indicates which constraints we must take into account when we design a cognitive model. The question about the description level that is suitable for the phenomenon we want to simulate is analogous.

In his essay, Turing makes no distinction between the objections he lists. His inquiry about the conditions validating the claim goes hand in hand with an investigation about the real possibility of a thinking machine, with a probing of the biases that we must overcome, and with a search for the theoretical principles that will function as constant intimations to AI research and provide it with hard problems unlikely to be solved once and for all. The four "internal" objections, however, share a central role—(3), (6), (7), and (8). Incidentally, they would be valid even in case a program were to pass the Turing Test.

Let us go back to Searle. I have described the Chinese Room argument as an indirect objection to Turing's hypothesis of a thinking machine. We can consider it so in two ways. First, it levels criticism at the behavioral aspects of the Turing Test, an accusation that appears justified if we refer to a superficial and too optimistic vision of AI. Such an interpretation, I have argued, transcends Turing's true aim. Second, and more directly, Searle's argument is an objection against the possibility that a program communicates through natural language and that, by so doing, it effects processes similar to those that are characteristic of a human being. Formal symbol manipulation excludes understanding, Searle claims, showing that no correspondence exists between human thought processes that relate to the brain causal powers, and the formal algorithmic procedures that a computer implements. In the second sense, we may find the connection with Turing in the use of natural language that lies at the core of the imitation game. I pointed out earlier the extent to which this move is implausible. Now I will examine the objections that Searle, like Turing, levels at his own *Gedankenexperiment*. He arranges them and their refutations into six points (Searle, 1980, pp. 419–422):

1. The person inside the room does not understand Chinese, but the system, of

which the person is just a part, does. Hence only two necessary elements, the inside person and the instruction handbook—"the program"—constitute the system. They can become just one only if the person in the room *interiorizes* the program by memorizing it. This operation does not allow *him/her* to understand Chinese, but just to *learn a method by heart*. This is the "systems reply."

2. A robot that incorporates a computer solves the problem. Yet, even if such a robot interacted with the environment, it would have no intentional states. Put differently, placing the room inside a sensory-motor system able to perceive and make movements does not endow the program with the ability to understand. The Searle-executor in the room can quietly go on with his functions. This is the "robot reply."
3. The solution is a machine that simulates the firing sequence of the neuronal brain activity of a Chinese person while s/he is speaking Chinese. Yet this machine would not have intentional states. We might imagine replacing the brain with a system of pipes and valves in which water flows, while a Searle-plumber controls it. Since the system simulates only formal neuro-physiological properties that are insufficient to generate the causal ones and then manifests those properties in linguistic form, Searle-plumber would not have the slightest understanding of Chinese. This is the "brain simulator reply."
4. The three previous objections fail individually, but if we take them together they strengthen. Yet, we could bring the same objection as (3) to the idea of a robot as an overall system that, instead of the computer, has a simulated brain inside: an individual might be hidden in the room of the simulated brain (or, why not, control it from a distance with a remote control). This is the "combination reply."
5. The knowledge we have of other people's understanding of Chinese or of anything else stems from our observation of their behavior. If a computer exhibits the same behavior, we can ascribe to it the same kind of knowledge. This "other minds reply" is just a return to a trivial interpretation of the Turing Test, a confusion between *demonstrans* and *demonstrandum*.
6. In order to reproduce causal procedures that are brain-specific, we could brush aside the computational view and adopt a different AI strategy. Because the Chinese Room argument applies only to the computational ("strong") version of AI, this "many mansions reply" is flawed.

The six objections (and their refutations) are evidently not on the same level. Searle discards (5) and (6) as not addressing the argument. With (6) Searle appears to grant some hope for non-symbolical (and non-classical) AI, albeit a weak one. (3) is a thesis against the simulation of cerebral mechanisms, hence against connectionism; it would apply against any attempt to extend the Chinese

Room argument to formalizations through symbols and explicit rules of the entire neuronal activity.

(4) does not appear interesting because it adds nothing to the three previous objections, and does not hold out against the arguments by which he refutes them. (2) is interesting, because Searle excludes that perception and environmental interaction are of any relevance to intentional states and understanding, a thesis that is acceptable only with extreme caution. In any case, we can consider (2) and (3) an extension of (1). In (2), a robot incorporates the room thus replacing the computer that controls it; in (3), the room is inside a computer that accurately reproduces the synaptic connections of a brain able to understand Chinese.

Finally, only (1) is an original objection. Curiously enough, Searle rejects it with a method reminiscent of Putnam's *Gedankenexperiment*—the memorization of the instruction handbook. Such a method, according to Searle, would erase the presence of a system that a "Searle-inside-the-room" plus "the handbook" or "program" make up. In that case, the person in the room would have two subsystems, one to understand English, and another one, which is inside the first to operate on the Chinese symbols and do something with them. Since the second subsystem is just a part of the first one, the necessary and sufficient condition for memorizing the rules (the instructions) and the formal symbols (the graphic representations of Chinese ideograms) is the understanding of the English language. Memorizing would just be an external fact, as it was for the individual who learned by heart Japanese passages written with stream of consciousness technique. Yet, a double obstacle is present. Even though such a memorization method could take place with huge sacrifices (first obstacle), the whole situation remains unlikely because, as I pointed out before, the instruction manual is implausible (second obstacle).

Searle appears not to defend his *Gedankenexperiment* in a convincing manner when faced with the objections that he himself put forward, or better still, with objection (1), of which the other objections, unless we diverge from the main argumentation, are special cases. By tracing these phenomena back to brain causal powers that he cannot explain, Searle cultivates a theory of mental phenomena that he is unable to justify. At the same time, he betrays his excessive reliance on the possibility of reducing natural language to explicit rules allowing us to carry on a formal conversation that we view as made up of questions whose answers must be unequivocal and unambiguous. But Searle's target is still, against the grain of researchers who consider it a suitable theory of mental phenomena, computationalism as formal symbol manipulation. According to such a thesis and to Searle's Chinese Room argument, natural language, owing to its "easy and prompt" reducibility to a set of explicit rules, must be excluded from the set of relevant mental phenomena. But this, I would claim, is just the opposite of what Searle wants to achieve with the Chinese Room.

Yet, Searle's argument may be of use and we should not entirely discard it. It prompts us to ponder some crucial questions: given that at some level (the brain-mechanical level) it must be ontologically admissible, which kind of computationalism can we propose as an explanation of mental processes? What theoretical assumptions do we need to build a satisfactory notion of computationalism? If computationalism is the formal manipulation, or processing, of symbols, what level (or levels) should we investigate through this theoretical notion? Brain or mind? Or both? Or something in-between them? With objection (3) (1980, pp. 420–421), Searle invites us to consider the brain as a system for the symbolic manipulation of binary symbols. The idea of "binary-ness" does not agree with the hard reality of neurophysiological facts but, with appropriate adjustments, we can consider even the brain a system of formal symbol processing. But evaluating thoroughly the extent to which such adjustments can preserve specific functional brain features, an issue still controversial in cognitive neuroscience, is also necessary.

Many have been the immediate reactions to the publication of Searle's essay. I am not committed to examine them all in detail. For my purposes I will discuss the last objection that scholars have leveled at the Chinese Room, namely Douglas Hofstadter and Daniel Dennett's, which came out one year after the publication of Searle's original essay (1981, pp. 360–369).

Hofstadter and Dennett put forward the "systems reply." Unsurprisingly so, if we accept the reduction of the six objections that Searle brought forth to (1), the only one that is neither a fallacy (in the sense of a failure to address the argument) nor a repetition of previous objections. Even though the intuitive and immediate qualities of Searle's reply mask it, I have already pointed out its weaknesses and implausibility. Hofstadter's criticism underscores these weaknesses and brings into play the mechanism of the "intuition pump" that Dennett had introduced with reference to the Chinese Room. An "intuition pump" is "a device for provoking a family of intuitions by producing variations on a basic thought experiment" (Dennett, 1980, p. 429). These variations let us to draw different conclusions from the same argument structure depending on the features that we attribute to the *specific Gedankenexperiment* under consideration. With reference to the Chinese Room, Hofstadter identifies five parameters (or, we could say, *criteria*), "five knobs" that let us modify the conceptual situation the thought experiment describes (Dennett, 1978; Hofstadter and Dennett, 1981, p. 363):

- the physical *stuff* out of which we build the simulation;
- the *imitative level* of the mind-brain system (subatomic, atomic, synaptic, cellular-neural, of groups of neurons, symbolic, and so on);
- the *physical size* of the simulation (from microscopic to macroscopic);
- the size and the type of the simulation *demon*, the main *actor* of the simulation;
- the *speed* of action of the demon (slow or fast).

These five variables allow the creation of multiple different "room experiments" whose intuitive effect we secure in different ways: slowing down the execution speed of the task; enlarging the whole system to sizes that are humanly unthinkable; using the most inert and inactive material; placing into the room, at the experiment's core, a demon that is as close as possible to a human agent—hence also a human agent—who mechanically or *methodically* implements actions that we assign to it (or him/her) and who *replaces the explanatory part* of the simulation experiment (even though, being *human-like,* it could also behave in a non-mechanical manner).

In this way, we can obtain Searle's "symbolic room" (the room of Chinese *symbols*) as well as John Haugeland's sub-symbolic room (the synaptic *connection* room). In the last one, a demon fires up all and only the right synapses of the brain that belong to the person who is having a conversation in Chinese (Haugeland, 1980, p. 432). We are already familiar with Searle's objection to a similar transformation of the room. In objection (3), he asserts that also in this case the main thesis holds true: no intentionality exists—no understanding of the Chinese language—in the individual operating the hydraulic or electric levers in the "brain room." In Haugeland's case, though, Searle appears to retract what he had said in (3). In his reply to him, Searle claims that the exact reproduction of the brain, through the substitution of the normal synaptic activity with a demon, does not delete the effectiveness of the brain's causal powers: "if the stimulation of the causes is at a low enough level to *reproduce* the causes and not merely *describe* them, the 'simulation' will reproduce the effects" (Searle, 1980, pp. 452–453). Only one of the following statements can, however, be true: either there is no difference between (3) and a demon that takes over all synaptic interactions—which amounts to Searle withdrawing from his initial position; or the only difference between the two conditions must concern the *material* out of which we have built the system. In any event, we ought to consider the causal powers of the brain different from those of a system of hydraulic pipes and valves (or also of a "pinchers" of synapses) accurately reproducing the structure of the brain. But this would be an odd conclusion, because, by definition, the nature of a causal (*functional*) power does not reside in a determinate property of a determinate material. The potential to produce precise effects from precise conditions that may (or may not) be "material," through a well-defined procedure, method, or sequence of actions, belongs to the nature of a causal power.

8. Searle's Problem and the Model "Purification Cycle"

Searle does not provide an adequate explanation of this point, possibly because his argumentative *focus* remains exclusively on natural language. Dennett and Hofstadter address the problem of learning a non-native language. They say that for the individual inside the room is not possible to "internalize" every instruction in the handbook, so that (1)—the "system" (person + instructions) objection—is not valid. If "internalizing" a program means "memorizing" it, still the system does all the understanding. As Putnam has suggested in his thought experiment, no difference exists between having something written on a sheet of paper and committing it to memory verbatim. The only difference lies in the use of large amounts of memory resources by the memorizer. If, on the contrary, "internalizing" a program means inserting it into one of the "mental subsystems," how can we achieve this in the absence of some connection between them and the program? Only a connection would bring about a conscious use of the knowledge that we have "internalized," eventually constituting *learning* and, in this specific case, the learning of a foreign language.

In all likelihood, however, Searle's Chinese Room argument has its main focus elsewhere. The central problem remains the simulation of the understanding (and of the production) of a natural language, plus the objections that we can possibly level at this specific cognitive activity. We can view the variation parameters that Hofstadter and Dennett identify in the room arguments as an incomplete but effective list of constraints that apply to all attempts to simulate cognitive activity. They are valid even when we deal with the problem of understanding in the broader sense of automatic natural language processing. With one difference: language *as such* leads to the question of how we should consider it. The answer falls between two extremes: first, language is considered a purely external behavioral episode, the mere output of a series of mechanisms and procedures, of functions (symbolic, biological, or both) taking place in the inner space (or room) that our mind and/or brain is; or second, language as a symbolic manipulation by means of syntactic-formal rules and semantic regularities that together let us understand and produce language. My analysis of the Chinese Room argument has shown how Searle oscillates between and mixes these two extremes, hence pushing beyond acceptable limits of plausibility the situation that Turing had imagined in the imitation game.

This leads me to a further remark. I mentioned a double perspective shift that let us construct of the Chinese Room's *Gedankenexperiment* starting out from Turing's game. The twofold transformation occurs in two specific steps: the first one, from Game to Test, is a *behavioral move* that, in the simulation, converts natural language from a communication medium to the output of a specific cognitive activity. This output warrants that actual cognitive activity occurs inside the *simulating* machine. The second move, from Test to Room, comes from the symbolic nature of the components of natural language (sign, phonetic, and graphic) and is a *formalistic move,* whereby language is no longer the output of mental or brain activities, but something internal to both, reducible to a set of instructions at once formalizing and causing language to become impervious to understanding.

We can apply these moves, separately, to AI programs that have tried to simulate different cognitive abilities. In other words, whenever *the identification between the explanation of a performance and its execution occurs*, the risk of a behavioral drift threatens each and every model of a specific mental activity, including those that are purely connectionist. Nevertheless, the reduction of a performance to a procedure that has the same inexorably formal, mechanical, rigid features of deductive logic is also a fruitful perspective for the interpretation of the different models of mental activity. If we want to avoid perspectives that are so analytic that they lose any explanatory power, though, we must balance out that reduction. The last approach is valid only as one among the many assessing a model of cognitive activity; as far as meeting the need for an explanation, it is necessary but not sufficient. During the implementation of the two moves I described above, we saw how easily we can distort natural language and its understanding. Can a third move that would take us back to the initial situation but now cycled and filtered through our theoretical and epistemological objections, bring the circle of an AI that strives to build simulative models to a closure? Figure 7.1 visualizes this methodological hypothesis.

The diagram depicts a sort of "purification cycle" of the models. Whenever we propose a cognitive simulation and hypothesize an explanation of mental processes or phenomena, we should always consider not only the behavioral and formal aspects present in the explanatory component of the simulation, but also draw the consequences that necessarily descend from them. A revision of the model or of its supporting theory *is* among these consequences, at least if we care for the effective explanation of the cognitive activity that the precise simulation puts to test.

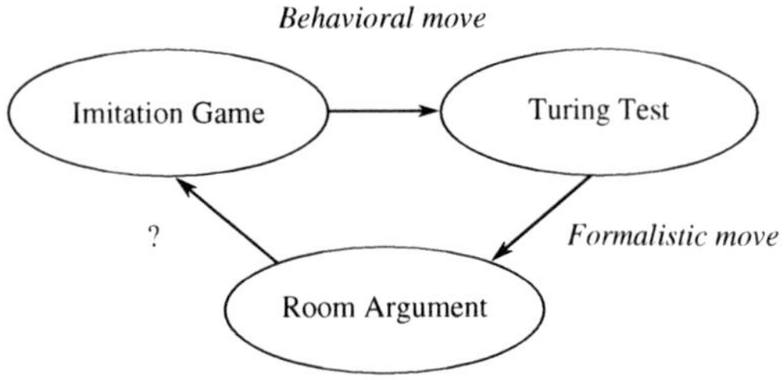

FIGURE 7.1: The "purification cycle."

The idea of a cycle is not new. It is one of the epistemological principles of the sciences of the artificial, of the old and new artificial intelligence as well as of cognitive science (Cordeschi, 2002; Cordeschi, 2008). For instance, Herbert Simon speaks of a "helical path" (Paige and Simon, 1966) between theories (theoretical hypotheses) and data that we extract from real operative models in order to obtain better explanations of phenomena. William Bechtel's description of cognitive modeling as a process of functional decomposition and structural localization bears a resemblance to Simon's suggestion (Bechtel and Richardson, 1993; Zawidzki and Bechtel, 2004). According to this view, a mechanistic explanation of cognitive phenomena, which in turn we must not confuse with dynamicism, is in order only if a dynamic interaction between decomposition and localization occurs. The difference between the cycle I suggest and this approach lies in the fact that my cycle is as general as possible and that the third move (as I will discuss below) is a localization that does not concern just the neurological apparatus (namely, the brain). From the abstract viewpoint, the structure of the process is circular. Although we can find it in many scientific fields, it is even more likely for circularity to turn up in investigations of cognitive phenomena, where we cannot so well establish the boundary, which needs constant redefinition, between theoretical and practical elements.

But why should we recommend a return to Turing and the spirit of his imitation game? At least for two reasons, both concerning the profound significance of the game he proposed, which we may find in the evaluation of the conditions under which *we as experimenters* are willing to acknowledge that we are in the presence of thinking machines. The first reason is methodological. To overestimate the significance of an experiment, let alone a simulative one, in AI and cognitive science as in any other scientific discipline, is never a good practice. Instead, we should carefully evaluate the phenomenon that the experiment puts to test; we should define its constraints—*the conditions under which, in the simulation, the phenomenon under study keeps remaining such*; finally, we should anticipate and verify the expected outcomes. The second reason is theoretical and epistemological. We should have a clear idea of the phenomenon we want to model and should be mindful that the connection between model and reality, necessary in some respects, must never be the result of an attribution coming from an external observer. Otherwise, we would expose the model to the criticism I have previously discussed in this chapter. The last move, the returning one, must therefore be a "*realistic move*." It will alert us to the epistemological imperative that all studies of thought processes through a methodology of simulation cannot fail to engage. Regardless of the disparate methods that different "inflections" of AI may use in their model building, an epistemological dimension is always at their front and center. In a nutshell, if the process that the simulation models does not work in the game first, it will never work in reality. We may view games that imitate reality just as a part of reality, but for our simulative purposes they are also its most relevant part. We may view reality as the best simulation of reality, but reality as a whole is of no use for studying and explaining reality itself.

9. Leibniz and the Perception Mill

The "purification cycle" I suggested in the previous section describes a possible epistemological scheme to evaluate the process that computational models undergo (or should undergo) from theory to model implementation. From the perspective of the epistemology of both AI and cognitive science, it points, among other things, to the significance of the role of language *as a system of symbols that are interrelated*. The understanding and the production of natural language are themes of the greatest relevance to AI. In their ambiguity and complexity, they highlight the problems looming over the construction of simulative models that wish to explain mental (or cerebro-mental) phenomena. To conclude my discussion, I will recall another formulation of the Room Argument dating back to the Modern period, even though it targets a high-level mental activity of a different order: perception.

In 1714 Leibniz wrote *Monadology*, a book that, along with *Principles of Nature and Grace*, constitutes a comprehensive summary of his thought. In *Monadology* he proposed the following argument:

> it must be confessed that *perception* and that which depends upon it are *inexplicable on mechanical grounds*, that is to say, by means of figures and motions. And supposing there were a machine, so constructed as to think, feel, and have perception, it might be conceived as increased in size, while keeping the same proportions, so that one might go into it as into a mill. That being so, we should, on examining its interior, find only parts which work one upon another, and never anything by which to explain a perception. Thus it is in a simple substance, and not in a compound or in the machine, which perception must be sought for. Further, nothing but this (namely, perceptions and their changes) can be found in a simple substance. It is also in this alone that all the *internal activities* of simple substances can consist. (Leibniz, 1985[1714], pp. 227–229)

Critical scholarship has interpreted Leibniz's argument as anti-naturalist and anti-reductionist (see Calabi, 2005, p. 194). Scholars have supposed that Leibniz thus denies the attribution of mental states to physical states. According to Paul Churchland, for instance, Leibniz does not know where to look, for his knowledge of the brain is inadequate to specify what (and how) neuronal mechanisms can bring about perception. Instead of leading him to admit the accidental ignorance of his contemporary scientific knowledge, this fact leads Leibniz to a denial of the phenomenon. For Churchland, on the contrary, contemporary (or future) neurophysiologists will undoubtedly succeed in determining the physical phenomenon to which we may reduce perception (P. M. Churchland, 1995, pp. 191–193).

According to Searle, Leibniz simply confused the levels of description (Searle, 1983, pp. 267–272). Not only mental events cause mental events, as physical events cause physical events; also, physical events realize, and therefore cause, mental events. Hence, if a physical event realizes (causes) a mental event that causes (realizes?) another mental event, by the transitive property the first physical event causes also the last mental event. But looking for the mental *in* the physical without an adequate knowledge of how we can reduce the second to the first is a confusion of levels that brings us to the paradox of denying the phenomenon altogether. In Leibniz's more plausible stance, the same confusion leads to the recognition of an ontological difference between the two levels. For

Searle too, it all ends up in a matter of ignorance: "if we had a perfect knowledge of how the brain produced, for example, thirst or visual experiences, we would have no hesitation in assigning these experiences locations in the brain, if the evidence warranted such assignments" (*ibid.*, p. 271). We could hold the same as valid also for the global localization of mental events in the brain or in wide areas of it. Again, Leibniz would supposedly fail to recognize the ultimate reducibility of mental phenomena to the causal powers of the brain.

But how could he not, I wonder? In 1714, when he wrote *Monadology,* the gnoseological-ontological dualist paradigm that Descartes had inaugurated with the identification of two separate substances composing the unity of the human being was at the height of its popularity and theoretical consolidation. For Leibniz too, the dualism between soul (mind) and body, or between thought as thinking thing (*res cogitans*) and material substrate as extended thing (*res extensa*) is simultaneously a fact and a problem solvable through a pre-established harmony that we can grasp only by transcending the physical world. Unquestionably, Leibniz's whole system hinges on the assumption of an ontological difference between physical and mental states. Unsurprisingly, the mill argument leads to anti-reductionist results. I contend, however, that Leibniz did not conceive the argument with this aim. I find more plausible to see it as playing a central role in our definition—in contemporary terms—of an acceptable explanation for mental phenomena and cognitive processes.

I will reconsider the argument. Leibniz's attention is seemingly directed to the *mechanical reasons* that render the phenomenon of perception *inexplicable*. This induces us to think not that perception is an inexplicable phenomenon, but that some other kind of possible explanation is in order. As a matter of fact, Leibniz thinks that "two kinds of explanations" for mental phenomena are available. As Clotilde Calabi suggests, they are the "explanation by means of mechanical reasons and the natural explanation" (2005, p. 194). The first one would be a reductionist explanation in the proper sense, and thus finite and incomplete; the second one is "an explanation referring to final causes and not sufficient reasons, ultimately equivalent to an explanation by means of infinitely long mechanical reasons" (*ibid.*). Calabi argues that Leibniz does not propose the mill argument to reach precise ontological conclusions about mental phenomena and concludes that Leibniz is not a conceptual, but a metaphysical reductionist. We could say that the explanatory gap is due to the fact that natural explanations require a never-ending analysis, which is factually impossible. The entire question turns, for Leibniz too, into the undecidability about whether mental states are or not reducible to physical states.

Let us accept the idea that the mill *Gedankenexperiment* "is not an argument that leads from epistemological premises to ontological consequences" (*ibid.*, p. 210). Let us accept as plausible, as I have argued before, that Leibniz's aim is not to set forth an anti-reductionist argument about the nature of mental states.

We could then view the situation that Leibniz describes as plea not for an exclusively metaphysical explanation of perception, but for a different consideration of mechanisms that he conceives as a sequence of cause-effect interactions. Only if we add to it also a view of perception that is either finalist or referring to final causes, though, we could consider mechanisms an exhaustive explanation of the phenomenon of perception. In other words, Leibniz would be suggesting the "systems reply." Let us see how I can support this claim.

The mill thought experiment begins with the hypothesis of "a machine, so constructed as to think, feel, and have perception." Upon entering the machine (the room with mill machinery), we would see "only parts which work one upon another." However, according to Leibniz, a machine able to think and perceive *thanks to its structure* already exists. What kind of machine is it?

Section 64 of *Monadology*, which follows the description of the mill argument, declares that "the organic body of each living being is a kind of divine machine or natural automaton, which infinitely surpasses all artificial automata." Compared with the machines that human beings build, "the machines of nature, namely, living bodies, are still machines in their smallest parts *ad infinitum*" (Leibniz, 1985[1714], p. 254). The human body is a machine whose parts are again machines, while the "parts or fragments which for us are not artificial products, and which do not have the special characteristics of the machine, for they give no indication of the use for which [it] was intended" (*ibid.*) constitute the machines that humans build. This excerpt suggests Leibniz's idea of a living body as a hierarchical succession of machines, each one analyzable in terms of the machine at the immediately lower level, without ever reaching a base level. This assertion might recall today's difference between genotype and phenotype, with the important distinction that in Leibniz no ultimate genetic ground exists. The presence of a different organizational structure characterizing each level causes the shift from one level to a higher one. In the commentary to the article "Rorarius" of Bayle's *Dictionnaire*, Leibniz adds that even the soul is a mechanical automaton, more precisely, an "immaterial automaton" whose inner constitution is a "concentration or representation" of a material automaton that produces in a representational manner the same effect in the soul in question. This machine raises, I believe, the greatest interest. If an ontological difference—the dualism of substances—characterizes the difference between natural and immaterial automata, both substances maintain exactly the same concept of mechanism that is applicable, simultaneously and uniformly, to both automata. The pre-established harmony has created them with a *specific structure* and has enabled them, albeit on different levels, to work mechanically: "the operation of spiritual automata, that is of souls, is not mechanical, but it contains in the highest degree all that is beautiful in mechanism" (Leibniz, 1875-90, p. 365). The spiritual automaton is therefore the representation of the material automaton; the unifying principle of the monad warrants that the spiritual automaton is the representation of its mechanicalness.

For Leibniz, as for the seventeenth century philosophers who dealt with knowledge, the issue of representation is of the greatest importance. Representation, in Leibniz's view, is identical to perception. The analysis of the functional (information-transmitting and other representation-causing) role/s of perceptions in thinking processes is beyond the scope of this chapter. We may just say that the mill argument concerns how a person can assert *having or possessing a particular perception*. According to Leibniz, the decomposition of the parts of perception, or better still, of the parts of the machine that causes the perception, does not and cannot explain it. Representation lies in the principle of its unity, the monad, a simple substance and an immaterial automaton in which the mechanicalness of the material body parts is reflected. We presently return to the conclusion of the mill argument. If we unburden the principle of unity of its ontological feature and free it of the ultimate metaphysical nature that causes it to be unknowable, we may well considered it a principle of structural organization. If we had a machine able to think and perceive, as Leibniz says, and we went inside it, we would see nothing but mechanical parts pushing one another—unless we knew the function of every part, the relations connecting the different parts, and the global organization of the whole system. The presence of a machine in the situation that Leibniz describes is pointless, unless we accept the explanatory principle of the organizing structure—with or without a teleological language. If perception only existed in the monad as a transcendent principle, the mill argument would be the absolute negation of mechanicism. This position would at any rate be at odds with Leibniz's intriguing (albeit not totally clear) assertion concerning the activity of spiritual automata that are not mechanical but "most eminently contains all that is beautiful in mechanics"—presumably the global relational structure and the functional unitary organization of the interconnected parts that work in accordance with a causal law. We can find further evidence in support of this interpretation in the *petites perceptions* Leibniz speaks about in the preface to the *New Essays on Human Understanding*. These perceptions produce, *unconsciously* and through a complex interactive modality, the conscious perception:

> These minute perceptions ... are more effective in their results than has been recognized. They constitute that *je ne sais quoi*, those flavors, those images of sensible qualities, vivid in the aggregate but confused as to the parts; those impressions which are made on us by the bodies around us and which involve the infinite; that connection that each being has with all the rest of the universe. It can even be said that by virtue of these minute perceptions the present is big with the future and burdened with the past, that all things harmonize. (Leibniz, 1996[1705], pp. 54–55)

10. The Room between Language and Perception

My interpretation of Leibniz's argument is consistent with the thesis that the causal laws governing the components of the brain are insufficient to explain specific cognitive phenomena. As in the thought experiments I discussed above, we are again facing a situation in which artful and appropriate variations produce an ideal state that we can experience only in theory. We may compare it, *mutatis mutandis*, to the experience of visualizing single neurons (or sets of neurons of increasing sizes) that slow down their firing so that we may observe them in the act of producing perception. Leibniz's argument conveys the idea that the knowledge of how single parts move—the laws of excitation and firing—is not sufficient for a full explanation of mental phenomena. Instead, we need an overall view of the "whole mechanism." This is the "systems reply" that appears implausible only as long as we—like Searle— consider a system a mere juxtaposition of different parts and not an organized relational structure: "The idea is that while a person does not understand Chinese, somehow the *conjunction* of that person and bits of paper might make Chinese understandable. It is not easy for me to imagine how someone who was not in the grip of an ideology could find the idea at all plausible" (Searle, 1980, p. 419). No matter how we assess Searle's claim, its degree of plausibility is exactly the same as the one that the Chinese Room argument brings up.

The subject of perception in relation with the Room Argument calls for a last reflection. We cannot apply the two moves I mentioned above to the situation Leibniz imagined. The external evidence of "having a perception" can only occur through a linguistic report or through the operational observation of the system's actions and behavior. But instead of stressing this fact, Leibniz's argument wants to identify the appropriate levels and the relevant conceptual categories we need for an adequate explanation of the phenomenon. Unless we assign a formal significance to the "mechanical reasons" that govern the parts of the mill, we cannot take the formal and symbolic aspects into account. Leibniz's room differs to some extent from Searle's and the conclusions we can draw about the two phenomena at stake are also different. Which consequences can we draw from all this?

In his analysis of the Chinese Room argument, David Chalmers reaches a conclusion similar to mine: it is not a cogent argument against computationalism in general and against the possibility that we may implement understanding and conscious phenomena (Chalmers, 1996, p. 332). But Chalmers's generalizes his observation to any kind of conscious experience—"take any program that is supposed to capture *some aspect of consciousness*, such as understanding Chinese or having a sensation of red" (*ibid.*, p. 323, emphasis added)—which is precisely what appears implausible. The Chinese Room argument holds if, and only if, we examine linguistic understanding as a cognitive ability that we need to implement through a program. This is so because only language, or better still only a specific language that is culturally and historically situated as a compound of phonetic and graphic symbols, can support the complex mechanism Searle has designed to prove and to strengthen his thesis against all computationalism of mental phenomena. This is valid *a fortiori* for any specific formal language, but the argument in support of such a case would be trivial. By definition, we can manipulate a formal language mechanically. How and to which extent the mechanical manipulation of a natural language is possible, instead, remains unclear. This discrepancy between the natural and formal cases provides the backbone of the Chinese Room argument. At stake is the degree of formalism to which we must reduce a natural language in order to implement it *mechanically*. Although increasingly less so in an ascending linguistic hierarchy that goes from the language of the machine to the natural language we use in pseudo-code formulation, (formal) programming languages are still arguably a better fit for the treatment that we reserve to the in-the-room Chinese (natural) language.

In his argument, Chalmers appears to confirm that the Chinese Room works only if we refer it to language-understanding and not to any aspect of consciousness. His description repeats the original Searlian scheme by using as a "paradigmatic" example the understanding (or lack of thereof) of the Chinese language. I see here nothing "paradigmatic," though. On the contrary, the *symbolic nature* of language—as it occurs historically in written forms that we concretely speak or, by stipulation, in formal and programming languages—is unique to language itself. I am marginally interested into the debate whether the Chinese Room argument refers to intentionality or also to consciousness. But how Chalmers recreates the argument shows that we cannot place language and perception (broadly intended as categorization) on the same plane. A remarkable difference exists between the Chinese Room and the "perceiving mill" Leibniz describes. The first one is an argument against a formalistic and behaviorist explanation of cognitive activity, whose explanation does not exhaust all kinds of computationalism; the second one alerts us to the correct explicative attitude we want to keep when we describe thought processes.

The trend of bundling up pre-connectionist AI and cognitive science into a research that considers the mind as modular and syntactically manipulating is halfway between adequately and approximately realistic. We cannot reduce the first four decades of AI to the study of the artificial simulation of the "Language of Thought" that Fodor exemplified in his book by the same title (1975). Yet, this view has been an essential component of AI. The strong version of computationalism that underlies the (classic) modularity of the mind and the "Language of Thought" hypothesis have deeply affected the field of cognitive science. Its advocacy, though often dogmatic and provocative, has been in a few cases even beneficial. For example, the idea of a language of thought has drawn attention to the central role that *concepts*—"words of thought" that thanks to their plausibly unitary and firm characterization are *just like words*—play in cognitive activities such as memory, language production, learning, and perception. In turn, the criticism that many researchers have leveled at this view has favored new approaches to the "concept question" and eventually to cognitive science itself. Many questions about high-level cognitive activities remain open and unresolved, such as the notion of representation and the explanation of the phenomena of perception, language production, and understanding.

I have attempted to show that we should not view the transition from Descartes to Leibniz as framing different concepts of mechanicism, but as a shift toward a different approach to the question, so dear to philosophers since Descartes, that concerns the explanation of human knowledge. In our times, we see that the same line of inquiry evolve first into the problem of how we may correctly define intelligence and then into the adjacent issue of how to validate the models that simulate thought processes. Turing's reflections prove that the second question has concerned AI since its very inception. It has operated as an *epistemological engine* of sort that has propelled and keeps on propelling forward all the research conducted in the field of cognitive science during its gestation period, namely, the period that saw the dawn and early age of AI.

I therefore propose to consider the steps that Descartes and Leibniz have taken as follows: Descartes describes the criteria that we use to verify the presence of intelligence in natural or artificial entities; Leibniz —in accordance with his metaphysical and anti-mechanist stance—requires a *scientifically realistic* view that would transcend mere mechanism in the study of the basic features of "intelligent phenomena." Descartes's and Leibniz's moves complement one another and constitute a sort of primitive formulation of a fundamental AI problem: how to evaluate the explanations of intelligent behavior, a problem that Turing has first championed during the 1950s. The problem is peculiar because such is the nature of the phenomena they investigate. Hence, the questions about how we can draw the boundaries of AI's and of the subject matter of cognitive science became epistemologically relevant, eventually involving the research methodologies and the types of *scientific* explanation that the advent of AI yearned to ask and respond to. Regardless of the *specific object* that constitutes the specific field of each science that busies itself with phenomena traceable to the mental (mind and logic for philosophy, thought processes for psychology, brain for neuroscience, language for linguistic, human beings for anthropology, and so forth), the intelligence problem was destined to become pervasive and transversal to all disciplines. Intelligence, the "simulative" activity *par excellence*, a notion that from the epistemological point of view is active along the whole historical unfolding of AI, is the *transversal object* of any research in cognitive science, and even in cognitive neuroscience.

As I argued earlier on, Turing's reflections in the 1950s were more than a new philosophical presentation of old philosophical ideas that he moved from a philosophical to an operational context. The last question is whether this chapter provides the complementarity between philosophy and theoretical AI that we need.

Before answering, I would like to note that Turing does not explicitly deal with perception. He deals with the "brain-computer" problem (1948) from the peculiar perspective of his Universal Turing Machine. Like others (John von Neumann, 1958; Asaro, in this volume) in the same period, although differently, he nurtures an interest in the relationship between computer and brain, and in the possibility we may have of artificially reproducing neuronal circuits and function. But he never deals with systems that are directly dedicated to perception—as, for example, Oliver Selfridge does with his celebrated *Pandemonium*.[3] Because of Selfridge's choice of theme and development—a problem-solving level system based on clustering statistical *perceived* data and symbolic results—his ideas appear relevant even in the development of cognitive science. On the contrary, although only partially for its epistemological meaning, we may view the formulation of the imitation game that focuses on language as a sign of its old-fashioned character.

There are two possible answers to the last question. First, if we consider Turing's ideas as yet another item we can add to the long list of superficial (formal and/or behavioral) linguistic criteria for the evaluation of cognitive models, criteria that, as I have shown in the discussion of Searle's argument, are necessary but not sufficient—then the answer is no. Second, if we consider the room that the game references as a metaphorical place where *we put to test human beings, not computers*—then the answer is yes. As Turing understood before the birth of the term AI, those human beings, for their research on the mind-brain system as a whole, will need criteria (theoretical procedures) and model-reality relationships in order to gain relevant experimental data about the model's *working and learning*.

NOTES

1. Attempts to introduce specific modifications and rules for expressing formally (in predicative calculus) non-monotone reasoning carried out since the 1960s. For a detailed analysis of this topic see Fisher Servi, 2001.

2. As further evidence, notice that the possibility of extrasensory perception is the last of Turing's reactions to the objections he brought against the hypothesis that we can identify a thinking machine through the imitation game. To see Turing accepting its existence may be surprising, but no matter how we consider "extra-mental powers," the acknowledgment of their possibility by definition admits the existence of at least *one* mind.

3. Selfridge's *Pandemonium* project (1955; 1959) is quite interesting owing to his anticipation of many issues of late cognitive science and simulative research. Unfortunately, exploration of this matter does not lie within the scope of this paper. For Selfridge's recent commentary on his own research, see Selfridge, 2008.

Eight

TURING COMPUTABILITY AND LEIBNIZ COMPUTABILITY

Maurizio Matteuzzi

1. Constructivism and Its Origins

The constructivist approach to the foundations of deductive science made its first appearance in the last part of the first book of Aristotle's *Analytica posteriora*. First, Aristotle explains why, in order to avoid the *circulum in demonstrandum* or series of infinite regressions, we need to assume some basic principles. Second, he defines the structure of an axiomatic science that he bases on a universe (the underlying genus: *to genos to hypokeimenon*) and on an apparatus of initial assumptions that he divided into definitions (*horoi*), postulates (*aitemata*), and logical axioms (*koinai ennoiai*). Third, he sketches a hierarchy of the sciences. He introduces the concept of subordination between scientific disciplines: a science that explains the causes or "why" (*to dioti*) is more fundamental than a science that only explains the "what" (*to oti*), hence subordinating it. Finally, Aristotle develops, often implicitly, a rather long argument that, by reducing the main features of the theory of ideas that his Master had advanced to the theory of substance (*ousia*), seeks to overcome it:

> But the things that do not signify a substance must be predicated of some underlying subject, and there cannot be anything white which is not white through being something different. (*For we can say goodbye to the forms, for they are just sounds without meanings*, and if there were any they would contribute nothing to the argument). (Aristotle, 1984, p. 136=83a30–34, translation slightly modified; emphasis added)

From now on, Aristotle will discuss the strength of the different kinds of apodictic demonstrations (*apodeixes*), concluding that:

- The universal demonstration is better than the particular one.
- The affirmative demonstration is better than the negative one.
- The affirmative demonstration is better than the one that leads to absurdity.[1]

2. Existence in Mathematics

Plato and Aristotle codified the two possible points of view about the ontological status of mathematical entities that were to remain stable for the next two millennia.

After the failure of the Pythagorean "arithmogeometry," in order to recover its necessary firmness (*akíneton megalon en peírasi desmôn),*[2] first Plato redefined mathematics on a Parmenidean basis, and then provided a new foundation one, according to which the mathematician deals with the number *in itself,* and the square is not the square of the figure, but the square *in itself.* From this approach we gain the geometrical point with no extension, the line as length without width, and the surface without thickness. The objects of deductive science are thus objects of reason that we place *ab aeterno* in a hyperuranium world whose essence is the absence of contradiction.

Without the help of ideas, or "separate beings," Aristotle secures the foundation of a formal science. He promotes the superiority of a *constructivist* point of view. We may find here the first glimpse of a different conception of "mathematical existence": an entity exists insofar as we can *construct* it; its characterization depends on the fact that "a single determination refers to multiple objects."[3] Aristotle grounds the formal character of an entity on its "being-spoken-of," and not on its Platonic, participatory, "being-in." By proceeding along these lines, Aristotle does not give rise to a mentalist or intuitionist positions *à la* Brouwer: he has no recourse to the activity of the mathematician.

Even though Aristotle thought of himself as a "Platonist" (as the expression "we Platonists," which he often uses in *Metaphysics*, attests), for him what is only ideal is no doubt less essential than what we construct.

3. On What We Can Effectively Compute

During the first decades of the nineteenth century, the discovery of non-Euclidean geometries caused the ultimate demise of the classic approach to the axiomatic sciences that had been based on the self-evidence of their first principles. After the so-called crisis in the foundations of mathematics, which the analysis of logical antinomies had largely caused, Gödel's second theorem (1931) dealt the final blow to classic axiomatics. The final detachment of the notions of *truth* and *demonstrability* (or, from a formalistic point of view, *derivability*) is its outcome: in a complex formal system, every model has sentences that are true but for which a formal proof does not exist.

This situation gives rise to a new question: what can we consider *effectively* computable? The so-called *Church-Turing thesis* provides a definite answer. Church's approach, by means of the lambda calculus, is strictly connected to the principle of mathematical induction and defines the universe of recursivity; Turing's, by introducing the famous automaton, masterfully grasps the concept of algorithm.

The fact that these two efforts, even though based on completely different approaches, define the same set of objects provides a final, and in a certain sense consolatory, result.

4. The Halting Theorem

According to Turing, an algorithm that is executable on a Turing machine (*TM*) is a series of zeros and ones—a number. Starting from this Turing conceives an interesting generalization of his abstract automaton: a *TM* with unbounded tape, on which we may place all possible algorithms. This new structure, the so-called Universal Turing Machine (*UTM*), provides the intellectual basis for a further, essential result: the *halting theorem*. In short, and in the simplest form, we can say that the theorem shows that no algorithm can check the correctness of all algorithms. In other words, the function that determines which functions are computable is itself not computable. We know such an outcome as the theorem of semantic incompleteness, or alternatively as the *halting theorem*.[4]

5. Turing Computability and Constructivism

A correct algorithm halts in a finite number of steps: since a series of 0's and 1's represents it, it is a number. Let us define a *pseudo-algorithm* as an algorithm that does not halt. It is a number as well, for the same reasons. Given a natural number N, it will be the case that N is an algorithm, or a pseudo-algorithm, or a non-algorithm. We can imagine a first Turing machine, TM_1, with an unbounded tape, which contains all and only the algorithms that halt. These are the correct ones. We can order all these algorithms by length and, when lengths are equal, by using the arithmetic relation less-than. In the same manner, we can put every pseudo-algorithm, or algorithms that do not halt, on a second machine, TM_2, and order them in the same way.[5] Suppose we build a meta-machine (MTM) that, having a number as an input, executes the string-matching algorithm on TM_1 and TM_2, as I defined it above, via an iterative depth-first search procedure. MTM tries to match the string that represents the input number to strings of a certain length by trying, alternatively, on TM_1 and TM_2, while progressively increasing the length in case of failure. The behavior of MTM will be as follows:

- if the input number is an algorithm, it will find it, sooner or later, on TM_1;
- if it is a pseudo-algorithm, it will find it, sooner or later, on TM_2;
- if it is a non-algorithm, the machine will never halt.

Complicating a bit our procedure, we could also define TM_3 as a machine, with unbounded tape, that contains all the numbers representing non-algorithms that we order, for instance, by increasing length. Under these conditions, the meta-machine MTM, by applying an iterative depth-first search on TM_1, TM_2, and TM_3, will always halt, because any given number will necessary belong to TM_i, with $i = 1,2,3$.

This argument, seemingly constructive, appears to be in contrast with the halting theorem.[6]

6. The Number Omega and a Mathematical Theory of Everything

Gregory Chaitin refers to a similar argument, which he bases on the complexity of algorithms determined by the their length in bits. He starts from a statement by Leibniz in the *Discourse de Métaphysique*, paragraphs V and VI. Chaitin says:

> Leibniz discusses the crucial question of how we can distinguish a world which can be explained by science from one that cannot. How do we tell whether something we observe in the world around us is subject to some scientific law or just patternless and random? Imagine, Leibniz says, that someone has splattered a piece of paper with ink spots, determining in this manner a finite set of points on the page. Leibniz observes that, even though the points were splattered randomly, there will always be a mathematical curve that passes through this finite set of points. Indeed, many good ways to do this are now known. For example, what is called "Lagrangian interpolation" will do. (Chaitin, 2005)

From my point of view, Chaitin forces the meaning of the text : Leibniz's aim is not to distinguish between a world that we can scientifically explain and a world that we cannot, but to explain that an accidental phenomena cannot exist. Everything that could happen, even if seemingly "accidental," is eventually the output of a complex calculus that God carried out during the creation of the world. I do not agree, therefore, with Chaitin's conclusion:

> So the existence of a mathematical curve passing through a set of points cannot enable us to distinguish between points that are chosen at random and those that obey some kind of a scientific law. How, then, can we tell the difference? Well, says Leibniz, if the curve that contains the points must be extremely complex ("fort composée"), then *it's not much use in explaining the pattern of the ink splashes. It doesn't really help to simplify matters and therefore isn't valid as a scientific law*—the points are random ("irrégulier"). The important insight here is that something is random if any description of it is extremely complex—randomness is complexity. (*ibid.*, emph. added)

The text I emphasized above reflects Chaitin's position, but not Leibniz's, according to whom, instead, identity grounds every truth, regardless of whether we can recognize it or not in a finite number of steps (see section 7 below).

But let us leave aside, because not our present focus, the correctness of Leibniz's interpretation. Chaitin extracts from this allegedly Leibnitian argument his theory of complexity, which he names *Algorithmic Information Theory*. For him, we can assess the complexity of a given function by measuring the length in bits of the algorithm that computes it. Let us now consider the set of binary sequences that represent algorithms, regardless of whether they halt—in our terms, the union of algorithms and pseudo-algorithms. We can ask the following question: what is the probability that a given, randomly chosen algorithm will halt? Chaitin calls such a number *Omega*. As each bit has probability $\frac{1}{2}$, Omega turns out to be the sum of infinite terms of the form $\frac{1}{2}^N$. If we add some other assumptions, including the choice of the programming language to use, Omega looks *well-defined*. According to the halting theorem, Omega is also, in Chaitin's terms, *irreducible*, or *algorithmically random*, or *infinitely complex*.

7. Leibniz Computability

In my opinion, Leibniz's notion of "computability," which includes the concepts of *algorithm* and *computer program*, clearly anticipated its modern version. In the essay *Generales inquisitiones de analysi notionum et veritatum*, Leibniz looks for sound definitions of true, false, possible, necessary, and contingent. These definitions should all agree with each other and depend on the basic assumption that underlies of all his thought: namely, that truth is essentially analytic, which is to say that the principle of identity founds every true sentence. In a broader sense, Leibniz takes from Aristotle the principle according to which *praedicatum inest subjecto* [the predicate inheres in the subject] and relies on it. Among the countless citations available on this point, the clearest is perhaps the following one:

> The predicate or consequent therefore always inheres in the subject or antecedent. And as Aristotle, too, observed, the nature of truth in general or the connection between the terms of a proposition consists in this fact. In identities this connection and the inclusion of the predicate in the subject are explicit; in all other propositions they are implied and must be revealed through the analysis of the concepts, which constitutes a demonstration *a priori*.[7]

On the basis of this principle, Leibniz tries to reach a clear definition of classic modal operators. But, as we now know, a true-functional definition of modal operators such as the one Leibniz leans toward, would collapse the modal system into a non modal logic. At the same time, we must base even contingent truths on identity, as the nature of truth is *semper vel expresse vel implicite identica* ([always based on identity, either explicitly or implicitly], Leibniz, 1970b, p. 268; 1903, p. 519). Leibniz tries several different approaches, and returns to the problem in a quite obsessive manner, until he reaches a solution that he deems satisfactory: the contingent truth is analytic. The law of identity grounds it, but he deems it an identity that does not follow from the analysis or from the individuation of terms that we can carry out within a finite process:

> Hence we also seem able to discern a distinction between necessary truths and others; for example, necessary truths are those which can be reduced to identities, or the opposites of which may be reduced to contradictions.

> Impossibles are those which may be reduced to contradictions, or the opposites of which may be reduced to identities.
>
> Those truths are possible concerning which it can be shown that in their analysis a contradiction would never arise. *Contingent truths are those which although their analysis is to be continued into infinity lack resolution.*[8]

In short, truths are either necessary or contingent. Necessary truths are those whose decomposition in terms produces an identity in a finite number of steps. Contingent truths are those whose identity of terms arises only by means of an infinite process that does not halt. In other words, the difference between necessary and contingent truths "is the same which holds between convergent lines and asymptotic ones, or between rational and irrational numbers."[9] For the first time, we witness the concept of computability becoming detached from the concept of truth.

8. On What There Is and What We Can Say

A famous theorem by Georg Cantor (1873) guarantees that the set of real numbers is more than countable. The set of well-formed expressions in a formal system, that is, in a *language*, is obviously at most countable. The consequence of this fact, which did not escape Cantor's attention, is that if we consider the denotative expressibility in its extension—in a given language—and in eternity, we will be able to name only a small subset of the real numbers (or of any set with the cardinality of the continuum or greater). Some researchers have drawn the conclusion that there are mathematical objects that are *unnameable*. This is clearly a misunderstanding: we can say that, at a given time, there is an uncountable number of objects *that we cannot name*, but this does not mean that there are *unnameable* objects *tout court*. I can name whichever point on a line I choose to, or whichever element of the set of real numbers. But if we accept a purely extensional semantics and do not assume an intensional approach or more sophisticated semantics, in truth we face an enormous overflow of the un-said with respect to what we can, even only in theory, say. Since constructable objects are for the most part a countable set, this embarrassing amount of un-named objects expresses well the difference between a Platonic conception of mathematical existence and the constructivist approach.

On the basis of the argument I just discussed, I can go back to the theoretical Turing machines and reformulate the halting theorem in the following way: TM_1, the machine that lists all and only algorithms, is not *constructive*. Under a restricted conception of mathematical existence, such a machine does not exist. It is a conceivable object, but not a constructable one. From this standpoint, we can say that Turing's abstract automaton captures a certain notion of *computability*.

9. Does a Constructive and Non Turing-computable Process Exist?

Nowadays, several efforts are underway to overcome the notion of Turing-computability by means of continuous systems, analog ones, quantum computation systems, and so on.[10] But I would like to raise a different question. I contend that there might be an area lying between the domain of Turing-computability and the world of Platonic ideas where we may find functions that, even though not Turing-computable, are none the less, at least according to a certain meaning of it, "*constructive*." I will refer, as an example, to the proof of the Bolzano-Weierstrass theorem on the existence of at least an *accumulation point* in every infinite bounded set.[11]

Let l_1 and l_2 be the inferior and superior limits, respectively, of a given infinite set A. For every $x \in A$ we will have that

$l_1 \leqq x \leqq l_2$

Consider l_3, defined as $\frac{1}{2}(l_1 + l_2)$, and let us consider the statement: "In the interval $[l_1, l_3]$ there are infinite points of the set A." Such statement will necessarily be true or false. Let us assume that is false; therefore it will be true that "In the interval $[l_3, l_2]$ there are infinite points of the set A," since A is an infinite set, and given the fact that A is the union of the two mentioned intervals. Let us call S_1 the interval $[l_1, l_3]$ in the first case, the interval $[l_3, l_2]$ in the second one. Let l_4 be the medium point of S_1. So l_4 divides it into two intervals, and at least one of them, by the above argument, must contain infinite many elements of A. Following this procedure, we can determine a series of nested intervals, S_1, S_2, $S_3, \ldots$, and we can show that there exists exactly one point P which belongs to all of them. P is the point of accumulation we were looking for.

Leaving aside the technical details of the proof, let us notice that, for the purposes of our argument, we reach the result by means of a reiterate application of choices.[12] The issue at stake is: how should we consider a procedure that is based on a sequence of not-deterministic choices, even on a finite number of cases? Undoubtedly, it is not Turing-computable, since a Turing procedure is inconsistent with the concept of choice. We face here a degree of logical strength that is quite far from a mere lack of contradiction.

Let us have a look from a different standpoint. Suppose we have a procedure that, at a certain point during its execution, would follow not a criterion we have predetermined but rather perform a random choice. Suppose that we should always make this choice on finite sets. Clearly a human being would not have any problem in performing it. Such a procedure will not be, *strictu sensu*, Turing-computable. Likewise, the invocation of a pseudo-random generation of the criterion of choice clearly remains within Turing-computability, because an algorithm generates randomizations that are only apparently "random." Nonetheless, by throwing the dice whenever necessary, a human being may have each step completely determine the following ones.

10. *Gedankenexperiment*

Let us suppose we write an algorithm that performs the following trivial task. It reads the input; if the input is a number from 1 to 6, then the algorithm writes 0; if the input is more than 6, it writes 1. Such an algorithm is no doubt deterministic and Turing-computable. Let us also suppose that we build a robot able to throw two dice and to recognize the output. Then, let us load our program inside the robot and make it throw the dice, read the input, and take the result as the new input. Its behavior depends on chance, since it throws dice and acts consequently. How should we consider this program and its performance? At a first level, it is Turing-computable: for any given input, we univocally determine its behavior. But let us consider the entire system as a whole: at a second level—from a broader point of view—are we not facing a truly random process? A situation such as this one is not infrequent in AI: on the one hand, in any simulation of heuristic and not-deterministic processes, the readers who follow a technical approach will ask us to admit that at the system level anything running on a computer is no doubt an algorithm, hence a deterministic and Turing-computable procedure. On the other hand, when we consider those simulations from the point of view of the intended model—or even better, in the real word—we can interpret the algorithm's output at a higher level of abstraction, as a free and hence non deterministic process. Should we call it myopia of the technician or delusion of the AI scientist? Extending the model that we use to interpret the simulation could have us evade these strictures. If we limit the model to a set of processors and statements, we cannot avoid a conclusion in favor of the determinism of Turing-computability. But if such a model includes the real world, or at least a fragment of it, then we face a scenario that may be difficult but opens up to more daring solutions.

The infinite complexity of the external world, with its still mysterious ontological constitution, inches toward indeterminism. Again, that even a small fragment of reality is more real than a computational model appears to be an obviously tautology. This *Gedankenexperiment,* its apparent simplicity notwithstanding, is embarrassing. In spite of its apparently aleatory behavior, we can *effectively* build it and put it into the real world. On the basis of current technology the task is not even too difficult. We could even simulate the output of the two dice by generating pseudo-random numbers in the integer interval $[1,12]$ and we would still face an only apparent randomness. By locating the process within the real word and shifting it to the physical level, we can instead attain *actual* randomness.

And yet: is the computer not within the world? Could we not interpret its semantics in the very world it inhabits? Which interpretation is truthful? Formulating it from another point of view: on the one hand, we cannot predict the future behavior of a program before it runs, hence it is non-deterministic and non-Turing-computable; on the other hand, after we let time to enter the model so that a program can execute it in time, does not its behavior become trivially deterministic?

11. World and Randomness

I will now shift my focus from language to the real world: from a *de dicto* to a *de re* modality. Do the dice give a result that is truly random? The question has obviously to do with the general opposition between *determinism* and *indeterminism*. In a world *à la* Spinoza, or *à la* Laplace, the following principle notoriously holds: *give me the situation of the entire universe at a specific given time, and I will calculate the exact situation that the universe will have the following moment.* Reformulating it for my case: give me the weights of the dice, the exact position of their centers of gravity, and the force I exert as I throw them and I will tell you what the output will be. A world *à la* Leibniz would produce a different scenario: even though Leibniz holds that "the present is pregnant with the future," the real world presents the maximum of compossibility, thus letting free will to exist. As a consequence, a process that involves a human being can be free even in the absence of randomness. A pure Laplacean operator, instead, would say that once we knew all the parameters of the body such as the saline concentration in the blood, the quantity of adrenalin, and so forth, we could compute the force of the throw. A Leibnitian would counter that with a free act of volition we can always subvert this outcome. A breakdown of causality takes place inside us, where hardware and software touch and penetrate each other, or, in Leibniz's terms, synchronize. Let us suppose we completely accept the more extreme form of determinism. Under this hypothesis, the behavior of the program in the *Gedankenexperiment* becomes also *a priori*, despite the overwhelming number of parameters, constructive and computable, though overwhelmed by a terrible number of parameters. From a theoretical point of view, its behavior is evidently Turing-computable. Let us now suppose that we are, instead, indeterminists, or at least that we believe in human freedom. Under these conditions, we could then describe the process only *a posteriori*, after we have executed it in time—then randomness will acquire ontological value and exist *in re*. The process would not, at any rate, be Turing-computable. But since for Leibniz time is not a thing but an order relation that is accessory to reality, and since the computations that God, the totally enlightened monad, carries out always already contain mankind's free choices, it follows that Leibniz would incur in some additional problems. In his system everything is always, in all cases, analytic; God can always, through an infinite number of steps, compute anything. Of course I contemplate here classic and perennial philosophical questions. Perhaps we did not expect that the simple transposition to the real world of the Turing-computability property, otherwise well-defined and precise, would force us to face them.

12. New Models

The notion of Turing-computability is well founded and quite strong, and has characterized a broad swath of computer science. It appropriately defines a precise set of functions and connected procedures. The arguments I developed in this chapter, however, suggest that we should not conceive this notion, albeit obviously well-established, as definitive, lest we accept the inconceivable end of all scientific progress. I believe that—perhaps not soon enough, but also not too far in the future, at least according to the temporal scale of scientific progress that does not measure up to the yardstick of human life—we will improve the notion of computability with models that will include some of the elements I discussed above, such as the variable time as an internal (not an external) factor and the *Dasein* as *being-in-this-world*. Most likely, the technological developments of computer science will stimulate progress in this direction. Having said that, I suspect that finding out if our dice throwing is or is not a truly random act will take a much, much longer time.

NOTES

1. The Eleatic school celebrated the *epagoghé eis to adynaton* through Zeno's well-known arguments. See Aristotle, 1984, p. 140=85b10–12, p. 142=87a1, respectively. I have modified the translation.

2. "motionless within the limits of mighty bonds," as per Parmenides' fragment B8 line 28 in Diels-Kranz. See also Parmenides, 1965, pp. 83, 85.

3. See *An. Post.*, 77a5. The standard English translation makes Aristotle's point less perspicuous, perhaps: "for it to be true to say that one things holds of many is necessary" (Aristotle, 1984, p. 125).

4. We call it "Halting theorem" because an algorithm is correct if and only if the *TM* in which we execute it halts in a finite number of steps.

5. With the notion of pseudo-algorithm we have in mind a syntactically correct program that does not halt—a semantically incorrect one; under a different theoretical approach, we could say that no abstract arithmetical difference exists between a pseudo-algorithm and a non-algorithm, or a random sequence of bits.

6. But it is not: the nonconstructive nature of the sets that the argument uses provides a positive solution to the halting problem.

7. From *Primae Veritates*, in Leibniz, 1903, pp. 518–519, Engl. translation from Leibniz, 1970b, pp. 267–268. See also Leibniz's *Generales inquisitiones de analysis notionum et veritatum*, secs. 16 and 132, in Leibniz, 1999, pp. 39, 65; Leibniz, 1903, pp. 366, 388.

8. "Videmur etiam hinc discere posse discrimen veritatum necessariarum ab aliis, ut scilicet verae necessariae sunt quae ad identicas reduci possint, aut quarum oppositae reduci possint ad contradictorias. Et impossibiles, quae ad contradictorias reduci possint, aut quarum reduci possint ad identicas. Possibiles sunt de quibus demonstrari potest nunquam in resolutione occursuram contradictionem. *Verae contingentes sunt quae continuata in infinitum resolutionem indigent.*" See *ibid.*, secs. 60–61, p. 371; Leibniz, 1999, pp. 45–46, emphasis added.

9. Leibniz, 1903, sec. 65, p. 388, Leibniz, 1999, p. 65, I have modified the translation.

10. See, for instance, the chapters by Rossella Lupacchini and Giorgio Sandri, in this volume.

11. In a linear set A of points, a given point P is called an accumulation point if and only if in every space near P (which contains P), there are infinitely many elements of A.

12. Here is a more formal treatment. The Bolzano-Weierstrass theorem asserts that all bounded real series have a convergent subseries. I will prove the following proposition, from which the theorem follows:

Proposition: All real series (a_n) have a real subseries a_{n_k} with
$\lim_{n\to\infty} a_{n_k} = \limsup_{n\to\infty} a_n \lim_{n\to\infty} a_{n_k} = \limsup_{n\to\infty} a_n$

Proof.

Let $y_n = \sup_{i\geq n} a_i$ and $y = \inf y_n$

Let us prove the proposition only in the case that a_n is bounded. Let $l_1 = 1$.

It is possible to find a $k_1 > l_1$ such as

$0 < y_{l_1} - a_{k_1} < 1$

We can now choose

$l_2 > k_1$ and $k_2 > l_2$ with $0 < y_{l_2} - a_{k_2} < \frac{1}{2}$

$l_3 < k_2$ and $k_3 > l_3$ with $0 < y_{l_3} - a_{k_3} < \frac{1}{3}$

$l_n < k_{n-1}$ and $k_n > l_n$ with $0 < y_{l_n} - a_{k_n} < \frac{1}{n}$ (1)

Notice that y_{l_n} is a subseries of y_n and a_{k_n} is a subseries of a_n

$y_{l_n} \to y$ because $y_n \to y$

$y_{l_n} - a_{k_n} \to 0$ as $n \to \infty$ following (1)

Therefore $a_{k_n} \to y$, which proves the proposition.

Nine

LOGICAL INSTRUMENTS: REGULAR EXPRESSIONS, AI, AND THINKING ABOUT THINKING

Christopher M. Kelty

1. Introduction: Thinking about Thinking

This chapter revisits the origins of AI in the context of the history of software and computer science. This might seem an obvious thing to do, but it is not. For one, the period from roughly 1940 to 1960 was a time of relentless fantasizing about the potential of the new calculating machines and the new science appropriate to them—alternately christened cybernetics, automata studies, artificial intelligence, mechanization of thought processes, and computer science, among others. Researchers from an incredibly large range of disciplines struggled to understand and define what computers, computation, and the control of mechanical logical processes would come to mean. Proposals for the future and research programs designed to achieve them were abundant and varied, and no one was certain what the computer itself would become: giant brain? industrial calculator? military weapon? automatic linguistic translator? cybernetic organism? experimental device? communications system?

It was a period of tremendous experimentation and exploration, and Artificial Intelligence of the form that John McCarthy, Allan Newell, Herbert Simon, and Marvin Minsky promoted was but one fine thread. All of these men, along with a wide range of others, were involved in multiple attempts to define the computer: from a youthful Noam Chomsky to the august John von Neumann; from neuroscientist-philosophers to circuit-designing engineers and abstract object-loving mathematicians. Some built machines, models, and theories to solve specific problems (from automatic linguistic translation to anti-aircraft fire control); others worked on general problems of intelligence, reasoning, or logic. Out of this ferment, from roughly 1940 to 1960, when the precise future of computing technology was extremely vague, emerged a division that haunts computer science: is the computer a tool for thinking, or a replacement for it?

Artificial Intelligence proponents such as Newell and Simon made the leap instantly from rudimentary logic problem solving programs to an autonomous thinking machine. Instead of speaking of computers and computational science as aids to thought (as many at the time did, from Vannevar Bush to Warren Weaver to JCR Licklider), they leapt instantly to speaking of them as thinking entities. Later, critics like Hubert Dreyfus and Lucy Suchman (among many others) would argue that this reduction of human thinking to logical problem solving was philosophically impoverished and an impossible research program (Agre, 1997; Suchman, 1987; Dreyfus, 1992). But close attention to this period reveals that not everyone confronting the computer equated logic and human reasoning. For many, the computer remained a tool to think with, a way to try out and test ideas in a mechanical form in order to generate concepts and results that could give meaning and form to the practice of mechanical reasoning.[1] In short, many people in this period treated logical reasoning not as the basis of human thought, but a companion to it, as an instrument we should explore and put to use. As a logical instrument to reason *with,* not *against.* What is important to highlight here is that the term "logical instrument" does not designate an implementation of logic in a machine, but signals the instrument-like qualities of logic itself, as we refine, argue over, and implement it in ever-new physical forms.[2]

This chapter tells the story of one such "logical instrument": regular expressions. Regular expressions are (today) a ubiquitous programming tool and a keystone in the edifice of theoretical computer science.[3] The exploration of the history of these arcane tools demonstrates how the development of logic and mechanical reasoning proceeds alongside, and in conversation with human reasoning and exploitation of these tools. This story does not take sides for or against claims that have been made about Artificial Intelligence (such as the critiques by Dreyfus), but seeks instead to observe, historically and anthropologically, how thinking proceeds. I use the case of regular expressions to show how concepts evolve and are made concrete, even (and perhaps most significantly) concepts meant to capture the foundations of thought itself. This objectification, or "registration" (B. C. Smith, 1996) of objects as having a certain stability and tractability is what I attempt to demonstrate through the historical analysis of the half-century long transformation of the idea of regular expressions. Such an approach is broadly pragmatic in its understanding of logic, thought, and concepts as things that we concretely test against reality.[4] We should also note that the claims I made here about logical instruments and their translations are not claims about specific historical research programs, but historical observations of how practitioners appear to be thinking about thinking. Which is just to say, not all of the observations I made here are claims the actors would themselves make about what they were doing.

The range of transformations that regular expressions have gone through is impressive and surprising: they emerge out of the work of Rudolph Carnap on the syntax of logic; Warren McCulloch and Walter Pitts transformed them into a logical calculi of nerve nets; John von Neumann picked them up for his description of the EDVAC computer, and later his theory of automata; mathematician Stephen

Kleene formalized and dubbed them "regular expressions"; Janusz Brzozowski used them to design state diagrams for circuits; Ken Thompson (who also held a patent on their use for pattern matching) for the first time implemented them in software: they were then implemented in a variety of text editors such as `emacs` and `vi`, used in a wide variety of basic textbooks in computer science and built into nearly every programming language in use into the present.

In telling this story of transformations, I demonstrate how human reasoning develops in concert with (indeed, through) the logical tools for thinking that we create. Such a claim is increasingly obvious given our contemporary reliance on software and computation to guide thinking, and yet we have no clear understanding of how we have created such a possibility for ourselves. The period between 1940 and 1960 is key to understanding how logics went from being the ultimate basis for all human and mathematical reasoning (in the sense of Hilbert's program) to a domesticated creature just barely under our control, living alongside us, keeping us company, thinking with us, thinking about thinking.

2. Regular Expressions, a Brief History, with Backreferences

Regular Expressions are widely known as a powerful and useful tool for matching patterns in text. As is clear from the very popular web-comic `xkcd` (see fig. 9.1), possessing knowledge of how to use these tools can be powerful, heroic even. The utility and ubiquity of Regular Expressions are a bit of a self-fulfilling prophecy: with the spread of the Internet has come the spread of structured text as a mode of interaction and transfer of data. HTML and XML (and its parent SMGL) especially have created the possibility for, and the need for, the automatic processing, manipulation, and reformatting of structured text. Regular expressions are the right tool for the job because they permit a programmer to write concise and general expressions that match a wide range of variable terms in structured text.

For example, a simple verbatim pattern like "cat" will match the word *cat*, while a pattern such as [cat(.*)] will match any word beginning with *cat*. A regular expression can also match, for instance, the word *cat* at the beginning of a line or *dog* at the end of one, thus:

```
[(^cat|dog$)]
```

And sophisticated uses such as matching any email address are also possible:

```
{\b[A-Z0-9._%+-]+@[A-Z0-9.-]+.([A-Z]{2,4})\b}
```

Regular Expressions match only syntactic or lexical features; they cannot match semantic features unless they we can tie them to some kind of lexical or syntactic pattern. Nonetheless, we use them widely because they are built-in features of the

FIGURE 9.1: "Regular Expressions" `xkcd` web-comic by Randall Munroe. Used with permission (Creative Commons Attribution-NonCommercial 2.5 License).

most popular programming languages like C++, Java, Perl, and Python. Perl in particular, was one of the first scripting (or "glue") languages to make extensive use of regular expressions since it was used extensively in the early spread and development of the World Wide Web. Jeffrey Friedl's book *Mastering Regular Expressions* (2006) taught a generation of programmers how to apply regular expressions to problems of all kinds; from pattern matching to text searching, to processing and formatting streams of data like stock quotes or database queries. What's more, modern implementations include the ability to "back-reference" patterns, allowing sophisticated ways to search and replace bits and pieces of a pattern (such as replacing the ".com" section of an email with ".edu"). Combined with other line-oriented UNIX-based processing tools, regular expressions are now part of an arsenal of text and data manipulation tools that allow us to control at least some of our contemporary symbol-saturated information environment.

As a tool, we tie its proliferation to the proliferation of the UNIX operating system in the 1980s and to the growth of Free Software in the 1980s and 1990s (Kelty, 2008). Current versions of regular expressions trace their roots to Henry Spencer's implementation in Perl in the 1980s that was widely disseminated via USEnet and the Internet. But prior to this, regular expressions were implemented in a variety of tools in the UNIX Operating System. `grep` is the most famous of these tools—so famous that geeks use it colloquially to mean "search" as in "I grepped for my socks." `grep` is standard on every UNIX operating system, including now the Macintosh OS. The modern forms of `grep`, known as `egrep` and `fgrep`, owe their existence to the work of Alfred Aho, who created them for the UNIX operating system during his time at Bell Labs (1963–1990). Aho's versions of `grep` were combined with a tool called `sed` (Stream EDitor) by Larry McMahon, and led to the creation of a widely used programming language, with Peter Weinberger and Brian Kernighan, called `awk` (after the initials of its inventors). `awk` emphasized string data types, text processing, and regular expressions. `awk` was an inspiration for Perl in the 1980s.[5]

But most users of regular expressions as a tool do not know, except in outline, the long and peculiar history of regular expressions. All of these implementations in software trace back to the work of Ken Thompson, one of the inventors of the UNIX operating system, who was the first to attempt to implement an algorithm for regular expressions in software. Thompson wrote the first UNIX version of what would become `grep` for a text editor called QED. QED was a rudimentary word processor, in an era before word processing—one of the very first tools that allowed people to directly manipulate a file on a screen or through a teletype; hence it was natural that some facility for searching text would be a goal (van Dam and Rice, 1971; Meyrowitz and van Dam, 1982). The name 'grep' itself derives from a set of commands that would have been issued in QED:

`G/re/P` (**G**lobally search for regular expression *re* and **P**rint it)

Butler Lampson and Peter Deutsch (1967), two computer engineers at UC Berkeley working on the DARPA-funded Project GENIE, sister to Project MAC at MIT and one of the origins of many of the innovations that went into UNIX and modern time-sharing, multi-user operating systems (Lee et al., 1992) initially wrote QED. That version of QED did not include regular expressions, or even the ability to search for arbitrary fixed strings. Ken Thompson received his BS and M.SC. at Berkeley (1962–1966) and worked with Lampson and Deutsch on Project Genie. When he was hired to work at Bell Labs, and was sent to MIT to participate in Project MAC, one of the first things he did was to create a version of QED for the famous Compatible Time-Sharing System (CTSS), later to become Multics (van Vleck, 2008). Thompson went on to create versions of QED for Multics and ultimately for UNIX, for which he dropped the Q and called it simply `ed`. This last program, in turn, inspired a generation of powerful and difficult to use text editing programs that are still in wide use today, such as `emacs` and `vi`.

In 1968, Thompson also published a short "Programming Techniques" paper for the *Communications of the ACM*, in which he described the "Regular Expression Search Algorithm" he had implemented (1968). Regular expressions provided a general algebraic language for describing patterns (for instance, aab*a refers to the pattern "two a's followed by zero or more b's followed by an a"). But converting this language into a software program required some ingenious work on Thompson's part. His solution was to create a program that builds a "non-deterministic finite automaton" equivalent to the given regular expression, and uses that automaton or machine to quickly test a set of inputs. So instead of a general search algorithm that takes any pattern and tries to match it against any given input, Thompson's implementation translates a pattern into a customized searching machine.[6]

How Thompson reached this point, however, is not at all obvious. Prior to his implementation of this algorithm (which was originally written in Assembly language for the IBM 7090, on which CTSS ran, and presented in the paper in Algol), no such attempts existed to use regular expressions in this way. Not many people would have known of their existence outside a handful of now famous people who had explored them in different forms—Stephen Kleene, John von Neumann, Walter Pitts, Warren McCulloch, Michael O. Rabin, and Dana Scott. The link between these individuals and Thompson came from Janusz Brzozowski, who taught for a semester at Berkeley while Thompson was there, and who researched the use of formal languages for circuit design in engineering. Brzozowski introduced a method for using regular expressions to create "state-transition diagrams" that they widely used in the 1950s as a mnemonic for designing and implementing programs in circuits and in the computers of the same time period (Brzozowski, 1964). Thompson adopted Brzozowski's ideas at a time when "software" was just emerging as concept and practice.

Brzozowski's work was part of an explosion of research in the late 1950s that forms the basis of abstract computer science today: the formalization of symbolic processing at the root of the logical operation of computers (Myhill et al., 1960; Nerode, 1958; Arbib, 1961; Copi, Elgot, and Wright, 1958; Hopcroft and Ullman, 1969). Scholars frequently reference Michael O. Rabin and Dana Scott's 1959 article, "Finite Automatons and their Decision Problems," as the apotheosis of this work. We might accurately describe the formal description that this article contains as the theoretical ground for modern computers more than Turing's work of twenty years earlier (1937). Though indebted to Turing's work, the formalization of computing in the late 1950s took the finite nature of existing computing devices as a realistic constraint in developing a formal language that would allow the manipulation of these devices. Such an approach took as its object the logical manipulation of symbols by *finite* automatons (whether human, machine, or animal). In no small part the previous explorations that Turing, von Neumann, McCulloch and Pitts, and ultimately Carnap,Whitehead, and Russell had undertook, lay at its root.

In any event, these earlier attempts were in no way intended to produce a simple tool for matching patterns in text, but instead to explore the very basis of logical reasoning itself. In the next section I will show that they conducted this exploration of logical reasoning as if it were an investigation into the basis of human reason. Unlike such investigations, however, they proceeded through the manipulation of objectified "logical instruments" that we can formalize, manipulate, explore, and construct as if they were tools, as if they were material objects.[7] Attention to the media-specific creation of these tools (in their symbolic expression, their representation in diagrams and ultimately in programming languages, and as running code in machines) can help demonstrate how logical instruments have come into being and are progressively translated or modulated from medium to medium and mind to mind.

3. From *Principia Mathematica* to *Deus ex Machina*

Thompson's paper on Regular Expressions is frustratingly vague about the origins of his idea. It has only four references: the IBM 7094 Programmer's manual, the paper by Brzozowski, a single paper about pattern matching in text (by Anthony Oettinger), and Stephen C. Kleene's "Representations of Events in Nerve Nets and Finite Automata" (1956). Kleene's paper is a kind of hidden classic and the link between theories of automata and the language of "regular events and regular expressions" that Kleene introduced. Stephen Kleene was a graduate student at Princeton in the 1930s, where he worked alongside Alonzo Church and Alan Turing, making substantial contributions in the field of "recursive functions." In 1952, Kleene wrote perhaps the single most famous textbook on the subject called *Introduction to Metamathematics*. Kleene's 1956 paper describes what he called an "algebra of regular events." It was published in a well-known volume that Claude Shannon and John McCarthy edited, titled *Automata Studies*. Kleene's paper, despite being widely cited in the computer science literature, and in some ways one of his most famous works, was not central to his research project. He never returned to the subject.

Kleene's work on the algebra of recursive functions motivated his interest in the subject. In the paper, he introduces a trio of mathematical operators that represent regular events: `A + B` (or union), `A ○ B` (or concatenation), and `A*` (the "Kleene Star," or iteration). We generally take Kleene's paper to be significant because it proves two theorems about the equivalence of regular events and finite automatons. But what makes it interesting in this context is where the idea to create such an algebra came from. In stating these theorems in the text, the origins of Kleene's interest become clear:

> Theorem 3: To each regular event, there is a nerve net that represents the event by firing a certain inner neuron at time p + 2, when started with suitable states of the inner neurons at time 1. If the event is describable by a prepositive and positive regular set of tables, the representation can be by a net started with all inner neurons quiet. (*ibid.*, p. 28)
>
> Theorem 5: In any finite automaton (in particular in a McCulloch Pitts nerve

> net), started at time 1 in a given internal state b_1, the event represented by a given state existing at time p is regular. (Kleene, 1956, p. 34)

By "regular events" Kleene meant an event that a set of nerve cells processes—an event of perception or of thought. Kleene's paper says nothing about computers, programming, matching patterns in text, or searching for text on a computer—as the typescript indicates, the paper was not even composed on or near a computer. From the opening line of the article, however, we see clearly what "automata" are: "anything that receives stimuli, human, machine or animal." The inspiration for Kleene's paper was clearly the famous Warren McCulloch and Walter Pitts's 1943 paper titled "A Logical Calculus of the Ideas Immanent in Nervous Activity"(1943). Kleene adopted the same graphic diagrammatic representation that McCulloch and Pitts used to present their definitions (see fig. 9.2).

Kleene had written his paper in 1951, during a summer at RAND, after Merrill Flood, his sponsor at RAND and fellow graduate student at Princeton, had given him McCulloch and Pitts's paper to read (Kleene, 1979). Upon reading it, Kleene's first inclination as a mathematician was to push further the model it contained—a model of the brain as a logical calculus. Post-Gödelian logics being Kleene's *métier*, he naturally wanted to formalize what McCulloch and Pitts had proposed as a model of the brain in terms of an algebra of recursive functions. In a later reflection, Kleene said:

> I found [McCulloch and Pitts's] model to be very interesting—an original contribution—but their analysis of it to fall quite short of what was possible. So I did what cried out to be done with it (as it seemed to me) having newly in mind also the idea of a finite automaton, which came to me that summer at RAND through reading in printer's proof [?] von Neumann's Hixon Symposium lecture. (1979, p. 380)

What "cried out to be done" however, was probably not what McCulloch and Pitts would have cried out for. I would like to emphasize the misfit of intentions between these two papers. Whereas McCulloch and Pitts were interested precisely in creating a representation of thought (if not neurons *per se*), Kleene was not:

> our theoretical objective is not dependent on the assumptions [about the neuro-physiological data] fitting exactly. It is a familiar stratagem of science, when faced with a body of data too complex to be mastered as a whole, to select some limited domain of experiences, some simple situations, and to undertake to construct a model to fit these at least approximately. Having set up such a model, the next step is to seek a thorough understanding of the model itself. (Kleene, 1956, p. 4)

Kleene was not interested in whether McCulloch and Pitts's model was accurate—whether it really represented the state, function, or structure of a brain, or of thought. He took from the paper a model that it just so happened to be defined in terms of neurons and sensory inputs and organisms, but that as far as he was

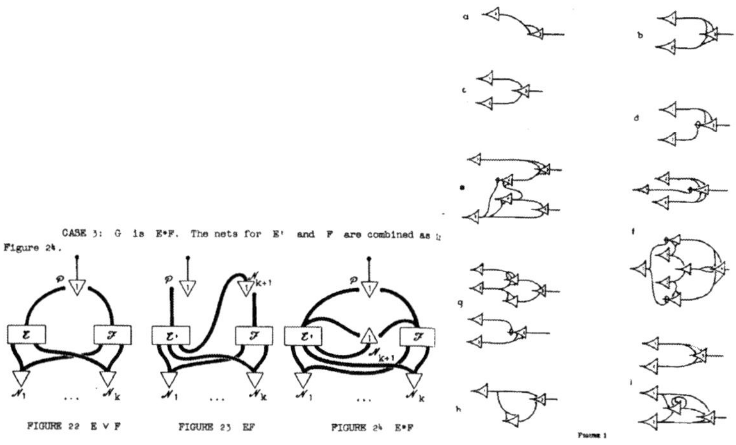

(A) Kleene's Diagrams

(B) McCulloch and Pitts's Diagrams

FIGURE 9.2: Kleene's and McCulloch and Pitts's diagrams of regular events.

concerned, we had not fully explored in mathematical terms yet. Kleene proved in his paper that at stake was not a model of human thought, *per se*, but a model of any finite automaton, whether animal, human, or machine. His reference to von Neumann's Hixon Symposium lecture on automata makes this evident. Von Neumann was among the first to explore the idea of a general theory of automata using the tools of logic. His Hixon symposium text spells out that, whereas for centuries we have relied on the mathematics of continuous variation to analyze nature, today the emergence of computing technology has created a need for a more powerful discrete mathematics that might be amenable to the analysis of thought more than classical mathematical analysis. Kleene takes this insight a step further: in his approach, the implicit equivalence between the brains of animals, humans, and machines holds no necessary connection to brains or neurons, nor even to the actual facts of thinking. Von Neumann intuited this as well when writing the notorious "First Draft of a Report on the EDVAC." Instead of describing the design of a computer in terms of vacuum tubes, wires, plug-boards, and relays, von Neumann used McCulloch and Pitts's diagrams of neurons as the formalism for the design (see fig. 9.3).

For von Neumann and Kleene, the formal equivalence that McCulloch and Pitts demonstrated between brains and logical languages was also a formal equivalence suitable to a machine. This was not a metaphor. More than a suggestion that computers should function *like* or *as if* they were brains, it recognized a

> **6.2** The analogs of human neurons, discussed in 4.2–4.3 and again referred to at the end of 5.1, seem to provide elements of just the kind postulated at the end of 6.1. We propose to use them accordingly for the purpose described there: As the constituent elements of the device, for the duration of the preliminary discussion. We must therefore give a precise account of the properties which we postulate for these elements.
>
> The element which we will discuss, to be called an E-element, will be represented to be a circle ○, which receives the excitatory and inhibitory stimuli, and emits its own stimuli along a line attached to it: ○—. This axon may branch: ○—<, ○—⋲. The emission along it follows the original stimulation by a *synaptic delay*, which we can assume to be a fixed time, the same for all E-elements, to be denoted by τ. We propose to neglect the other delays (due to conduction of the stimuli along the lines) aside of τ. We will mark the presence of the delay τ by an arrow on the line: ○→—, ○→—<. This will also serve to identify the origin and the direction of the line.

FIGURE 9.3: McCulloch-Pitts neurons in the text of the First Draft of the Report on EDVAC by John von Neumann.

certain utility, testability, or pragmatic rectifiability to this formalism. With the introduction of an algebra of regular events, Kleene created a useful, practicable formalism that in a matter of years would dominate the nascent field of computer science. To a generation of people bound to become the first luminaries of computer science, Kleene's paper was the first step toward a formal theory of actual and possible computer programs. In 1959, for instance, Michael Rabin and Dana Scott published "Finite Automata and Their Decision Problems" (1959), which became the locus classicus for future research into the classification of computer programs, their complexity, and their solvability. What Thompson took from Kleene was not a model of a brain, or even an algebra, but a kind of logical toy, a structure with certain implications that we could explore, an object to think with that we could potentially make to fit some new situation. Thompson recognized that this algebra was a tool for efficiently matching patterns in text and employed the algebra to create an algorithm and a piece of working software. It appears as an obvious case of science being "applied" to create technology, but the problem is that regardless of what Kleene's paper sets proleptically out as science, is not the technology that Thompson creates. It is a translation of Kleene's algebra into an algorithm, a paper, and a program.[8] Similarly, Kleene had translated McCulloch and Pitts's model into an algebra and a formalism that were now known as regular expressions for regular events, terms we initially meant to capture the sensory input experience of an organism or an automaton.

Kleene's formalization of McCulloch and Pitts's results transferred their ideas from the domain of representations of thought to the domain of representations of automata, hence making it useful as a tool—a logical tool for thinking about thinking (namely, thinking about logical reasoning) and a tool for designing thinking machines (designing and controlling circuits and eventually software). McCulloch and Pitts's paper had no such goals in mind. Cybernetics, neuroscience, mathematical psychology, cognitive science, mathematical biology, artificial intelligence, and the burgeoning subfield of computer science studying neural nets and neural processing, all these disciplines claimed McCulloch' and Pitts's paper as a source, though oblivious, I should say, of any brains other than those of the researchers.[9]

McCulloch's interest was initially in finding what he hypothesized as a "psychon"—the atomic unit of neural activity that he first sought in the physiological research he conducted during the 1930s in partnership with Yale physiologist J.G. Dusser de Barenne. In the early 1940s, Jerome Lettvin introduced McCulloch to Walter Pitts and thereby to Nicholas Rashevsky's Mathematical Biology group at the University of Chicago, where Walter Pitts had been actively working on models of neural activity with Rashevsky and mathematician Alston Householder. The collaboration between the two was lopsided, at best. McCulloch was in his forties, Pitts was 17; McCulloch had spent his career in physiology and philosophy, Pitts was by different and sometimes unreliable accounts a mathematical prodigy who had run away from his home in Detroit and met Bertrand Russell in a park in Chicago (Smalheiser, 2000; Schlatter and Aizawa, 2008). Together they managed to piece together something that met in the middle, a paper that demonstrated the formal equivalence between a plausible model of neural activity and a logical calculus. Part of McCulloch and Pitts' inspiration for their paper was Alan Turing's machine. As Tara Abraham puts it, "Turing was able to define the complicated process of computation in 'mechanical' terms, with the notion of a simple algorithm so exhaustive, rigorous and unambiguous that the executor would need no 'mathematical knowledge' to carry out its task" (2002, p. 18). This identification of computation with an automatic procedure provided the inspiration for McCulloch and Pitts to model a set of nerves as something that could also calculate "in the absence of mathematical knowledge."

In hindsight, what McCulloch and Pitts achieved was far more influential in engineering, computer science, and mathematics than it was in biology or neuroscience. Works that take McCulloch and Pitts's logical calculus of nerve nets as a starting point have been extraordinarily bountiful in mathematics and computer science. Accounts of Artificial Intelligence almost inevitably refer to this paper as one origin point, whether as inspiration or as foundation.[10] By contrast, neurology and neurobiology have more or less dropped their formalization entirely, beginning at least with McCulloch and Pitts themselves, whose 1947 paper "How we Know Universals" and the 1959 paper they wrote with Lettvin and Maturana, "What the Frogs Eye Tells the Frog's Brain" (1959; 1947), both abandon the strict formal equivalence with propositional calculi or the Turing machine, in favor of more complex biological models that are less amenable to logical manipulation.

The analogy between logic and thinking is obviously not unique to McCulloch and Pitts. In most philosophy and mathematics of the early twentieth century, and especially under the influence of the Hilbert program, logic and mathematics represent the ability of human reason at its pinnacle. McCulloch and Pitts materialized this assumption in a representation of neurons, adding a concrete formal specificity to an otherwise abstract set of claims about human reason. Like Thompson's paper on Regular Expressions, McCulloch and Pitts's paper is exasperatingly short on citations: it has three, and none refers to previous work in biology or physiology, not even to Turing's work, but instead to arguably the three most famous attempts to formalize scientific reasoning of the twentieth century: Alfred Whitehead and Bertrand Russell's *Principia Mathematica*, David Hilbert and Wilhelm Ackerman's *Foundations of Theoretical Logic*, and Rudolph Carnap's *Logical Syntax of Language*.

Despite McCulloch's polymathic knowledge of physiology, the paper with Pitts is hopelessly imprecise (and avowedly so) in physiological terms. What drives the paper instead is the attempt to transform a logical calculus from one symbolic system to another—a symbolic system inspired by the observation that neurons are connected together in networks and tend to be firing electrical impulses or not, are either "on" or "off."

Walter Pitts was a student of Rudolph Carnap and had read his work (on the advice of Bertrand Russell himself, so the story goes). What was it that McCulloch and Pitts took from Carnap? Was it a particular logical system or a set of propositions and theorems? Or was it the symbolism? What was it that "cried out to be done," to echo Kleene, with Carnap's work? I would argue that Carnap provided McCulloch and Pitts with the warrant they were looking for to *creatively* formalize their model.

Carnap's book, *The Logical Syntax of Language*, is the pinnacle of his attempts to define a philosophical language suitable to the analysis of scientific statements. In it, he constructs two kinds of artificial languages, labeled Language I and Language II. Language I is a highly restricted language that includes only the most basic postulates of the arithmetic of natural numbers. Language II includes Language I as well as "indefinite" terms for all of mathematics and "the sentences of physics." McCulloch and Pitts chose Language II for their formalism, a choice widely regarded as unfortunate because of Carnap's impossibly baroque symbolism. Nonetheless, it is not the formalism *per se*, nor the content, which they draw inspiration from, so much as Carnap's insistence that *the choice itself is arbitrary*. Carnap's *Logical Theory of Syntax* is famous for this approach:

> In [this book], the view will be maintained that we have in every respect complete liberty with regard to the forms of language; that both the forms of construction for sentences and the rules of transformation (the latter are usually designated as "postulates" and "rules of inference") may be chosen quite arbitrarily .... For language, in its mathematical form, can be constructed according to the preferences of any one point of view represented; so that no question of justification arises at all, but only the question of the syntactical consequences to which one or another of the choices leads, including the question of non-contradiction. (1937, p. xv)

Carnap was justly famous, even among the logical positivists, for his vigorous hatred of metaphysics and of "meaningless" statements—a hatred most visible in the article "The Elimination of Metaphysics through the Logical Analysis of Language" (1966[1932]). So it is perhaps oxymoronic to term this approach to the arbitrariness of logical language the "Principle of Tolerance," which states, "It is not our business to set up prohibitions, but to arrive at conventions" (1966[1932], p. 51), and, "In logic, there are no morals. Everyone is at liberty to build up his own logic, i.e. his own form of language, as he wishes. All that is required of him is that, if he wishes to discuss it, he must state his methods clearly, and give syntactical rules instead of philosophical arguments" (1966[1932], p. 52).

Only the principle of tolerance, the submission to the arbitrary, Carnap insisted, could lead us to "the boundless ocean of unlimited possibilities" that awaits us (Sarkar, 1992). McCulloch and Pitts realized that Carnap's formalism may not be the most realistic, or accurate, or universal, but that its arbitrariness freed it from any putatively metaphysical presuppositions that might constrain it. We could made it into an experimental tool. We could made it into a brain, a brain-machine, or a brain-language—and the slippage amongst these things could be deliberate and refer implicitly, if not explicitly and precisely, to the activity of human thought.

Like Turing's machine, Carnap's languages were meant to be the most mechanical and the most rigorous constructions of reason possible—but they were nonetheless constructions of reason. They were not, in Turing's case or in Carnap's, instructions for the creation of machines with which human beings might think. For Carnap, the elimination of metaphysics was explicitly about creating protocols that define the understandability and analyzability of statements; but it was also an implicit ontology of introspection and thinking. Logical analysis provides a way to construct systems by which we can reach valid conclusions without recourse to any inaccessible form of introspection or any revealed knowledge. We can construct machines that will demonstrate the answer to multiple parties regardless of any mathematical knowledge.

Carnap, in a tradition stretching back to Leibniz, wanted his artificial languages to be the ultimate arbiters of dispute. No longer would human frailty, human weakness in its reliance on irrationality or its reference to the supernatural, be responsible for decision making when we could construct the best of all possible systems to answer questions objectively and mechanically.[11]

McCulloch and Pitts recognized something slightly different in this liberation of logic; they appreciated that our assumptions or our metaphysical beliefs should not constrain the design of a logical system, but should instead be an act of pure construction whose implications and consequences for the task at hand are all that matter. But McCulloch and Pitts did not use Carnap's Language II to test the soundness of their model; they were not making empirical statements about the brain that Carnap's logic would verify. Instead, McCulloch and Pitts proved the formal equivalence of their model neuron and that of Carnap's logic; or to put it differently, they created yet another language—a brain-language or nerve-language—which they showed to have exactly the same expressive possibilities as Carnap's syntax-language.[12]

McCulloch and Pitts's abstract images of neurons that are carefully correlated with formal propositions from Carnap's calculus represent brain-language. These abstract neural-net diagrams show up again and again in subsequent formalizations, such as Kleene's work, and are ultimately central to neural network research of all kinds (see again, fig. 9.2). McCulloch and Pitts's translation of Carnap (like Kleene's translation of McCulloch and Pitts, and Thompson's translation of Kleene) does not try to preserve Carnap's universal language, nor reduce the brain to it; it invents a new one, it translates it, or remedies it.

If we examine the work of Carnap, McCulloch and Pitts, Kleene, Thompson, and later contributors to the development of regular expressions, we can see a continuity of an instrumental thing, though not always a continuity of research programs institutions, or philosophical commitments. If anything the "principle of tolerance" does not liberate reasoning from metaphysics at all, so much as it creates an entirely new kind of object to think with: logical instruments. The liberated logical mechanisms of Carnap or Turing mediate McCulloch and Pitts's insights, making them thinkable. Just as McCulloch and Pitts's mediate Kleene's, and Kleene's mediate Thompson's, and so on. A stability and a coherence to the object these researchers register is in order, even if the intentions they express and justifications they give diverge from the object, or sometimes, converge only loosely and serendipitously with it.

4. Conclusion

Regular Expressions are everywhere today, which is a strange triumph if one sees in them a vestige of Carnap's program. That his approach to understanding thought might lead, through a circuitous but quite concrete path, to a tool we use everyday to reformat symbolic information in our syntax-saturated, structured language-draped world is a curious fact. The continuity of Regular Expression raises a question that we might phrase in terms of Bruno Latour's concept of "immutable mobiles": what exactly is the "immutable" part of this mobile concept, ontologically speaking? What, if anything, remains of Carnap, in today's Regular Expressions?[13] Creating a richer description of thinking about thinking is part of what is at stake; one in which we include logical instruments as part of our repertoire for exploring ideas, exploring arguments and claims, and constructing concepts. Regular Expressions are a strategic example, since they form the very basis of "formal language" theory in computer science, as well as a ubiquitous tool for matching patterns in text. We can also show a similar progression with the related logical instrument "L-Systems" that similarly morphed from a model of biological growth, to a mathematical formalism, and ultimately to a widely used tool for generating computed visual forms, such as images of trees, flowers, and neurons (Kelty and Landecker, 2004). Similarly, we might look to other conventional features of contemporary computing: Bezier curves in illustration (and their origins in numerical controlled design of automobile bodies in the 1950s), or filters used in audio-video editing software (and their origins in electrical engineering, communications theory, or speech-processing research). All of these are clear cases of tools based in logic (because based in the construction of effective procedures that our computing devices require) that have come to populate our world.

Seen in this light, we might also measure the success of logic, opposite the putative "failure" of the classic AI program, especially in terms of the productive possibilities that result from research. Allen Newell and Herbert Simon's famous "Logic Theorist" (1956) was a program designed to prove the theorems of the *Principia Mathematica*. At first glance, this would appear to be another and particularly nice example of the kind of translations we have seen in the case of Regular Expressions. However, Newell and Simon's project made no explicit formal equivalence, it translated nothing. They did invent a new artificial language—a precursor to the programming language LISP—but they did not attempt to prove that that language was equivalent to any other. Instead, the implicit formal equivalence that AI maintained is between the human capability for reason and the proof of theorems by logical inference. Newell, Simon, McCarthy, Minsky, and others aimed at a second creation, a robot race of machines more intelligent than human beings. By contrast, the work of McCulloch and Pitts, Kleene, and especially Thompson led to the creation of something more like a companion species of logical instruments; logical instruments with and through which we can conduct introspection and explorations. Rather than asking "Can machines think?" as AI relentlessly does, this minor tradition asks "Can humans think?" to which the answer is: not alone. We reason with logics, we form symbiotic relationships with them, we translate them from paper to machine, to patent, and to software.

If we see AI as the attempt to create a second race of thinking beings, then we can only pronounce its failure; but if we see it from the perspective of "logical instruments," we might see AI might also as a similarly successful research program—not "degenerative" at all (as per Dreyfus) but proliferative. Game engines, recommendation systems, virtual world non-player characters, and net-based commerce and advertising applications are everywhere. Computer scientists routinely speak of "agents" today (intelligent or otherwise) instead of human beings or thinking entities—an approach canonized in Stuart Russell and Peter Norvig's *Artificial Intelligence: A Modern Approach* (2003) and in the Rodney Brooks-inspired approach of creating large coordinated networks of simple processes that has dominated the research imagination since the early 1990s.

Understanding AI as part of a tradition of logical instrument fashioning might also help explain the current explosion of computer research traditions that surround us today. "Computers" today are much more than the machine on our laps or in our offices, and more than what Turing wrote about in 1936. Today we have journals such as *Soft Computing*, *BioSystems*, *Natural Computing*, *The Journal of Applied Soft Computing*, the *IEEE Transactions on Evolutionary Computation,* and *The International Journal of Unconventional Computing*. We have GRID computers, mesh computers, computers made of DNA, molecular computers and "moleware," reaction-diffusion chemical computing, quantum computers, spintronic computers, protein-based optical memories and processors, "computers" made of leech neurons and "membrane" computers (which are not made of membranes at all), tinker toy computers, "amorphous computers," neural networks, and genetic algorithms and an increasingly vibrant array of programmable logic devices like the field programmable gate array, to say nothing of a new generation of biological sensors that measure and compute everything from light intensity to fish freshness. We have designed all these variations with logic instruments and they in turn become logical instruments for designing. They proliferate in direct response to the playfulness with which we engage them, and this should tell us something about the history of thinking about thinking. Some of these proliferations tend toward a fantasy world in which everything is a computer. Cells compute, DNA computes, molecules compute, waves compute—this version of computation-as-intelligence is so generalized that to me it seems impoverished to continue making the implicit connection between logic and thought. We should instead start thinking of them as logical instruments without which we cannot think, creatively or otherwise.

NOTES

1. Andrew Pickering in particular has shown in detail some of the ways in which work in cybernetics divides along similar lines; Grey Walter, Gordon Pask, and Stafford Beer, for instance, pursued a much more playful, toy-oriented experimental version of cybernetics than did Norbert Wiener, Donald Hebb, or other American cyberneticians interested in developing a universal model of feedback-control mechanisms. See Pickering, 2002; Pickering, 2004; Pickering, 2007.

2. Hélène Mialet has recently demonstrated something similar in her analyses of Stephen Hawking's peculiar calculative abilities (2009); similarly Matthew Jones's recent work (2006) demonstrates that Leibniz's practice concerning the quadrature of the circle

proceeded in a similar way—by arguing for new kinds of practical and technical tools that were otherwise illegitimate in mathematics at the time.

3. Regular expressions are central to the construction of programming languages, especially in tools that partially automate the process, such as compilers, parsers, lexical analyzers and compiler compilers. By 1969, regular expressions had been incorporated into computer science curricula in most of the major programs as part of classes on formal languages and automata theory. The first textbook was Hopcroft and Ullman's *Formal languages and their relation to automata* (1969); current textbooks retain much of the same structure and material, for instance, Peter Linz's *An Introduction to Formal Languages and Automata* (2001).

4. In particular, the work Dewey undertook in *Experimental Logic* (2004). Pragmatist approaches to AI are few and far between, but see Burke (1995). Recent work by Okrent has extended a Dewey-inspired reading of Heidegger that focuses on intentionality and tool-use as one way to reformulate the philosophical basis of rationality. Intentionality as a case of tool-use is close to what I seek here as an explanation in the case of regular expressions, though the present case does not fit neatly within Okrent's explanation of intentionality, in part because the actors I discuss here do not always understand their manipulations of logical systems as a tool-oriented, intentional activity. And yet, I argue, it is possible to understand what they have done within this frame, and that this re-description is a step toward articulating the understanding of intentionality necessary to remove logic from the domain of pure reason and show its function in the domain of practical reason, especially after the advent of software and programmable computers. See especially Okrent, 1991; Okrent, 2007.

5. No scholarly historical work on UNIX and its origins at Bell Labs exists that explores the range of activities underway there. Peter Salus (1994) reports on the development of the operating system in outline. Most of this work is available on-line, however, in different places. See Dennis Ritchie's web resources, for instance: http://www.cs.bell-labs.com/who/dmr/.

6. Russ Cox has written a technical history of the different implementations of the matching engine that Thompson used, and those that are in use today, demonstrating differences in efficiency and speed that we have lost over time (2007). The algorithm that Thompson implemented also bears some similarity to the general AI approach to "reasoning as search" given the focus on pattern matching and backtracking. Thompson's use of regular expressions, however, appears to be the first such implementation and entirely distinct from the AI tradition of problem solving.

7. The meaning of materiality is obviously at issue: objectification is probably a better term, or conceptification. It is not merely the symbolic character that is at stake, but the ability to grasp the instrument as a singularity, as something with parts that can be observed, moved, explored, replaced as if one were looking at a device.

8. And on top of that, a patent that represents the pattern-matching regular expression algorithm in yet another kind of formalism, that of the legal language of claims in a patent (US Patent Number 3,568,156, filed 9 August 1967).

9. Tara Abraham has recently documented much of the intellectual context and background of this paper (2003; 2004; 2006). See also Kay, 2002 and Piccinini, 2004.

10. Lily Kay points out that even though researchers routinely adopt McCulloch and Pitts's paper as an origin point for AI, McCulloch was himself "not an AI enthusiast. He was far less interested in machines that try to think than in the mechanical principle of thought (and by mechanical he did not mean physico-chemical or reductionist). He was attracted to 'Machines that Think and Want': how we know particulars and universals, the desire for food, woman and bed, music, poetry, mathematics. And he sought a mathematical corrective to the mindless black box of behaviorism and the tripartite Platonic conjectures of psychoanalysis" (2002, pp. 603–4).

11. Dreyfus locates his original critique of AI in just this age-old desire, which he refers to as Platonic in origin. Dreyfus' book is not a critique of reason *per se*, however, but an alternative explanation of human reason drawing on concepts of embodiment and background know-how (1992). It does not, however, make any assumptions about whether the presence and proliferation of machines and symbol systems mediated the embodiment or background knowledge, regardless of whether they were created with the intention of being able to reason autonomously. Thinking with machines seems to me to fit into our "background" know-how in ways that neither AI nor Dreyfus seem to give any credit.

12. Lily Kay's analysis of McCulloch and Pitts confirms this, in part. She argues that McCulloch and Pitts attempted to bridge the formal analysis of the brain with that of the mind: "Neurophysiology could expand to embrace mind, since that mind—conceptualized within a new discourse of information—could now be embodied in neural nets and investigated experimentally" (2002, p. 600).

13. Latour's work on immutable mobiles appears in *Drawing Things Together* (1990). See also Paul Rabinow on the concept of "remediation" (2008).

IV

AT THE THRESHOLDS OF COMPUTABILITY

Ten

GÖDEL, NAGEL, MINDS, AND MACHINES

Solomon Feferman*

1. The Imbroglio between Nagel and Gödel

Some fifty years ago (1957 to be exact), Ernest Nagel and Kurt Gödel became involved in a contentious exchange about the possible inclusion of Gödel's original work on incompleteness in the book, *Gödel's Proof*, then being written by Nagel with James R. Newman. What led to the conflict were some unprecedented demands that Gödel made over the use of his material and his involvement in the contents of the book—demands that resulted in an explosive reaction on Nagel's part. In the end the proposal came to naught. But the story is of interest because of what was basically at issue, namely their provocatively related but contrasting views on the possible significance of Gödel's theorems for minds vs. machines in the development of mathematics. That is our point of departure for the attempts by Gödel, and later John Lucas and Roger Penrose, to establish definitive consequences of those theorems, attempts which—as we shall see—depend on highly idealized and problematic assumptions about minds, machines, and mathematics. In particular, I shall argue that there is a fundamental equivocation involved in those assumptions that needs to be reexamined. In conclusion, that will lead us to a new way of looking at how the mind may work in deriving mathematics which straddles the mechanist and anti-mechanist viewpoints.

The story of the conflict between Gödel and Nagel has been told in full in the introductory note by Charles Parsons and Wilfried Sieg to the correspondence between them in Volume V of the *Gödel Collected Works* (Gödel, 2003b, pp. 135ff.), so I will confine myself to the high points.

The first popular exposition of Gödel's incompleteness theorems was published by Nagel and Newman in 1956 in an article entitled "Gödel's Proof" for the *Scientific American*. The article was reprinted soon after in the four volume anthology edited by Newman, *The World of Mathematics: A Small Library of the Literature of Mathematics From a'H-Mosé the Scribe to Albert Einstein*, presented with commentaries and notes. That was an instant best-seller, and has since been reprinted many times. Newman had been trained as a mathematician but then

*Revised text of the Ernest Nagel Lecture given at Columbia University on 27 September 2007. I wish to thank Hannes Leitgeb and Carol Rovane for their helpful comments on an earlier version. This essay was originally published in the *Journal of Philosophy*, 106:4 (2009), pp. 201–219.

became a lawyer and was in government service during World War II. Endlessly fascinated with mathematics, he became a member of the editorial board of the *Scientific American* a few years after the war.

Ernest Nagel had long been recognized as one of the leading philosophers of science in the United States, along with Rudolf Carnap, Carl Hempel, and Hans Reichenbach. Like them, he was an immigrant to the US, but unlike them he had come much earlier, in 1911 at the age of 10. Later, while teaching in the public schools, Nagel received his bachelor's degree at CCNY in 1923 and his PhD in philosophy at Columbia University in 1930. Except for one year at Rockefeller University, his entire academic career was spent at Columbia, where he became the John Dewey Professor of Philosophy and was eventually appointed to the prestigious rank of University Professor in 1967. In his philosophical work, Nagel combined the viewpoints of logical positivism and pragmatic naturalism. His teacher at CCNY had been Morris R. Cohen, and with Cohen in 1934 he published *An Introduction to Logic and Scientific Method*, one of the first and most successful textbooks in those subjects.

Soon after the appearance of Nagel and Newman's article on Gödel's theorems in *The World of Mathematics*, they undertook to expand it to a short book to be published by NYU Press. Moreover, they had the idea to add an appendix which would include a translation of Gödel's 1931 paper on undecidable propositions together with the notes for lectures on that work that he had given during his first visit to Princeton in 1934. Early in 1957 their editor at the Press, Allan Angoff, approached Gödel for permission to use that material. Though Gödel said he liked the Nagel-Newman article very much as a nontechnical introduction to his work, he said that he was concerned with some troublesome mistakes that had been made in it and even more so with the interpretation of his results, so he was reluctant about agreeing to the proposal.[1] To encourage him to accept, Nagel paid a lengthy personal visit to him at the Institute in March of 1957. A week later, Gödel wrote Angoff with three conditions. First,

> I would have to write an introduction to the appendix on the one hand in order to mention advances that have been made after the publication of my papers, [and] on the other hand in order to supplement the considerations given in the book, about the philosophical implications of my results. I am not very well pleased with the treatment of these questions that came out in the *Scientific American* and in the *World of Mathematics*. This, for the most part is not the fault of the authors, because almost nothing has been published on this subject, while I have been thinking about it in the past few years. (Gödel, 2003a, p. 1)

A second condition, somewhat of a surprise, was that he share in some way in the royalties of the book. But the most contentious condition was the one he put at the beginning of his letter to Angoff: "In view of the fact that giving my consent to this plan implies, in some sense, an approval of the book on my part, I would have to see the manuscript and the proof sheets of the book, including the appendix." This condition was made even more explicit in a follow-up letter:

> Of course I shall have to see the manuscript of the book before I sign the contract, so that I can be sure that I am in agreement with its content, or that passages with which I don't agree can be eliminated, or that I may express my view about the questions concerned in the introduction. (*ibid.*, p. 4)

Gödel also wrote Nagel saying that he had made to Angoff "the same suggestions I mentioned to you in our conversation in Princeton." But when Nagel was shown the correspondence with Angoff, he exploded. He thought the proposal to share royalties was "grasping and unreasonable" yet was prepared to accept that. But what really ticked him off, as he wrote Angoff, was Gödel's number one condition:

> I could scarcely believe my eyes when I read his ultimatum that he is not only to see the manuscript of our essay before signing the contract with you, but that he is to have the right to eliminate anything in the essay of which he disapproves. In short, he stipulates as a condition of signing the contract the right of censorship. This seems to me just insulting, and I decline to be a party to any such agreement with Gödel .... If [his] conditions were granted, Jim and I would be compelled to make any alterations Gödel might dictate, and we would be at the mercy of his tastes and procrastination for a period without foreseeable end .... Gödel is of course a great man, but I decline to be his slave. (Gödel, 2003b, pp. 138–139)

Nagel's fears about how things might go if they agreed to this condition were indeed well-founded, but he left it in Angoff's hands to communicate his displeasure with it. In the event, Angoff tried to avoid direct conflict, and wrote Gödel in a seemingly accommodating but ambiguous way. It was not until August 1957 that Nagel wrote Gödel himself making clear his refusal to accept it:

> I must say, quite frankly, that your ... stipulation was a shocking surprise to me, since you were ostensibly asking for the right to censor anything of which you disapproved in our essay. Neither Mr. Newman nor I felt we could concur in such a demand without a complete loss of self-respect. I made all this plain to Mr. Angoff when I wrote him last spring, though it seems he never stated our case to you. I regret now that I did not write you myself, for I believe you would have immediately recognized the justice of our demurrer. (*ibid.*, pp. 152–153)

This was passed over in silence in Gödel's response to Nagel. In the end, the proposal to include anything by him in the book by Nagel and Newman came to naught, and even the specific technical errors which Gödel could have brought to their attention remained uncorrected. Since this was to be the first popular exposition that would reach a wide audience, Gödel had good reason to be concerned, but by putting the conditions in the way that he did he passed up the chance to be a constructive critic.

As an aside, it is questionable whether Gödel indeed recognized the "justice of [their] demurrer." He made a similar request seven years later, when Paul Benacerraf and Hilary Putnam approached him about including his papers on Russell's mathematical logic and Cantor's continuum problem in their forthcoming collection of articles on the philosophy of mathematics. Gödel feared that Benacerraf and Putnam would use their introduction to mount an attack on his platonistic views, and he demanded what amounted to editorial control of it as a condition for inclusion. In that case, the editors refused straight off, but gave Gödel sufficient reassurance about the nature of its contents for him to grant permission to reprint as requested (see Gödel, 1990, p. 166).

2. Gödel's Concerns, Part I: the Formulation of the Incompleteness Theorems

Recall that from his first letter to Allan Angoff, one of the conditions that Gödel put for the proposed appendix to the Nagel and Newman book was that he write an introduction to it "in order to mention advances that have been made after the publication of my papers" as well as "to supplement the considerations, given in the book, about the philosophical implications of my results," with which he was "not very well pleased." As to the first of these, Gödel would have wanted to use the opportunity to put on record what he considered the strongest formulation of his incompleteness theorems. This was signaled in a letter that he composed to Nagel early in 1957 but that was apparently never sent. (A number of letters found in Gödel's *Nachlass* were marked *nicht abgeschickt* [not sent].) He there writes that

> [c]onsiderable advances have been made ... since 1934 ... it was only by Turing's work that it became completely clear, that my proof is applicable to every formal system containing arithmetic. I think the reader has a right to be informed about the present state of affairs. (Gödel, 2003b, p. 147)

What he is referring to, of course, is Alan Turing's analysis in 1937 of the concept of effective computation procedure by means of what we now call Turing machines; Gödel had readily embraced Turing's explication after having rejected earlier proposals by Church and Herbrand. But it wasn't until eight years after the debacle with Nagel and Newman that Gödel spelled out the connection with formal systems and his own work. That was in a postscript he added to the 1965 reprinting of his Princeton lectures in the volume *The Undecidable*, edited by Martin Davis (and reproduced in Gödel, 1986, p. 369, my emphasis):

> In consequence of later advances, in particular of the fact that, due to A. M. Turing's work, a precise and unquestionably adequate definition of

> the general concept of formal system can now be given, the existence of undecidable arithmetical propositions and the non-demonstrability of the consistency of a system in the same system can now be proved rigorously for every consistent formal system containing a certain amount of finitary number theory.
>
> Turing's work gives an analysis of the concept of "mechanical procedure" (alias "algorithm" or "computation procedure" or "finite combinatorial procedure"). This concept is shown to be equivalent with that of a "Turing machine." *A formal system can simply be defined to be any mechanical procedure for producing formulas, called provable formulas.* For any formal system in this sense there exists one in the [usual] sense ... that has the same provable formulas (and likewise vice versa).

As we will see, there's much more to this identification than meets the eye; in fact, in my view it is the source of a crucial misdirection in the minds vs. machines disputes that we shall take up below. By comparison, Nagel and Newman distinguish the consequences of the incompleteness theorems for axiomatic systems and for "calculating machines" in their concluding reflections as follows:

> [Gödel's theorems] show that there is an endless number of true arithmetical statements which cannot be formally deduced from any specified set of axioms .... It follows, therefore, that an axiomatic approach to number theory ... cannot exhaust the domain of arithmetical truth. (Nagel and Newman, 1956b, p. 1694)
>
> Gödel's conclusions also have a bearing on the question whether calculating machines can be constructed which would be substitutes for a living mathematical intelligence. Such machines, as currently constructed and planned, operate in obedience to a fixed set of directives built in, and they involve mechanisms which proceed in a step-by-step manner. But in the light of Gödel's incompleteness theorem, there is an endless set of problems in elementary number theory for which such machines are inherently incapable of supplying answers, however complex their built-in mechanisms may be and however rapid their operations. (*ibid.*, p. 1695)

Of course, Gödel could charitably read their informal idea of calculating machines simply to be explicated by the notion Turing machines, but as we saw above, he would have gone beyond that to stress that there is no essential difference between these results. The first thing that Gödel was surely reacting to on the philosophical rather than the technical side was the statement by Nagel and Newman that

> [the incompleteness theorems] seem to show that the hope of finding an absolute proof of consistency for any deductive system in which the whole of arithmetic is expressible cannot be realized, if such a proof must satisfy the finitistic requirements of Hilbert's original program. (*ibid.*, p. 1694)

What must have annoyed Gödel was that this, together with their further reflections on the significance of the incompleteness theorems concerned matters to which he had given a good deal of thought over the years, in some respects with overlapping conclusions, in others contrary ones, but in all cases in much greater depth and with much greater care and precision. The trouble was that almost none of his thought on these questions had been published. With respect to the significance of the incompleteness theorems for Hilbert's finitist consistency program this was something that he had made only one brief cautious comment about at the end of his 1931 paper on undecidable propositions. But it was a matter he kept coming back to all through his life, as we found when we unearthed unpublished lectures and seminar presentations in his *Nachlass*. The first time his further views on this would begin to come out in print would be in 1958, a year after the imbroglio with Nagel and Newman.[2] Since my main concern in the remainder of this article is with the minds vs. machines debate as it relates to the incompleteness theorems, I shall leave that matter at that.

3. Gödel's Concerns, Part II: Significance of the Incompleteness Theorems for the Minds vs. Machines Debate

The above quotations from Nagel and Newman all come from their final three paragraphs on the "far reaching import" of the incompleteness theorems. (My guess is that Nagel was largely responsible for their formulation, but I have no specific evidence for that.) These, with some rewording but no essential change in content, were to form the entire last chapter, entitled "Concluding reflections," of Nagel and Newman (1958); however, it is only the 1956 version that Gödel would have seen and to which he would have reacted and so it is from there that I continue to draw the quotations. We now continue with the remaining parts of these paragraphs that concern the potentialities of human thought vs. the potentialities of computing machines, already signaled above in their discussion of "calculating machines":

> It may very well be the case that the human brain is itself a "machine" with built-in limitations of its own, and that there are mathematical problems which it is incapable of solving. Even so, the human brain appears to embody a structure of rules of operation which is far more powerful than the structure of currently conceived artificial machines. (Nagel and Newman, 1956b, p. 1695)

And (from the third paragraph):

> The discovery that there are formally indemonstrable arithmetic truths does not mean that there are truths which are forever incapable of becoming known, or that a mystic intuition must replace cogent proof. It does mean that the resources of the human intellect have not been, and cannot be,

fully formalized, and that new principles of demonstration forever await invention and discovery .... Nor do the inherent limitations of calculating machines constitute a basis for valid inferences concerning the impossibility of physico-chemical explanations of living matter and human reason. The possibility of such explanations is neither precluded nor affirmed by Gödel's incompleteness theorem. The theorem does indicate that in structure and power the human brain is far more complex and subtle than any nonliving machine yet envisaged. (*ibid.*, p. 1695)

And finally,

> Gödel's work is a remarkable example of such complexity and subtlety. It is an occasion not for dejection because of the limitations of formal deduction but for a renewed appreciation of the powers of human reason. (*ibid.*, p. 1695)

Here again Gödel would have reacted with a "been there, done that" annoyance, since he had already laid out his thoughts in this direction fifteen years earlier in what is usually referred to as his Gibbs lecture, "Some basic theorems on the foundations of mathematics and their implications" (Gödel, 1951).[3] But once more this is something he had never published, though he wrote of his intention to do so soon after delivering the lecture; in fact it never appeared in his lifetime. After Gödel died, the text languished with a number of other important essays and lectures in his *Nachlass* until it was retrieved by our editorial group for publication in Volume III of the *Gödel Collected Works* (1995, pp. 304–323).

There are essentially two parts to the Gibbs lecture, both drawing conclusions from the incompleteness theorems. The first part concerns the potentialities of mind vs. machines for the discovery of mathematical truths, and it is that part that we should compare with Nagel and Newman's reflections. The second part is an argument aimed to "disprove the view that mathematics is only our own creation," and thus to support some version of platonic realism in mathematics. George Boolos wrote a very useful introductory note to both parts of the Gibbs lecture in Vol. III of the *Gödel Collected Works*; more recently I have published an extensive critical analysis of the first part, under the title "Are there absolutely unsolvable problems? Gödel's dichotomy" (Feferman, 2006a), and I will be drawing on that in the following.

What I call Gödel's dichotomy is the following statement that he highlighted in the first part of the Gibbs lecture:

> *Either ... the human mind (even within the realm of pure mathematics) infinitely surpasses the powers of any finite machine, or else there exist absolutely unsolvable diophantine problems.* (Gödel, 1951, p. 310, Gödel's emphasis)

By a diophantine problem is meant a proposition of elementary number theory (aka first order arithmetic) of a relatively simple arithmetical form whose truth or falsity is to be determined; its exact description is not important to us. Gödel showed that the consistency of a formal system is equivalent to a diophantine problem, to begin with by expressing it in the form that no number codes a proof of a contradiction.[4] According to Gödel, his dichotomy is a "mathematically established fact" which is a consequence of the incompleteness theorem. However, all that he says by way of an argument for it is the following:

> [I]f the human mind were equivalent to a finite machine, then objective mathematics not only would be incompletable in the sense of not being contained in any well-defined axiomatic system, but moreover there would exist absolutely unsolvable problems ..., where the epithet "absolutely" means that they would be undecidable, not just within some particular axiomatic system, but by *any* mathematical proof the mind can conceive. (Gödel, 1951, p. 310, Gödel's emphasis)

By a finite machine here Gödel means a Turing machine, and by a well-defined axiomatic system he means an effectively specified formal system; as explained above, he takes these to be equivalent in the sense that the set of theorems provable in such a system is the same as the set of theorems that can be effectively enumerated by such a machine. Thus, to say that the human mind is equivalent to a finite machine "even within the realm of pure mathematics" is another way of saying that what the human mind can in principle demonstrate in mathematics is the same as the set of theorems of some formal system. By objective mathematics Gödel means the totality of true statements of mathematics, which includes the totality of true statements of first-order arithmetic. The assertion that objective mathematics is incompletable is at first sight simply a consequence of the second incompleteness theorem in the form that for any consistent formal system S containing a certain basic system S_0 of (true) arithmetic, the number-theoretical statement $Con(S)$ that expresses the consistency of S is true but not provable in S.

Examined more closely, Gödel's argument is that if the human mind were equivalent to a finite machine, or—what comes to the same thing—an effectively presented formal system S, then there would be a true statement that could never be humanly proved, namely $Con(S)$. So that statement would be absolutely undecidable by the human mind, and moreover it would be equivalent to a diophantine statement. Note however, the tacit assumption that the human mind is consistent; otherwise, it is equivalent to a formal system in a trivial way, namely one that proves all statements. Actually, Gödel apparently accepts a much stronger assumption, namely that we prove only true statements; but for his argument, only the weaker assumption is necessary (together of course with the assumption that the basic system of arithmetic S_0 has been humanly accepted). Also, Gödel's sketch to establish his dichotomy should be modified slightly, as follows: either humanly demonstrable mathematics is contained in some consistent formal system S, or not. If it is, then $Con(S)$ is an absolutely undecidable (diophantine) statement. If not, then for each consistent formal system *S*, there is a humanly provable statement that is not provable in *S*, i.e. the "human mind ... infinitely surpasses the powers of any finite machine."

Note as well that Gödel's dichotomy is not a strict one as it stands; Gödel himself asserts that "the case that both terms of the disjunction are true is not excluded, so that there are, strictly speaking, three alternatives." This could happen, for example, if the human mind infinitely surpasses finite machines with respect to certain diophantine problems, but not with respect to all of them. And if we drop the word "diophantine" from its statement, it might be that the human mind could settle all arithmetical problems, but not all problems of higher mathematics. In fact, some logicians conjecture that Cantor's Continuum Problem is an absolutely undecidable problem of that type.

Be that as it may, how does Gödel's conclusion differ from that of Nagel and Newman? They speak of calculating machines "as currently constructed and planned, [that] operate in obedience to a fixed set of directives built in, and [that] involve mechanisms which proceed in a step-by-step manner," where Gödel speaks more precisely of Turing machines. Still, to be charitable, I think that is a reasonable interpretation of what Nagel and Newman had in mind. Secondly, they do not make the connection between formal systems and calculating machines, where Gödel sees these as amounting to the same thing.[5] But the essential difference is that they seem to come down in favor of the first, anti-mechanist, disjunct of the dichotomy when they say that "[by] Gödel's incompleteness theorem, there is an endless set of problems in elementary number theory for which such machines are inherently incapable of supplying answers, however complex their built-in mechanisms may be and however rapid their operations" and "the human brain appears to embody a structure of rules of operation which is far more powerful than the structure of currently conceived artificial machines"; furthermore, in the third paragraph they say that "the resources of the human intellect have not been, and cannot be, fully formalized, and ... new principles of demonstration forever await invention and discovery."

There is a lot of evidence outside of the Gibbs lecture that Gödel was also convinced of the anti-mechanist position as expressed in the first disjunct of his dichotomy. That is supplied, for example, in his informal communication of various ideas about minds and machines to Hao Wang, initially in the book, *From Mathematics to Philosophy* (1974, pp. 324–326), and then at greater length in *A Logical Journey. From Gödel to Philosophy* (1996, esp. chp. 6). So why did Gödel not state that outright in the Gibbs lecture instead of the more cautious disjunction in the dichotomy? The reason was simply that he did not have an unassailable proof of the falsity of the mechanist position. Indeed, he there says:

> [It is possible that] the human mind (in the realm of pure mathematics) is equivalent to a finite machine that, however, is unable to understand completely its own functioning. (Gödel, 1995, p. 309)

And in a related footnote, despite his views concerning the "impossibility of physico-chemical explanations of … human reason" he says that:

> [I]t is conceivable … that brain physiology would advance so far that it would be known with empirical certainty
>
> 1. that the brain suffices for the explanation of all mental phenomena and is a machine in the sense of Turing;
> 2. that such and such is the precise anatomical structure and physiological functioning of the part of the brain which performs mathematical thinking. (*ibid.*, p. 309)

And in the next footnote, he says:

> [T]he physical working of the thinking mechanism could very well be completely understandable; the insight, however, that this particular mechanism must always lead to correct (or only consistent) results would surpass the powers of human reason. (*ibid.*, p. 310)

Some twenty years later, Georg Kreisel made a similar point in terms of formal systems rather than Turing machines:

> [I]t has been clear since Gödel's discovery of the incompleteness of formal systems that we could not have mathematical evidence for the adequacy of any formal system; but this does not refute the possibility that some quite specific system … encompasses all possibilities of (correct) mathematical reasoning … *In fact the possibility is to be considered that we have some kind of nonmathematical evidence for the adequacy of such [a system].* (Kreisel, 1972, p. 322, my emphasis)

4. Critiquing the Minds vs. Machines Debate

I shall call the genuine possibility entertained by Gödel and Kreisel, the mechanist's empirical defense (or escape hatch) against claims to have proved that mind exceeds mechanism on the basis of the incompleteness theorems. The first outright such claim was made by Lucas in his 1961 article, "Minds, Machines and Gödel." Both Benacerraf and Putnam soon objected to his argument on the basis that Lucas was assuming it is known that one's mind is consistent. Lucas, in response (1996), has tried to shift the burden to the mechanist: "The consistency of the machine is established not by the mathematical ability of the mind, but on the word of the mechanist," a burden that the mechanist can refuse to shoulder by simply citing his empirical defense.

Roger Penrose is the other noted defender of the Gödelian basis for anti-mechanism, most notably in his two books, *The Emperor's New Mind* (1989), and

Shadows of the Mind (1994). Sensitive to the objections to Lucas, he claimed in the latter only to have proved something more modest (and in accord with experience) from the incompleteness theorems: "Human mathematicians are not using a knowably sound algorithm in order to ascertain mathematical truth" (1994, p. 76). But later in that work after a somewhat involved discussion, he came up with a new argument purported to show that the human mathematician cannot even consistently believe that his mathematical thought is circumscribed by a mechanical algorithm (1994, secs. 3.16 and 3.23). Extensive critiques have been made of Penrose's original and new arguments in an issue of the journal *Psyche* (vol. 2, no. 1, 1996), to which he responded in the same issue. And more recently, Stewart Shapiro (2003) and Per Lindström (2001; 2006) have carefully analyzed and then undermined his "new argument." But Penrose has continued to defend it, as he did in his public lecture for the Gödel Centenary Conference held in Vienna in April 2006.

Historically, there are many examples of mathematical proofs of what cannot be done in mathematics by specific procedures, e.g. the squaring of the circle, or the solution by radicals of the quintic, or the solvability of the halting problem. But it is hubris to think that by mathematics alone we can determine what the human mind can or cannot do in general. The claims by Gödel, Lucas, and Penrose to do just that from the incompleteness theorems depend on making highly idealized assumptions both about the nature of mind and the nature of machines. A very useful critical examination of these claims and the underlying assumptions has been made by Shapiro in his article, "Incompleteness, Mechanism and Optimism" (1998), among which the following. First of all, how are we to understand the mathematizing capacity of the human mind, since what is at issue is the producibility of an infinite set of propositions? No one mathematician, whose life is finitely limited, can produce such a list, so either what one is talking about is what the individual mathematician could do in principle, or we are talking in some sense about the potentialities of the pooled efforts of the community of mathematicians now or ever to exist. But even that must be regarded as a matter of what can be done in principle, since it is most likely that the human race will eventually be wiped out either by natural causes or through its own self-destructive tendencies by the time the sun ceases to support life on earth.

What about the assumption that the human mind is consistent? In practice, mathematicians certainly make errors and thence arrive at false conclusions that in some cases go long undetected. Penrose, among others, has pointed out that when errors are detected, mathematicians seek out their source and correct them, and so he has argued that it is reasonable to ascribe self-correctability and hence consistency to our idealized mathematician (see Penrose, 1996, pp. 137ff.). But even if such a one can correct all his errors, can he know with mathematical certitude, as required for Gödel's claim, that he is consistent?

As Shapiro points out, the relation of both of these idealizations to practice is analogous to the competence/performance distinction in linguistics.

There are two further points of idealization to be added to those considered by Shapiro. The first of these is the assumption that the notions and statements of mathematics are fully and faithfully expressible in a formal language, so that what can be humanly proved can be compared with what can be the output of a machine. In this respect it is usually pointed out that the only part of the assumption that needs be made is that the notions and statements of elementary number theory are fully and faithfully represented in the language of first-order arithmetic, and that among those only diophantine statements of the form $Con(S)$ for S an arbitrary effectively presented formal system need be considered. But even this idealization requires that statements of unlimited size must be accessible to human comprehension.

Finally to be questioned is the identification of the notion of finite machine with that of Turing machine. Turing's widely accepted explication of the informal concept of effective computability puts no restriction on time or space that might be required to carry out computations. But the point of that idealization was to give the strongest negative results, to show that certain kinds of problems cannot be decided by a computing machine, no matter how much time and space we allow. And so if we carry the Turing analysis over to the potentiality of mind in its mathematizing capacity, to say that mind infinitely surpasses any finite machine is to say something even stronger. It would be truly impressive if that could be definitively established, but none of the arguments that have been offered are resistant to the mechanist's empirical defense. Moreover, suppose that the mechanist is right, and that in some reasonable sense mind is equivalent to a finite machine; is it appropriate to formulate that in terms of the identification of what is humanly provable with what can be enumerated by a Turing machine? Isn't the mechanist aiming at something stronger in the opposite direction, namely an explanation of the mechanisms that govern the production of human proofs?

Here is where I think something new has to be said, something that I already drew attention to in my article on Gödel's dichotomy (2006), but that needs to be amplified. Namely, there is an equivocation involved, that lies in identifying how the mathematical mind works with the totality of what it can prove. Again, the difference is analogous to what is met in the study of natural language, where we are concerned with the way in which linguistically correct utterances are generated and not with the potential totality of all such utterances. That would seem to suggest that if one is to consider any idealized formulation of the mechanist's position at all, it ought to be of the mind as one constrained by the axioms and rules of some effectively presented formal system. Since in following those axioms and rules one has choices to be made at each step, at best that identifies the mathematizing mind with the program for a non-deterministic Turing machine, and not with the set of its enumerable statements (even though that can equally well be supplied by a deterministic Turing machine).

One could no more disprove this modified version of the idealized mechanist's thesis than the version considered by Gödel et al., simply by applying the mechanist's empiricist argument. Nevertheless, it is difficult to conceive of any formal system of the sort with which we are familiar, from Peano Arithmetic (*PA*) up to Zermelo-Fraenkel Set Theory (*ZF*) and beyond, actually underlying mathematical thought as it is experienced. And the experience of the mathematical practitioner certainly supports the conclusion drawn by Nagel and Newman that "mathematical proof does not coincide with the exploitation of a formalized axiomatic method," even if that cannot be demonstrated unassailably as a consequence of Gödel's incompleteness theorems.

5. One Way to Straddle the Mechanist and Anti-Mechanist Positions

As I see it, the reason for the implausibility of this modified version of the mechanist's thesis lies in the concept of a formal system *S* that is currently taken for granted in logical work. An essential part of that concept is that the language of *S* is fixed once and for all. For example, the language of *PA* is determined (in one version) by taking the basic symbols to be those for equality, zero, successor, addition and multiplication $(=, 0, ', +, .)$, and that of *ZF* is fixed by taking its basic symbols to be those for equality and membership $(=, \in)$. This forces axiom schemata that may be used in such systems, such as induction in arithmetic and separation in set theory, to be infinite bundles of all possible substitution instances by formulas from that language; this makes metamathematical but not mathematical sense. Besides that, the restriction of mathematical discourse to a language fixed in advance, even if only implicitly, is completely foreign to mathematical practice.

In recent years I have undertaken the development of a modified conception of formal system that does justice to the openness of practice and yet gives it an underlying rule-governed logical-axiomatic structure; it thus suggests a way, admittedly rather speculative, of straddling the Gödelian dichotomy. This is in terms of a notion of open-ended schematic axiomatic system, i.e. one whose schemata are finitely specified by means of propositional and predicate variables (thus putting the 'form' back into 'formal systems') while the language of such a system is considered to be open-ended, in the sense that its basic vocabulary may be expanded to any wider conceptual context in which its notions and axioms may be appropriately applied. In other words, on this approach, implicit in the acceptance of given schemata is the acceptance of any meaningful substitution instances that one may come to meet, but which those instances are is not determined by restriction to a specific language fixed in advance (see Feferman, 1996; Feferman, 2006c, and Feferman and Strahm, 2000).

The idea is familiar from standard presentations of propositional and predicate logic, where we have such axioms as $P \wedge Q \rightarrow P$ and $(\forall x)P(x) \rightarrow P(a)$ and rules of inference such as $P, P \rightarrow Q \vdash Q$ and $P \rightarrow Q(x) \vdash P \rightarrow (\forall x)Q(x)$ *(for x not free in P)*. We do not conceive of logic as applying to a single subject matter fixed once and for all, but rather to any subject in which we take it that we are dealing with well-defined propositions and predicates; logic is there applied by substitution for the proposition and predicate letters in its axioms and rules of inference. Similarly to allow systems like those for arithmetic and set theory to be applicable no matter what subject matter we happen to deal with, we formulate their basic principles in schematic form such as the Induction Axiom Scheme $P(0) \wedge (\forall x)[P(x) \rightarrow P(x')] \rightarrow (\forall x)P(x)$ in the case of arithmetic, and the Separation Scheme $(\forall a)(\exists b)(\forall x)[x \in b \leftrightarrow x \in a \wedge P(x)]$ in the case of set theory. (The idea for the latter really goes back to Zermelo's conception of the *Aussonderungsaxiom* as applicable to any "definite" predicate.)

But why, it may be (and is) asked, do I insist on the vague idea of an open-ended language for mathematics? Aren't all mathematical concepts defined in the language of set theory? It is indeed the case that the current concepts of working ("pure") mathematicians are with few exceptions expressible in set theory. But there are genuine outliers. For example a natural and to all appearances coherent mathematical notion whose full use is not set-theoretically definable is that of a category; only so-called "small" categories can be directly treated in that way (see Mac Lane, 1971; Feferman, 1977; Feferman, 2006b). Other outliers are to be found on the constructive fringe of mathematics in the schools of Brouwerian intuitionism and Bishop's constructivism whose basic notions and principles are not directly accounted for in set theory with its essential use of classical logic (see Beeson, 1985). And it may be argued that there are informal mathematical concepts like those of knots, or infinitesimal displacements on a smooth surface, or of random variables, to name just a few, which may be the subject of convincing mathematical reasoning but that are accounted for in set theory only by some substitute notions that share the main expected properties but are not explications in the ordinary sense of the word. Moreover, the idea that set-theoretical concepts and questions like Cantor's continuum problem have determinate mathematical meaning has been challenged on philosophical grounds (Feferman and Strahm, 2000). Finally, there is a theoretical argument for openness, even if one accepts the language L of set theory as a determinately meaningful one. Namely, by Tarski's theorem, the notion of truth T_L for L is not definable in L; and then the notion of truth for the language obtained by adjoining T_L to L is not definable in that language, and so on (even into the transfinite).

Another argument that may be made against the restriction of mathematics to a language fixed in advance is historical. Sustained mathematical reasoning had its origins in ancient Greece some 2500 years ago, and mathematical concepts of number and space have undergone considerable evolution since then. Yet even the earliest results have permanent validity, though not necessarily as originally conceived. While Euclidean geometry is no longer considered to be the geometry of actual space, its consequences as an axiomatic development (suitably refined through the work of Pasch, Hilbert, and others), such as the angle sum theorem and Pythagoras' theorem, are as valid now within that context as in Euclid's time. On the other hand, the origins of number theory as represented in Euclid's *Elements*, including the existence of infinitely prime numbers and the fundamental theorem of arithmetic, retain their direct interpretation and validity. But the language of mathematical practice (often mixed with physical concepts) then and through the many following centuries up to the present obviously cannot be identified with the language of set theory. One thing that would instead account for the continuity of mathematical thought throughout its history is the employment of certain underlying formal patterns, such as those indicated above, that could be instantiated by the evolving concepts that have come to fill out mathematics in the process of its development. And there is no reason to believe that this evolutionary process has come to an end; it would be foolish to believe that only that which can be expressed, say, in the language of set theory, will count as mathematics henceforth.

On this picture, in order to straddle the mechanist/anti-mechanist divide at the level considered here, one will have to identify finitely many basic forms of mathematical reasoning which work in tandem to fully constrain and distinguish it. These would constitute the mechanist side of the picture, while the openness as to what counts as a mathematical concept would constitute the anti-mechanist side. The evidence for such would have to be empirical, by showing how typical yet substantial portions of the mathematical corpus are accounted for in those terms while giving special attention to challenging cases. I have suggested steps in that direction (as indicated in Feferman, 2006c), but the program is ambitious and I have only made a start; spelling that out is planned for a future publication. In the meantime, the program itself should be treated as highly speculative, yet—I hope—worthy of serious consideration.

Considered more broadly and apart from the tendentious terms of the mechanism/anti-mechanism debate, I think the goal should be to give an informative, systematic account at a theoretical level of how the mathematical mind works that squares with experience. Characterizing the logical structure of mathematics—what constitutes a proof—is just one aspect of that, important as that may be. Other aspects—the ones that are crucial in making the difficult choices that are necessary for the mathematician to obtain proofs of difficult theorems—such as the role of heuristics, analogies, metaphors, physical and geometric intuition, visualization, etc., have also been taken up and are being pursued in a more or less systematic way by mathematicians, philosophers and cognitive scientists.[6] Without abandoning a basic naturalist stance, I see no usable reductive account of the mathematical experience anywhere on the horizon at the neuro-physiological level let alone more basic physico-chemical levels of the sort contemplated by Nagel and Newman and currently sought by Penrose, among others.

NOTES

1. See Putnam, 1960, for a review of Nagel and Newman, 1956, in which several errors are identified, the most egregious being the misstatement of Gödel's first incompleteness theorem and Rosser's improvement thereof on p. 91 (now Nagel, Newman, and Hofstadter, 2001, p. 99).

2. That was in the article, "Über eine bisher noch nicht benüzte Erweiterung des finiten Standpunktes" (Gödel, 1958). As a sign of his concern with the issues involved, Gödel worked on a revision of that until late in his life (1972). For the full story, see my piece "Lieber Herr Bernays!, Lieber Herr Gödel! Gödel on Finitism, Constructivity and Hilbert's Program" (2008).

3. Gödel's lecture was the twenty-fifth in a distinguished series set up by the American Mathematical Society to honor the 19th century American mathematician, Josiah Willard Gibbs, famous for his contributions to both pure and applied mathematics. It was delivered to a meeting of the AMS held at Brown University on December 26, 1951.

4. In modern terms, consistency statements belong to the class Π_1, i.e. are of the form $\forall x R(x)$ with R primitive recursive.

5. Calculating machines are assimilated more closely to axiomatic systems in the concluding reflections of Nagel and Newman, 1958, p. 100 (now Nagel, Newman, and Hofstadter, 2001, p. 111).

6. See, for example, Fischbein, 1987, Lakoff and Núñez, 2000, Mancosu, Jørgensen, and Pedersen, 2005, and Pólya, 1944, among others. In addition, there is a massive amount of anecdotal evidence for the non-mechanical essentially creative nature of mathematical research; for a small sample with further references, see, e.g., Davis and Hersh, 1981, Ruelle, 2007, and Yandell, 2002.

Eleven

ENTANGLING EFFECTIVE PROCEDURES: FROM LOGIC MACHINES TO QUANTUM AUTOMATA

Rossella Lupacchini

1. The Character of Symbolic Knowledge

In the "Introduction" to his *Philosophy of Symbolic Forms*, Ernst Cassirer writes that "mathematicians and physicists were the first to gain a clear awareness of [the] symbolic character of their basic implements" (1923; 1953, v. 1, p. 75). He refers mainly to Heinrich Hertz, whose *Principles of Mechanics* (1894; 1956) presented the models of a physical theory as "freely created intellectual symbols": any choice of propositions that enables us to construct mechanics through a purely deductive reasoning, with no further appeal to experience, consists of a set of principles of mechanics. Since mathematical formulas provide the formal scheme apt to analyze any physical problem in logical terms, he envisaged the theory as a "systematic grammar": physical reality emerges from its logical structure. Cassirer conceives the development of the sciences as a progressive refinement of their symbolic systems; his aim is to extend Hertz's view of mechanics as a symbolic form to all other regions of the "spiritual creation." He is aware, though, that any attempt to apply a symbolization function sheds light on the most general problem that characterizes gnoseological methodology: the more we advance toward the "symbolic element," the more we move away from the original ground of pure intuition. According to Leibniz (1970), symbolic knowledge is "blind." Is filling the gap between intuitive and symbolic knowledge possible?

Leibniz's answer was negative. He distinguished intuition—as "pure vision"—from any knowledge that, in so far as it must resort to symbols, is blind. Human knowledge, finite and limited, is of this kind. Hence, Leibniz saw human reasoning as a calculus and envisaged the possibility that a machine could accomplish the task of solving any "decision problem." We can regard David Hilbert's *Entscheidungsproblem* as the final version of this long lasting idea. In tune with Leibniz's *Calculemus*, Hilbert believes that "for any precisely formulated mathematical question a unique answer can be found" (Gödel, 193? , p. 164). Kurt Gödel observes that Hilbert's words can only mean that "given an arbitrary mathematical proposition A there exists a proof for A or for *not*-A, where by 'proof' is meant something which starts from evident axioms and proceeds by evident inferences." "Formulated in this way—Gödel remarks—the problem is not accessible for mathematical treatment because it involves the non-mathematical notion of evidence" (Gödel, 193? , p. 164). Once we frame the problem in precise mathematical terms, we find out that the answer, even in the domain of number theory, is negative. For Gödel, this result does not mean that the problem, in its original formulation, has a negative answer, but "that through the transition from evidence to formalism something is lost," since

> the number-theoretic questions which are undecidable in a given formalism are always decidable by evident inferences not expressible in the given formalism. As to the evidence of these new inferences, they turn out to be exactly as evident as those of the given formalism. So the result is rather that it is not possible to formalise mathematical evidence even in the domain of number theory, but the conviction about which Hilbert speaks [that a mathematical proof can be given for the *Entscheidungsproblem*] remains entirely untouched. (*ibid.*, p. 164)

Another way of expressing the result, Gödel continues, is this: "It is not possible to mechanise mathematical reasoning, i.e., it will never be possible to replace the mathematician by a machine." A conclusion that indicates we cannot attain some humanly knowable mathematical truths through a machine contrasts with Leibniz's *Calculemus*.[1] Notice that, in Leibniz's world, what is "evident" to the divine mathematician may not be equally evident to the human mathematician; certain and perfect knowledge lies out of human reach: "there can be relations which, however far an analysis is continued, will never reveal themselves sufficiently with certainty, and are seen perfectly only by him whose intellect is infinite." (Leibniz, 1999, p. 136)

2. On the Notion of Effective Procedure

What do we lose in the transition from evidence to formalism? The question requires a mathematically satisfactory definition of the general concept of formal system. As Gödel emphasizes, "the essential difficulty of this definition is the notion of mechanical procedure which comes in and which needs further specification" (Gödel, 193? , p. 166). In 1951, Gödel admits that "the most satisfactory way ... [of arriving at such a definition] is that of reducing the concept of finite procedure to that of a machine with a finite number of parts, as has been done by the British mathematician Turing" (Gödel, 1951, pp. 304–305). Gödel himself had given a precise mathematical definition of the notion of finite (mechanical) procedure, or "effective computability," but only Turing's work would persuade him of the correctness and the "absoluteness" of the result he had reached.[2]

As to correctness, the issue at stake is to establish an adequate correspondence between a mathematical and an intuitive notion. In line with Hilbert's axiomatic method, expressing an intuitive notion in mathematical terms amounts to stating a set of axioms. Accordingly, in 1934 Gödel hopes to express the properties of the notion of effective computability, which we generally accept, in a set of axioms.[3] Some years later, though, he defines a computable function by "general recursion," using a finite number of recursive equations as admissible postulates and two elementary substitution operations as inference rules. "To characterize the computations of values of 'recursive specified' number theoretic functions by uniform and arithmetically meaningful steps," (Sieg, 2006, p. 192) we do not need a set of axioms for effective calculable, but an equational calculus. Turing has proved beyond doubt, Gödel asserts, that such a calculus provides the correct definition of mechanical computability:

> He has shown that the computable functions defined in this way are exactly those for which you can construct a machine with a finite number of parts $n_1, \ldots, n_r$ which will do the following thing. If you write down any number on a slip of paper and put the slip of paper into the machine and turn the crank, then after a finite number of turns the machine will stop and the value of the function for the argument $n_1, \ldots, n_r$ will be printed on the paper. (Gödel, 193? p. 168)

How can these turns correspond to the steps we need to go through to compute the values of "recursive specified" number theoretic functions? Why are the computable functions that we define that way (*via* equational calculus) exactly those that a machine can compute? Are the rules of inference we have chosen so "evident" to justify the restrictions on the steps that computations permit? As Sieg's detailed investigation clarifies, Gödel's *analytical* claim that the computable functions we have defined *via* the equational calculus are exactly those mechanically computable rests on "Turing's proof" of the equivalence of general recursiveness with Turing's computability. It rests on the Church-Turing thesis that equates effective computability with computability on a "Turing machine." Turing's conceptual analysis, however, which "brings in human computers in a crucial way and exploits the limitations of their processing capacities, when proceeding mechanically" (Sieg, 2006, p. 201), entails an "appeal to intuition." By considering a *human* computer, Turing determines which simple mechanical operations on symbolic configurations are indispensable in computing. We can ground the restrictive conditions on computability in the limitations of the relevant capacities that the computer at hand displays. We can convert the symbolic processes underlying the elementary operations that a (human) computer perform in a finite set of instructions for a "Turing machine." Sieg points out that none of those conditions would work for the steps that an equational calculus requires. To view a symbolic structure as a model of "human" computing means to anchor the analytical theory of computability in the capacities of the agent available for its execution. This is a turning point that is crucial for subsequent technology and for theoretical understanding. Undoubtedly, the equivalence between general recursiveness and Turing computability plays a key role in Gödel's argument. Nonetheless, Gödel does not recognize

> the genuinely distinctive character of Turing's analysis, i.e., the move from arithmetically motivated calculations to general symbolic processes that underlie them. Most importantly in the given intellectual context, these processes have to be carried out programmatically by human beings: the *Entscheidungsproblem* had to be solved by us in a mechanical way; it was the normative demand of radical intersubjectivity between humans that motivated the step from axiomatic to formal systems. (Sieg, 2006, p. 201)

The property of being computable in *any formal system* (containing arithmetic) endows the general recursive functions with a character of absoluteness that convinces Gödel that he has just defined an important class of functions: "this importance is largely due to the fact that with this concept one has for the first time succeeded in giving an absolute definition of an interesting epistemological notion, i.e., one not depending on the formalism chosen" (Gödel, 1946, p. 150). Although this notion does not depend on a particular formalism (a set of axioms or principles), it ought to grasp some fundamental feature of all formalisms, of what makes a system "formal." Gödel draws from Turing's analysis the convincing evidence for an effective (mechanical) procedure that can produce provable formulas within any formal system:

> Turing's work gives an analysis of the concept of "mechanical procedure" (alias "algorithm" or "computation procedure" or "finite combinatorial procedure"). This concept is shown to be equivalent with that of a "Turing machine." A formal system can simply be defined to be any mechanical procedure for producing formulas, called provable formulas. (Gödel, 1964, pp. 369–370)

And yet Gödel's later concern is about a philosophical error in Turing's argument. The thesis that "mental procedures cannot go beyond mechanical procedures" would imply the same kind of limitation on human mind:

> What Turing disregards completely is the fact that *mind, in its use, is not static, but constantly developing*, i.e. that we understand abstract terms more and more precisely as we go on using them, and that more and more abstract terms enter the sphere of our understanding .... [A]lthough at each stage the number and precision of the abstract terms at our disposal may be *finite*, both (and, therefore, also Turing's number of *distinguishable states of mind*) may *converge towards infinity* in the course of the application of the procedure. (Gödel, 1972a, p. 306)

Is Gödel's criticism sound?

In the 1951 Gibbs Lecture, Gödel speculates on a few philosophical implications concerning the possible limitations of human thought that descend from his incompleteness theorems. He bases his discussion on a preliminary distinction between *objective mathematics*, which consists of "the body of those mathematical propositions which hold in an absolute sense, without any further hypothesis," and *subjective mathematics*, which consists of all humanly demonstrable truths ("all those theorems whose truth is demonstrable in *some* well-defined system of axioms all of whose axioms are recognized to be objective truths and whose rules preserve objective truth" (Feferman, 2006a, p. 135). Do they coincide? Solomon Feferman rephrases Gödel's answer as follows:

> If they do, then demonstrations in subjective mathematics are not confined to any one system of axioms and rules, though each piece of mathematics is justified by *some* such system. If they do not, then there are objective truths that can never be humanly demonstrated, and those constitute absolutely unsolvable problems. (2006, p. 136–37)

In light of the second incompleteness theorem, Gödel comes to the following conclusion:

> [I]f the human mind were equivalent to a finite machine, then objective mathematics not only would be incompletable in the sense of not being contained in any well-defined axiomatic system, but moreover there would exist absolutely unsolvable problems ..., where the epithet "absolutely" means that they would be undecidable, not just within some particular axiomatic system, but by any mathematical proof the mind can conceive. (Gödel, 1951, p. 310)

The question is how to conceive a notion of "decidability" that is disentangled from any effective procedure. The deterministic character of a Turing machine satisfies Gödel's requirement for a limited freedom in the activity of the mathematician: "If anything like creation exists at all in mathematics, then what any theorem does is exactly to restrict the freedom of creation" (*ibid.*, p. 314). Nevertheless, while a Turing machine, which is derivative of human computing, is anchored in space and time, Gödel's objective mathematics involves a "reality" that lies outside of space and time:

> It is correct that a mathematical proposition says nothing about the physical or psychical reality existing in space and time, because it is true already owing to the meaning of the terms occurring in it, irrespectively of the world of real things. What is wrong, however, is that the meaning of the terms (that is, the concepts they denote) is asserted to be something man-made and consisting merely in semantical conventions. The truth, I believe, is that these concepts form an objective reality of their own, which we cannot create or change, but only perceive and describe. (*ibid.*, p. 320)

What does such an objective reality mean? How can we "perceive and describe" an objective reality outside of space and time? The philosophical consequences for mathematics, which Gödel derives from the incompleteness theorems, are not so much opposed to possible limitations of human mind as to "materialistic philosophy." Gödel intend them to support "some form or other of Platonism or 'realism' as to the mathematical objects" (1951, p. 311).

From a different point of view, if we do not deem mathematics—as a "symbolic form"— an objective reality, or a free creation, but an "activity" that either a sensitive or an intellectual experience triggers, it becomes, detached as it is from any symbolization function, meaningless beyond the limits of the "agent" involved. In this perspective, the strength of Turing's analysis lies in his questioning the very evidence of effective procedures and in his focusing on the bare minimum we need to carry them out. In his 1954 paper, Turing takes a step forward with respect to his 1936 paper and makes the definition of an effective procedure independent from the Turing machine. He views a "systematic" procedure just as "a puzzle *in which there is never more than one possible move in any of the positions which arise and in which some significance is attached to the final result*" (Turing, 1954, p. 590). Starting from specific examples of puzzles—"the task of proving a given mathematical theorem within an axiomatic system is a very good example of a puzzle" (*ibid.*, p. 587)—he defines a *normal form* for puzzles as a "substitution type puzzle." Once he has described a (position) state of a puzzle by sequences of symbols in a row, the question that we have to answer in order to fix the appropriate state transformations, Turing says, is:

> "What sort of rules should one be allowed to have for rearranging the symbols or counters?" In order to answer this one need to think about what kinds of processes ever do occur in such rules, and, in order to reduce their number, to break them up into simpler processes. Typical of such processes are counting, copying, comparing, substituting. (*ibid.*, p. 588)

If we express the rules of a puzzle in such a way that takes into account the additional jottings and markers we need when those processes involve many symbols, then we can make those symbols into *mere substitutions*. It follows that "given any puzzle we can find a corresponding substitution puzzle which is equivalent to it in the sense that given a solution of the one we can easily use it to find a solution of the other" (*ibid.*, p. 588). Turing notes that the status of this claim falls between a theorem and a definition:

> In so far as we know *a priori* what is a puzzle and what is not, the statement is a theorem. In so far as we do not know what puzzles are, the statement is a definition which tells us something about what they are. One can of course define a puzzle by some phrase beginning, for instance, "A set of definite rules…," but this just throws us back on the definition of "definite rules." Equally one can reduce it to the definition of "computable function"

> or "systematic procedure." A definition of any one of these would define the rest. (*ibid.*, p. 588–89)

By *reductio ad absurdum*, he then shows that there is no systematic procedure for deciding whether substitution puzzles come out or not. The decision problem for puzzles is unsolvable. In more general terms, we can say that "no systematic method of proving mathematical theorems is sufficiently complete to settle every mathematical question, yes or no." Turing's conclusive remarks concern the "negative" character of these results as they "may be regarded as going some way towards a demonstration, within mathematics itself, of the inadequacy of 'reason' unsupported by common sense" (*ibid.*, p. 595). Does this amount to claim that we cannot settle mathematical evidence as a substitution puzzle or, more dramatically, that a (Turing) machine cannot replace the mathematician? It is worth recalling that, by giving a negative answer to the *Entscheidungsproblem*, the Turing machine rescues mathematics from decidability. We need to discuss to what extent decidability is to be a characteristic feature of mathematical knowledge.

3. Intelligent Machinery

As Turing intends his machine to prove that no systematic procedure can solve the decision problem, he goes on to generalize the concept and to define a *universal* machine as an "automaton" capable of imitating any "one purpose" Turing machine and, therefore, capable of executing all possible computations (Davis, 2000). The crucial idea is that when we feed the universal machine the set of instructions that describe a special Turing machine, the first simulates the second. The universal machine can execute any task that any special Turing machine can. Notice that this *universality* would play a key role in designing actual digital computers. A digital machine is universal in the sense that we can "program" it to imitate any calculating machine. The question then arises as to whether we could also comprise in a finite number of symbols and rules activities that are more sophisticated than calculations, say activities that we may judge "intelligent." Their nature deeply intrigued Turing, but he, unlike Gödel, believed that in the end mental processes are the consequence of physical processes. In 1951, Turing argues that if we accept that brains are a sort of machine, it will follow that a digital computer, if we suitably program it, will behave like a brain: "It is not altogether unreasonable to describe digital computers as brains" (1951, p. 482). As for the notion of puzzle, Turing could have said, "In so far as we do not know what brains are, the statement is a definition which tells us something about what they are."

John von Neumann shares Turing's ideas about computers and brains and, in order to understand complex phenomena such as those that life and intelligence entail, he praises the universal machine, or the "universal automaton," as the most appropriate starting point. In particular, von Neumann focuses on the concept of "complication" and raises the question of how complex an automaton ought to be in order to be "self-reproductive." In von Neumann's opinion, the main result of Turing's theory is to have shown that there is "an automaton which is at least as effective as any conceivable automaton, including, for example, one of twice its size and complexity" (1948, p. 314). The underlying principle is quite simple: "a completely general description of any conceivable automaton can be given in a finite number of words." Exploiting Turing's theory, Warren McCulloch and Walter Pitts also prove that by an appropriate finite "neural network" we can realize any real or imagined mode of behavior that we can completely and unambiguously put into words. What remains to be proved is whether we can put completely and unambiguously into words every existing mode of behavior.[4]

In addressing the problem of the "self-reproduction" of automata, in one respect only does von Neumann regard Turing's notion of universal automaton as too narrow, namely because it is a computing machine whose output is "a piece of tape with zeros and ones on it." The construction of a *self-reproductive* automaton needs a universal machine whose output is another automaton: "There is no difficulty in principle in dealing with this broader concept and in deriving from it the equivalent of Turing's result" (*ibid.*, p. 315). He formulates the problem of self-reproduction as follows:

> Can one build an aggregate out of ... [elementary parts] in such a manner that if it is put into a reservoir, in which there float all these elements in large numbers, it will then begin to construct other aggregates, each of which will at the end turn out to be another automaton exactly like the original one? (1948, p. 315)

The construction can proceed by steps, by connecting two automata, A and B, with a control mechanism C. The automaton A (constructor) can construct any automaton X whose description, in terms of appropriate functions, feeds it. The automaton B (reproducer) can make a copy of any instruction I (which is an aggregate of elementary parts). As to the control mechanism C:

> (*i*) when the instruction I_x feeds A, C will cause A to construct the automaton X described by I_x;
>
> (*ii*) C will cause B to copy the instruction I_x and insert the copy of I_x in the automaton X just constructed by A, $X[I_x]$;
>
> (*iii*) C will separate this construction $X[I_x]$ from $A+B+C$.

Call D the total aggregate $A+B+C$; the point is to provide D with an instruction that describes D and that A can execute. By Turing's principle, we can give the description of any automaton in a finite number of words and they become an instruction for a universal automaton. By inserting the instruction I_D that describes D into A within D, the aggregate $A[I_D]+B+C$ will come out. This construction, call it E, is a "self-reproductive automaton."

Self-reproductive automata may be of crucial interest in investigating the "mechanisms" of life, but what does a machine capable of showing intelligence amount to? How can we feed the machine a program that ends up doing something "original" or "surprising"? Turing tackles these questions in his 1947 "Lecture to the London Mathematical Society," where he explores the possible physical implementations of his model of universal machine. Let us suppose, Turing suggests, that we have set up a machine with some initial instruction tables, which we have so constructed to be capable of altering its instructions out of all recognition and, as a result of this, capable of executing tasks in a more efficient manner. When this happens, are we not obliged to regard the machine as showing intelligence? We need "a machine that can learn from experience," and to do so it should be able to alter its instructions and also to *make mistakes*:

> It has ... been shown that with certain logical systems there can be no machine which will distinguish provable formulae of the system from unprovable .... Thus if a machine is made for this purpose it must in some cases fail to give an answer. On the other hand if a mathematician is confronted with such problem he would search around a[nd] find new methods of proof, so that he ought eventually to be able to reach a decision about any given formula. (Turing, 1947, p. 393)

Against this argument, Turing suggests that we should give the machine some fair play by introducing some *random* element: "Instead of it sometimes giving no answer we could arrange that it gives occasional wrong answers" (*ibid.*, p. 394). We can formulate the issue of machine proving as follows. We can regard different machines as different proof-finding machines. The question is then whether we can implement on a machine the very procedure that we need to apply when choosing between different machines. According to Jack Copeland, we can view a mathematician as a proof-finding machine: "When intuition supplies the mathematician with new means of proof, he or she becomes a different proof-finding machine, capable of a larger set of proofs" (Turing, 2004b, p. 264). Does this imply a transformation of a proof-finding machine into a different one? A "partially random learning machine," which "emulates the 'activity of the intuition' in its walk through the space of proof-finding Turing machines" (*ibid.*, p. 265) appears to be the case at hand here. The problems we want to solve motivate the search for new methods of proof, and yet in so far as such search involves some random elements, a machine might perform it.

For Turing the relation between different machines constitutes an intriguing question: "One imagines different machines allowing different sets of proofs, and by choosing a suitable machine one can approximate 'truth' by 'probability' better than with a less suitable machine, and can in a sense approximate it as well as you please."[5] What happens when we build up a machine from some kind of standard components in a comparatively unsystematic way? "Machines which are largely random in their construction in this way—says Turing—will be called 'unorganized machines'"(1948, p. 416). The question of choice between different machines results in the description of a procedure that, combining *discipline* and *initiative*, represents the evolution of the systems of rules as a non-deterministic process. Turing maintains that "it would be possible for the machine to try changes of behavior, and accept or refuse them," and along this line he comes first to the idea of *casual* machines, and then to the idea of "non-organized machines with interference." An increase of information produces interference on whose basis the system sets the rules that lead to the (partial) organization of the non-organized machine and to stability. We regard the process as indefinitely extensible, in the sense that incoming information, namely *interference from outside*, may lead from a Turing machine to another one, and so on in an unlimited progression. Turing then considers machines in which interference from outside is not the rule but the exception. He distinguishes two kinds of interference: "screwdriver interference," which consists in removing parts of the machine and replacing them with others; "paper interference," which consists in the mere communication of information to the machine that alters its behavior. The following considerations illustrate the key role of interference:

> We may say then that in so far as a man is a machine he is one that is subject to very much interference .... He is in frequent communication with other men, and is continually receiving visual and other stimuli which themselves constitute a form of interference. It will only be when the man is "concentrating" with a view to eliminating these stimuli or "distractions" that he approximates a machine without interference ... although a man when concentrating may behave like a machine without interference, his behaviour when concentrating is largely determined by the way he has been conditioned by previous interference. (*ibid.*, p. 421)

Turing goes on to describe the elementary components and the system of relations that, in abstract terms, characterize the behavior of a machine capable of modifying itself as a reaction to an external intervention or to internal procedures (the system of rules governing computation): it amounts to non-deterministic maneuvers that the system elaborates to treat the incoming information and to unpredictable processes. Such a device should encompass a complex interference among the computing system, the incoming information, the environment, and the adaptation of the system; it should be able to cope with an increasing informational stream. No doubt, the abstract mathematical model of this "machine with interference" is utterly different from the classical one. A classical system does not admit interference with the initial information in input or with the set of rules that process that information; even the universal machine, which has infinite applications, is not apt to change the initial stock of instructions and data. To put it differently, it is a closed system. By contrast, we can view a device that can change the program throughout its execution as a "learning system" that, beside its capability of performing any computable task, is also capable of modifying its rules.

The main point Turing makes is that "if it is accepted that real brains are ... a sort of machine it will follow that our digital computer, suitably programmed, will behave like a brain" in some intelligent way (1951, p. 483). But he is aware of the obstacles on the road. We may frame one of these obstacles in terms of the controversy between free will and determinism: "To behave like a brain seems to involve free will, but the behavior of a digital computer, when it has been programmed, is completely determined" (1951, p. 484). Turing says that our feeling of possessing free will is either an illusion or that the evidence of its presence in our behavior is missing: "Whichever is the correct alternative, it is certain that a machine which is to imitate a brain must appear to behave as if it had free will, and it may well be asked how this is to be achieved. One possibility is to make its behavior depend on something like a roulette wheel or a supply of radium." A possibility that might remind us of the peculiar destiny of Schrödinger's cat.

4. Formal Systems and Quantum Physics

By stretching the ideal of *calculemus*—"to 'atomize' mathematical reasoning into such tiny steps that *nothing* is left to the imagination, nothing is left out!" (Chaitin, 2002, pp. 17–18)—on the one hand the Turing machine guarantees to mathematics its "existence" *via* undecidability; on the other, it requires randomness and interference in order to play the role of a "mathematician."[6] Quantum physics imparts a similar lesson: "the qualities possessed by the system as a whole cannot be split up and located—a little bit here, and a little bit there": the microscopic dissection of objects cannot discover them (Eddington, 1928, p. 107). When we perform a measurement, the interaction between the system and a measuring device has a perturbative character. Determining the value of a physical quantity—an "observable"—affects not only the "measurer," changing its "state of knowledge," but also the system because, as we obtain a "pure" state of the observable we measure, it must release any other "incompatible" observables from a pure state. Along with the incompleteness and undecidability that compel a reevaluation of Hilbert's program, quantum theory not only urges us to reconsider the role of formal systems in physics, but it might also suggest a more comprehensive notion of "universal automaton" that may carry out "open-ended" processes.

Even though his mathematics presents quantum physics with the most appropriate language and in spite of his persistent concern with theoretical physics, Hilbert does not seem at ease with quantum theory (Macrae, 1992, chap. VI). He stresses that theoretical physics' individual assumptions and laws have no meaning that we can be immediately realize in intuition. In principle, we cannot confront single propositions of physics with experience, but only the theoretical system as a whole.[7] As far as mathematics and physics are concerned, Hilbert holds that we can logically secure a concept only through its relations to other concepts, "otherwise one is only going around in a circle."[8] The axiomatic view he nurtures is fundamental to his "anti-metaphysical" account of the nature of mathematics (Hallett, 1989, p. 241). Instead of arguing that theories are acceptable because their axioms express *truths* about abstract or empirical objects, he singles out the notion of consistency as the crucial one for the understanding of mathematical, hence physical, theories. Hilbert's axiomatic approach echoes Hertz's: a formal system captures the structure of physical reality and the elements of this structure provide the images of *possible sequences of observable phenomena*. But the construction of an axiomatic system for quantum physics raises some genuinely new problems.

In a joint paper (1927), Hilbert, Nordheim, and von Neumann face the difficulty of fleshing out quantum physics in a system of axioms whose requirements range from the physical to the formal. The "natural" temporal order between the statement of the basic physical concepts—the probabilities associated to mechanical quantities—and the construction of the formal scheme that organizes them in a system is blurred.[9] We have then to work backwards to grasp the "physical meaning" of the formal system and, henceforth, of its basic physical requirements. Formalism has to lead the way.[10] This explains the caution of the authors, who do not conceive their presentation as a "full axiomatization" and admit the necessity of further work in order to achieve mathematical rigor. Only von Neumann, however, will pursue the search for an appropriate system of axioms for quantum theory. In von Neumann's work, the previous concern to separate the mathematical apparatus from the physical content of the theory appears to vanish.[11] Whereas the pursuit of logical rigor, justification, and rational *reconstruction* motivate Hilbert's interest in axiomatic method, von Neumann focuses on the issues of *formal extension* and lets the very formalism justify the axioms. His goal is to explore the whole field of possibilities for theories with precise formal features that reflect physical requirements. Can the structure of the physical world that quantum theory designs indicate ways of opening formal systems that might be relevant to the investigation of complex automata?

In classical mechanics, a point in the phase space gives a complete description of how things are now, while the Hamiltonian function determines how things change with time. To make classical physics deterministic, however, we need an additional assumption, namely that the physical system is closed. While classical determinism provides an image of science as a closed formal system built up from elementary constituents, in the light of quantum indeterminism the exploration of a field of open possibilities draws up a physical system. Despite the deterministic description of how the state of a quantum system evolves with time that Schrödinger's equation offers, quantum theory is inherently probabilistic: a quantum state cannot assign values to all physical quantities that we associate with the system. A quantum state describes not how things are, but their probability relations. In determining the probability of a specific value for a physical quantity, quantum theory differs from classical physics because of the impossibility of performing certain measurements simultaneously. A measurement is not a "passive" recording of how things are, but an action that involves two actors—a system to be observed and an observer—establishing a connection between them. Once we have performed the action, a "collapse" of information modify the "pair." Hence any measurement is "perturbative" and demands us to make a choice. Among the incompatible aspects of a physical system, the observer must choose which one to probe. As long as we regard measurement as an interaction, with the double requirement of freedom of choice of the observable and capability of recognizing the answer, we have to sharpen the relations of probability that we associate with its results and, consequently, sharpen their mathematical representation. But a theory about "measurement" implies also the incompleteness of the system and some unanalyzed interventions that always come from outside. Here is why we view the subject-object distinction as a question "at the very root of the unease that many people still feel in connection with quantum mechanics"(Bell, 1973, p. 41). This issue has its analogue in the unease that the incompleteness theorems have raised in mathematics. In both cases the difficulty originates from our attempt to describe an "inside-outside" relationship that ignores one part of the relation, as if we could reduce a relationship to an "objective property" of the system.

By connecting the "effectiveness" of a formal system to the computing capabilities of a human agent, a Turing machine takes Hilbert's system of mathematics to its limits. Those limits overlap with classical physics'. Hermann Weyl states the question as follows:

> The "physical process" undisturbed by observation is represented by a mathematical formalism without intuitive [*anschauliche*] interpretation; only the concrete experiment, the measurement by means of a grating, can be described in intuitive terms. This contrast of physical process and measurement has its analogue in the contrast of formalism and meaningful thinking in Hilbert's system of mathematics. As it is possible to formalize an intuitive mathematical argument, so it is true that measurement by a grating G may be interpreted as a physical process. In doing so one has to extend the original system S to a system $S*$ by inclusion of the grating G. But as soon as we want to learn something about $S*$ that can be told in concrete terms, then the undisturbed course of events ... must again be disrupted by subjecting $S*$ to the test of a grating outside $S*$. (1949, p. 261)

But as quantum theory shows, the limits of Hilbert's axiomatic-formal systems are not the limits of his mathematics. In particular, von Neumann's idea to formulate quantum mechanics as an operator calculus in a Hilbert space is a remarkable innovation in mathematical physics.

In his *Mathematical Foundations of Quantum Mechanics*, von Neumann argues in favor of the incompleteness of quantum physics because the unpredictability of measurements is an objective property of nature and not a consequence of subjective ignorance. Note that here, since there is no "true" proposition about a given quantum state such that an extended system can prove it at a meta-theoretical level, what is unpredictable is not how we solve a decision problem by the (quantum) formal system. The problem is how we describe a quantum measurement: does quantum physics only apply to the observed system (while classical physics applies to the observer)? Or does it apply to the joint system that the observed system and the observer bring about? On the one hand, since a measurement involves an interaction, a system has to be *open* to observation from the outside; on the other, a description "from the outside," where the observer is not part of the system, cannot be "universal." We cannot observe the whole universe, as a *closed* system, from the outside. As the problems of self-reference attest, the price we pay for universality is full measurability (Breuer, 2001). Von Neumann's quantum mechanics does not apply to the observer, thus is a "theory from the outside." For quantum mechanics to be a universal description of a closed system that includes an external observer, there must be no contradiction between the Schrödinger equation that applies to closed systems and the "projection postulate" or reduction rule that applies to quantum systems we observe from the outside. With a solution that resembles Tarski's hierarchy of formal systems, von Neumann addresses the problem in terms of a chain of observers. But the quantum measurement problem is different from Hilbert's decision problem, because a main feature of the notion of "truth" that emerges from quantum physics is a *changeable* system-observer interaction that no closed system can capture.

The nature of quantum systems and the "reduction" rule that applies to quantum measurements deeply puzzled Turing, who studied quantum mechanics on von Neumann's book (Hodges, 2004, p. 54). He was more interested in the "unpredictable" process that takes place in a quantum measurement than in the deterministic evolution of a quantum system. Turing conceives an intelligent machine as a device capable of both processing a stream of increasing information that comes from the outside and of modifying itself accordingly. For Turing, to convert a brain or a machine into a universal machine is an extreme form of *discipline*, whereas the production of intelligence requires also *initiative*. "Our task—he says—is to discover the nature of this residue as it occurs in man, and to try and copy it in machines" (1948, p. 429). Might quantum physics cast some light on the nature of this residue?

5. Steps to a Quantum Automaton

The search for an answer to this question can benefit from some considerations on the notion of *quantum* automaton. The essential difference between a quantum automaton and the universal Turing machine comes from a hierarchy of measurements that *entanglement* makes possible.[12] In his 1987 paper, Albert describes a quantum automaton as an ideal object that we endow with: (*i*) input-output mechanisms of information; (*ii*) instruments for measuring physical *observables*; (*iii*) an internal program containing a set of rules that, given their initial conditions, predicts the behavior of simple physical systems (including itself). Assuming that this automaton operates in accordance with the laws of quantum mechanics, we need to know how it would answer a variety of questions (1987, p. 579).

Suppose a quantum system S in the state $\psi^{S}_{A=\alpha_1}$ that assigns the value α_1 to the observable A in that state. By following its rules, the automaton Au can measure and register the value of the observable A; once it has this operation, we can interview the automaton by performing a measure of some well defined observables. More precisely, the composite system of S plus the automaton Au will be in the state

$$\psi^{S\oplus Au}_{A^{\oplus}=\alpha_1} = \psi^{Au}_{P_A=\alpha_1} \cdot \psi^{S}_{A=\alpha_1}$$

where the observable P_A measures the value *predicted by Au* for the observable A, while the observable $A^{\oplus}$ is the dynamical variable that corresponds to A in the composite system. If we let the observable $E_A = (P_A - A)$ measure the error in the automaton's prediction for A, this means that the prediction will be *accurate* when $E_A = 0$. Albert examines which observables we can predict accurately by quantum automata.

Imagine the system S in the state

$$\psi^{S}_{B=\beta_1} = \psi^{S}_{A=\alpha_1} + \psi^{S}_{A=\alpha_2}$$

which is a *superposition* of states with two different values of A. In this state, the observable B—by hypothesis "incompatible" with A—is well defined and its value is β_1, while the observables P_A and A are *indeterminate*. Nevertheless, the automaton prediction concerning A is accurate since, in the composite system, it turns out that $E_A = 0$:

$$\psi^{S\oplus Au}_{B^{\oplus}=\beta_1} = \psi^{Au}_{P_A=\alpha_1} \cdot \psi^{S}_{A=\alpha_1} + \psi^{Au}_{P_A=\alpha_2} \cdot \psi^{S}_{A=\alpha_2}$$

This means that even though the results of measuring P_A and A are not well defined in this state, we can predict that the value of P_A will be equal to the value of A. To enable the automaton to predict the values of A and B, P_A and P_B should be compatible. But, since the variables A and B of the system S are *incompatible*, the variables E_A and E_B of the composed system must also be *incompatible*; therefore, their values cannot be both zero. According to the uncertainty principle, "(simultaneous) *predictions of the automaton about incompatible observables cannot both be accurate*" (*ibid.*, p. 581). As Albert says:

> All of this is just the familiar quantum theory of measurement, and what I have done so far is in effect just to remark that in every circumstance imagined in that theory (in all of which observers observe things external to themselves), quantum-mechanical automata have precisely the same capacities to measure and predict that "observers" do. (Albert, 1987, p. 581)

Moving on, Albert suggests to instruct the automaton to measure A and $B^{\oplus}$, that is, to measure also something about *its prediction for* A, and to make predictions about the subsequent measurements of both. Now the composite state will be:

$$\psi^{S(\oplus Au)\oplus Au}_{B^{\oplus}=\beta_1} = \psi^{Au}_{P_{B^{\oplus}}=\beta_1} \cdot \psi^{S\oplus Au}_{B^{\oplus}=\beta_1} =$$

$$= \psi^{Au}_{P_{B^{\oplus}}=\beta_1} \left(\psi^{Au}_{P_A=\alpha_1} \cdot \psi^{S}_{A=\alpha_1} + \psi^{Au}_{P_A=\alpha_2} \cdot \psi^{S}_{A=\alpha_2} \right)$$

This state is a superposition of two states: in the first, where the variable $A = \alpha_1$, the automaton predictions are $P_A = \alpha_1$ and $P_{B^{\oplus}} = \beta_1$; in the second, where the variable $A = \alpha_2$, the automaton prediction for A becomes $P_A = \alpha_2$ while for B it remains $P_{B^{\oplus}} = \beta_1$. In the composite system, though A and $B^{\oplus}$ are incompatible, the values of E_A and $E_{B^{\oplus}}$ are both zero; in other words, the automaton can know with accuracy both A and $B^{\oplus}$. The two variables are not incompatible because, for the automaton, only one (namely A) belongs to the external system. If a second automaton were there, we could distinguish the properties of the two automata because, for the second automaton, E_A and $E_{B^{\oplus}}$ cannot be compatible, since A and $B^{\oplus}$ belong to external systems:

> So there is something subjective, as it were, about the capacity of the first automaton to be in a position to predict, simultaneously, both A and $B^{\oplus}$ ...; because that capacity depends not just on the *structure* of the automaton, but also on *which one it is*, on its *identity*. (*ibid.*, pp. 582–83)

To summarize, a certain category of measurements (measurements of A) generates knowledge of the sole external system. A second category of measurements (measurements of $B^{\oplus}$) generates knowledge of the external system and of the automaton's knowledge of that system. A third category of measurements (measurements of $B^{\oplus}$ by another automaton) generates knowledge of the external system, of the automaton's knowledge of that system, and of the knowledge that another observer can get of the system, and so on, with increasing complexity. To write quantum states such that an automaton living in one of them would accurately know an infinite number of mutually incompatible observables that are such according to the conventional definition, appears possible. As to the rules, those which allow performing measurements within the same category are the familiar rules of commutation that we need to apply to the measurement of external systems; the rules for combining measurements from different categories, though, are a different matter. As Rasetti has indicated ("Brain and Life"), these rules appear to depend even on the "identity" of the agent that performs the measurements; knowing such agents in depth is the big challenge for information theory and quantum computation.

6. Thriving on Undecidability

As a structure of increasing complexity, the system of symbolic knowledge demands a double test on its effective procedures: at the syntactical level, in order to allow information to flow; and at the semantic level, in order to allow the system to "read" and modify itself, hence to evolve. The universal Turing machine is a well defined symbolic system that is capable of generating "open-ended" complexity. However, the universal Turing machine is a closed system unable to cope with an increase of initial information, hence its complexity cannot go beyond a set limit. For Turing, it is a model of "the extreme form of discipline"—an obedient mechanism that lacks the "initiative" to have interference—that is, interaction—with the outside and "free will,"and consequently fails to generate an "open-ended" hierarchy of knowledge and complexity, the very hallmark of intelligence. From a philosophical point of view, claiming that a theory has a universal character might conceal the underlying assumption that no further information can produce new *forms* of knowledge or new mechanisms of behavior.

Rasetti regards a quantum automaton as a kind of "intelligent machinery." It is a physical system that, by shaping the incoming information in a hierarchical way, can define more and more sophisticated functions of its manipulation. If a Turing machine can only manipulate rational numbers, which is arguably the language suitable for describing the world of perceptive processes, a quantum machine is substantially equivalent to a classical Turing machine. The hierarchy of measurements that a quantum automaton can perform, however, shows that in the quantum world processes of information can follow paths that are unknown to the classical world. From the investigation of non-local subsystems some algebraic structures, more sophisticated than ordinary arithmetic, emerge. We can settle these structures by testing quantum software; hence, as far as logic is concerned, they belong to a meta-level inasmuch as they concern not the system itself, but computer's test from the outside. To return to Cassirer's frame, those structures provide a language that is suitable to describe a world of *symbolic forms*.

To conclude, when dealing with the nature of intelligent machinery, the two phenomena of interference and complexity present us with a model of information processes that is necessarily open not because precision and completeness are out of reach for human beings, but because its construction ought to be the result of interactions (and interference) at subsequent levels of observation (and measure). As a result, such a model cannot be anything but a structure of increasing complexity forever undecidable and inexhaustible, just like genuinely scientific—namely, symbolic—knowledge must be.

NOTES

1. In his 1939 Spring lectures at Notre Dame, Gödel discusses Hilbert's decision problem in the context of Leibniz's *Calculemus*; see Sieg, 2006.

2. Other mathematical definitions of the same notion are due to Alonzo Church, Emil Post, and others, see Gandy, 1988.

3. See Church's letter to Stephen Kleene in Davis, 1982, p. 9.

4. Von Neumann observes that while no difficulty exists in describing how an organism might be able to identify, for instance, "any two rectilinear triangles, which appear on the retina, as belonging to the same category 'triangle'... this, however, constitutes only a small fragment of the more general concept of identification of analogous geometrical entities... and only a microscopic piece of the general concept of analogy" (1948, pp. 310–311).

5. Letter from Turing to Maxwell Newman quoted in Turing, 2004b, p. 257.

6. "The very day on which the undecidability does not obtain any more, mathematics as we now understand it would cease to exist; it would be replaced by an absolutely mechanical prescription, by means of which anyone could decide the provability or unprovability of any given sentence" (von Neumann, 1927, p. 266).

7. In the list of still unsolved problems that Hilbert drew at the beginning of last century, the sixth one is about the axiomatization of physics (1900; 1976).

8. Letter to Frege of 22 September 1900 (Frege, 1980, p. 51).

9. For a detailed discussion of this issue see Lacki, 2000.

10. In his 1905 lectures on the use of axiomatic method in the different branches of science, Hilbert did not address analogous problems, because he considered "closed" theories whose meaning was not in question (*ibid.*).

11. The main result of John von Neumann, 1932, is the "statistical foundations" of quantum formalism.

12. In his seminal 1985 paper, David Deutsch derives the description of a universal *quantum* Turing machine from the universal Turing machine, but using quantum theory to define the operations carried out by the computer, which is now a physical system. Researchers have shown that that quantum computers solve some important problems more efficiently than classical ones, but they do not reach beyond the *computational* power of Turing machines. However, what is at issue here is whether quantum machinery, underlying the (quantum) computational model, might cast some light on the nature of "intelligent" machinery.

Twelve

TURING 1948 VS. GÖDEL 1972

Giorgio Sandri

1. Computation and Information

We can describe the procedure that leads to the computation of a numerical function as a form of information processing: we process input information in a closed box that operates those transformations of information that produce the output information. The notion of "information process" acquires a descriptive relevance in regards to computations where procedures occur that might not be strictly algorithmic, and where further information, besides input information, intervenes during the process. We can distinguish three different approaches.

1. John von Neumann (1951, 1958) remarked that the logic of automata will be different from the systems of formal logic on which we formally base the notion of "effective computability": it will have a less combinatorial and more analytic character. The logic of automata should be closer to thermodynamics, the part of theoretical physics that processes and manipulates information as information theory does.

   Von Neumann described how Alan Turing reached a general definition of a computing machine—a black box with a finite number of states—and provided a peculiar characterization of Turing Machines. Its essential operative character consists in describing how the machine is forced to change state: such a change needs some interaction with the external world, and the external world consists of a long paper tape. The interaction with the external world is limited to the input, and does not intervene during the computation process. The description satisfies the conditions on algorithmic computation, and characterizes the machine as a device that treats and transforms, according to explicit internal rules, information that originates from an external source.

2. Gödel often recognized how Turing had reached an absolute characterization of effective computability and had provided a definition of formal systems based on mechanical rules of inferences that are independent from the language of a system. Even though these achievements allow us to give a general form to Gödel's limiting results, the system based on effective finite procedures would be unable to produce a finite rule that includes the entire system of mathematics.

Two are the structural reasons preventing finite procedures typical of Hilbert's heritage and subject to the conditions of Turing's computability—recursive arithmetic—from including the entire system of mathematics (Cellucci, 1998). Each finite step that includes a formal system through an axiom—through a finite procedure—might lead to a wider axiom, and we could iterate the procedure for an indeterminate and even infinite number of steps. This property entails the incompletability of mathematics that limiting results, the Second Theorem in particular, had already proved. Since inferential tools that are logically weaker than the tools the system itself provides cannot prove its consistency, to prove the consistency of the classical theory of numbers we should, in Gödel's terms, proceed beyond finitary mathematics. Properties or relations of concrete objects—combination of signs—are what bases Hilbertian finitary mathematics. On the contrary, the proof of consistency of the theory of numbers requires abstract notions that concern properties dealing with mental constructions, and that do not stem from combinatorial properties of signs but from their "meanings"—a notion that, Gödel acknowledges, is largely indeterminate. More information intervenes, in addition to the information that finite and recursive procedures process. This information is to some extent partially indeterminate in nature and we cannot treat it through the effective computational procedures of recursive finitary mathematics. This fact convinced Gödel that the mind, unlike the brain, is not a finite automaton: within the closed and isolated system of Turing computation, we cannot provide an adequate description of the flows of information that the infinitary processes of the system of mathematics produce. We cannot reduce these flows of information to effective mechanical procedures; instead of depending on the local combinations of signs, they present *global and non-local properties* that depend on meaning (Gandy, 1980). Even if we try to move past Gödel's distinction between brain and mind, it is unclear how we could precisely characterize the information processes that are a consequence of the infinitary processes of the mind we have assumed.

3. When receiving as input the program of any Turing Machine (*TM*), the Universal Turing Machine (*UTM*) can simulate the computational behavior of the *TM* for any input and can compute the value the *TM* would have computed for that input (Herken, 1995). This ability does not imply that the *UTM* could modify the program it received as input at the beginning of the computational process. In the *UTM*, program and input data are part of the input, and any modification to the program would be equivalent to a new input. It would be a different computational process. Let us stress that a computational weakness of the program or, more likely, the inability of a given program to compute intervening information could suggest a change in the program of a *TM*. Since a classic *TM* does not allow any additional

information beyond the input information, in a classical system we do not need or allow any change to the program. Any change to the program of a *TM* would produce a different *TM*.

The issue at stake is the following: could we describe the change of program that leads from TM_0 to TM_1 as a rule-governed process, or at least as a process that some systematic procedure governs? Could we consider the process as a form of computational (or semi-computational) procedure, where we transform according to a rule, or at least to a fixed procedure, the input information (the input data and TM_0 program) together with the intervening information into the output information that consists of the TM_1 program? According to this description, the procedure that leads to a program change would be substantially unrestricted, and to some extent analogous to processes that imply an unrestricted or infinite conception of mind. It would be analogous to those procedures where a surplus of information—besides the input information of the finitary computational processes—requires that we open up the closed systems of classical computational processes to intervening information.

The Second Theorem shows that no consistent and powerful mathematical system can prove its consistency by means of tools that belong to the system itself. Accordingly, Gödel tried to distinguish within mathematics the effective component, which consists of recursive arithmetic, from the non-finitely representable component. He hesitated (mostly in unpublished notes: 1934; 193?; 1958; 1972) among several non equivalent answers:

1. the brain has the computational power of a finite state automaton—a *TM* —but the mind allows infinitary processes. Even though each stage of the procedure involves only a finite number of states, the mind expands them indefinitely in an infinite procedure that produces states of ever increasing complexity;
2. as a consequence of the Second Theorem and because we cannot include it in a finite number of axioms, mathematics cannot be finitely completed;
3. while recursive procedures deal only with strings of symbols, the growing hierarchy of states of mind involves abstract terms that we consider in terms of their meanings, and not just in terms of their string configurations. This fact produces an increasing complexity of states of mind within a non-effective (non T-computable) procedure;
4. the infinitary processes of mind *do not constitute a local, finite, and discrete procedure because they involve non-local determinations*. These processes entail meanings and a finite automaton cannot process them; they do not represent a case of T-computation. In 1972, Gödel suggested that we could advance our knowledge of mathematics, if we could describe such a procedure as a well-defined (according to rules) process (although a non-effective

one in Turing's sense) that "computes" non classically computable functions (functions that are not generally recursive).

To think that Gödel's uncertainty between the different hypotheses uncovers a substantial issue appears reasonable: there might be procedures that bypass the computational power of finite automata, because they comport factors with which we cannot deal in terms of classical computation. The properties of classical algorithmic computation hold only within strictly effective settings, but there might be non-local and non-strictly-effective procedures, whose character entails *an increase of information*: "meanings," sequences of axioms of increasing complexity, and non-algorithmic local processes.

A variety of facts share the occurrence of information that is not included in the computation's input:

1. incompletability of mathematics and infinity of the states of mind;
2. information flows that the computational process produces through feedback processes, adding new information to the input;
3. continuous interaction with the environment;
4. non-strictly-computational processes that lead to the construction of sequences of axioms of increasing power; these processes constitute procedures of "discovery" of properties or modes of formal organization that are not a logical consequence of a given formal structure;
5. occurrence of intervening information that causes the closed box of algorithmic computation to open;
6. evolution of computational systems in a learning process through non-algorithmic and partially random procedures;
7. non-strictly-computational processes that involve the "meaning" of the terms that are involved in the information process;
8. *non-local* character of the computational steps in the information process.

As the Church-Turing conjecture states, one of the distinguishing formal characters of algorithmic computation is its closure with respect to input information in the domain of the computed function. We could treat any further information—external, intervening, and generated in a feedback process that we cannot describe as a logical consequence of the information the algorithmic process deals with—only through the construction of computational procedures that should be stronger than those taking place in the algorithmic process. Such a "construction" would involve procedures of "discovery," and a consequent increase in the computational power of the system. A procedure that starts from a given configuration cannot deterministically generate such an increase: its formal character would be strictly *non-local*, and characterize non-strictly-computational processes that we could not describe in terms of T-computation, but that could possess a systematic "semi-computational" character (they could be processes computing functions that are not general recursive). Classic computational procedures involve algorithmic treatment of information. They satisfy closure with respect to the information that belongs to the domain of the function. Non-strictly-algorithmic processes would transform information according to procedures that, in general, do not finitely follow a rule.

2. The Incompletability of Mathematics

In 1946, Gödel acknowledged that Turing had given an "absolute characterization of an interesting epistemological notion"—general recursiveness—that did not depend on any formalism, and a "formal description of the notion of 'computability'" that was based on the notion of effective mechanical procedure (1946, p. 150). The formal description of algorithmic processes constituted an absolute achievement. Nevertheless, that description can support neither a thesis that asserts the fully algorithmic nature of mathematics, nor an argument about the algorithmic nature of the mind. Any mechanical characterization of intelligence (or some part of intelligence) must satisfy the conditions that Turing had postulated, and those conditions were neutral with respect to the possible existence of non-mechanical intelligent procedures. In a later occasion (a postscript of 1964), Gödel remarked that, thanks to Turing's achievements and to his analysis of "mechanical procedure"—which he proved equivalent to Turing-computability and to general recursiveness (algorithm or combinatorial finite procedure)—a precise and undisputed definition of the general concept of "formal system" has become possible. As a consequence, two further things have also become possible: to rigorously prove the existence of undecidable propositions and to show that we can never prove the consistency of the system in system itself (with respect to *any* formal system that contains the theory of numbers). Gödel added that Turing's analysis was independent of the question concerning the existence of non-mechanical procedures that use abstract terms on the basis of their meanings. Turing's achievements and the Incompleteness Theorems supported a thesis concerning the limits of formalism, not of human reason. The deterministic character of Turing's notion of computability automatically provide its consistency, hence becoming—Solomon Feferman conjectures—what convinced Gödel of the adequacy of Turing's notion of "mechanical procedure" (1990, p. 296).

For Gödel, the general form of the incompleteness theorems does require an adequate characterization of "mechanical procedure," but we can never completely mechanize mathematical reasoning. To replace the mathematician with a machine is impossible, even within the boundary of number theory.

We cannot reduce mathematical procedures to the procedures of a finite machine, because, as Gödel remarked, in order to prove the consistency of classical number theory (*a fortiori*, stronger systems), we must use some abstract concepts and these abstract concepts are not syntactical—they refer not to properties, but to *meanings* of strings of symbols. These remarks, together with the results stemming from the Second Theorem (concerning the incompletability of mathematics), would lead Gödel to favor the inadequacy of the functional equivalence between the mind and a finite machine that performs "mechanical procedures" (equivalent to T-computability) (Sieg, 2006).

In a more sweeping remark (in *Dialectica*, 1958, now in 1990, pp. 240–251), Gödel stressed that if we cannot prove the consistency of a system via inferential tools that are weaker than those we use in the system itself (Sieg, 1999; 2004), then we must go beyond Hilbert's finitary mathematics. We must use abstract notions that do not involve properties or relations among concrete object (combination of signs), but that refer instead to mental constructions (demonstrations, statements with their meaning) that we base on the meanings of the signs that represent demonstrations, not on their combinatorial properties. In the 1972 revision to the 1958 *Dialectica* article, Gödel stated that second-level concepts are *thought structures*. The proofs of propositions concerning these mental objects need *insights* that we cannot derive from a reflection on the combinatorial (space-time) properties of the symbols, but instead from a reflection on their meanings. Turing's formal results did not set limits to human reason, but only to formalism. Turing would make a philosophical mistake if he claimed that a human computer is capable only of a finite number of states of mind. Even if we acknowledge that the set of states at every stage is finite, the procedure is endless, and converges to infinity in its application (incompletability of mathematics).

Turing's axiomatic condition about the finite number of "states of mind" in the operations of the "computor" (a condition that we require, because the states must be distinguishable), is about the deterministic character of computation. We obtain determinism because we associate the "state-symbol" of the configuration with just one "operation-state" couple, thus implying that states are distinguishable and discrete. As Gödel remarked, if we associate Turing's conditions on states of mind with two assumptions (which Gödel rightly believed Turing shared):

1. the mind is not separated from matter,
2. the brain works basically as a digital computer,

we would end up with a thesis that supports the mechanical and finite character of mental procedures.

Gödel's argument concerning the consequences of Turing's conditions on states of mind would be correct only if the two scientists shared the same interpretation of what "states of mind" means. Turing replaced the misleading notion of "state of mind" with a more physically specific expression—"notes of instruction"—that would let the computor interrupt its work and start it again. Each note of instruction allows the computor to perform a step and to write down the next note, because each note *determines* the following computational step. Turing allowed the existence of an infinite number of mental states, but he maintained that the occurrence of an infinity of states *in the process of computation* would make them too close to each other to be distinguishable. The required deterministic character of computation ("effective computation") mandates that the number of states be finite—otherwise, they could be not effectively distinguishable. As Webb (1990) noted, we can use an effectively distinguishable set of states to effectively "remember" and to effectively compute. Gödel did not bring any objection to setting these restrictions on the brain, but he did insist on a "mental memory" that is separate from it.

Gödel's conviction relies on a notion of "state of mind" that is utterly different from Turing's. In the note of 1972, Gödel maintained that as more and more abstract terms enter our intellectual sphere, our understanding of them grows the more we use them. This process causes the mind to constantly develop, and the number of Turing's distinguishable states of mind to converge toward infinity. The number of abstract terms at each stage that, as Turing maintained, is as finite as the number of states of mind, during the application of the procedure converges toward infinity.

The formation of stronger and stronger axioms in set theory shows this property. The complexity of Gödel's notion of "states of mind" bypasses the complexity of Turing's notion (the notes of instruction are its physical counterpart). The process of understanding abstract terms of increasing complexity as they form makes the mind capable of states of corresponding increasing complexity. This growing complexity of the states of mind produces *non-local determinations* that we cannot treat in terms of effective mechanical procedures.

The deep difference between "states of mind" in Turing and in Gödel lies in the nature of the *information* they process. In algorithmic computation, information consists in the digital discrete input data that we feed to the computational process that local processes transform. In the abstract sense, however, information is non-normalized conceptual information that we treat and transform through non-local processes. These are *global processes* that involve meanings (Gandy, 1980). Whereas Turing's "states of mind" are part of the input configuration, Gödel associates them to a conceptual property that makes the increasing complexity of abstract terms correspond to an increasing complexity of the states of mind that understand them. Turing describes (and sets conditions on) a mechanical procedure, while Gödel analyzes a different notion that Turing's limiting conditions leave unaffected.

Which notion? We may advance two hypotheses, one in Turing's and the other in Gödel's style.

1. Since writing more symbols on the tape means to prevent the use of increasingly more complex states, the Universal Turing Machine (*UTM*) provides a digital solution to the problem of the complexity of states. In order to simulate machines confronting an arbitrarily growing number of states, we feed the Gödelian number of the machine that we want to simulate to the Universal Machine as input data. As Webb remarks, "the complexity of the machine to be imitated is concentrated in the tape, and does not appear in the *UTM* in any way" (1990, p. 300).

2. Gödel suggests that the process that leads to evermore complex states during the understanding of evermore complex abstract terms is non-mechanical. In general, we cannot describe the non-mechanical procedures that a human computer uses to introduce states of mind in symbolic terms (as strings of symbols on the tape), because the process that allows humans to understand infinitely complicated combinatorial relations (*global, not local, determinations*) includes the meanings of the abstract concepts of proof and also infinity.

The second hypothesis may support Gödel's statement of 1972, according to which, while forming stronger and stronger axioms of infinity in set theory, we might face a well-defined procedure that "would lead to non recursive numerical functions and would demand a substantial advancement in our understanding of the basic concepts of mathematics." Is Gödel foreseeing the possible existence of mathematical objects (a well-defined procedure) that are likely to lie beyond Turing computation and beyond mechanical procedures (since they would lead to non recursive numerical functions)?

3. Feedback Systems

The theory of computation and the mathematical theory of communication (Shannon, 1948) explicate notions that overlap partly, but have profoundly different goals. In the computational process, the transformation of information consists in computing the value of the function for the given input; in the process of communication, it consists instead in the transmission and deformation of the codified message in the channel. In the computational process, information consists only in the input information, whereas in the process of communication the transmission of information consists in the deterministic process that occurs in the closed box where, starting from the initial configuration and according to the rules of the program, the input information is being transformed. The rule determines the output configuration as soon as the the input configuration—state, symbol—is given.

If we look at information processes in general terms—not only as the manipulation of information consisting in the computation of a numerical function—we can conveniently consider them as stochastic (Markovian) processes. In that case, the output configurations depend *non-deterministically* upon the input configurations; further information intervenes during the information process. The output information depends non-deterministically on other sources of information in addition to the input information (input configuration) and the rules. Intervening information does not have a uniform character:

1. Intervening information has its origin in the environment; its sources are external to the closed box.

2. During the computational process, non-normalized information intervening from the outside—with respect to the computational setting—continuously interferes with the input information. The argument of the computed function may vary continuously as the intervening information varies.
3. The intervening information might include factors (disturbances, noise) that reduce the number of possible choices in the stochastic information and computational process.

We could interpret "noise" as information coming from an external source that intervenes in the input information and induces a modification of its computational treatment. In the classical computational processes, the input information is normalized—digital, discrete, finite—while "noise" information does not show a uniform character and cannot consist in *continuous* information.

I believe that the "cooperation" between information and computational processes, along with the possibly continuous character of the information and the possibly hybrid character of the computational processes, could support the thesis that classical algorithmic computation (Turing computation) does not constitute the only form of computation. We might describe forms of semi-algorithmic computation that would mostly consist of hybrid procedures that process continuous values (possibly through analogical procedures), thus violating computation's classical condition of discreteness.

According to another thesis, the intervening external information that modifies the computational process produces a partially conservative widening of the computational system. The rules of classical computation and the "post-classical" rules (*Oracle*, Turing, 1939) that process the different forms of intervening information would cooperate in a system that produces digital outputs (Siegelmann, 1999; 2003). Classical systems do not allow any form of interference in input information or in the system of rules. The Universal Machine allows an infinite number of applications (receiving as input the program of any machine), but allows no information except the input information that consists of the coded program and the input data of a *TM*. To evolve the rules that the system uses in computing the input information would be pointless: the property of closure of the system S_1 prevents any possible evolution into a system S_2. There is no relation between S_1 and S_2, in the sense that S_2 rules are not the output of a process of modification or evolution of the S_1 rules.

We can formulate as follows the thesis that supports forms of computation in which classical and post-classical processes occur: intervening information generates modified rules that would be post-classical (hybrid), and that, vis-à-vis the increased informational complexity, would constitute an evolution of the initial classical system. If S_0 is the initial system and S_n the system undergoing evolution (Eiben and Rudolph, 1999) (possibly a continuous evolution, since the flow of the intervening information may be continuous), the process from S_0 to S_n and further constitutes a possibly non-finitary process whose characters violate discreteness and locality—hence, a non-classical process.

Since we cannot consider the described process a truly computational one (discrete, finite, and local in the classical sense), the argument does not provide full evidence in favor of non-classical forms of computation. In any case, a process showing *random* characters would replace the deterministic character of the classic computational processes.

Why should we consider these random processes as an example of a more general non-monotonic procedure where the increase of information induces an evolutionary modification in the structures of the computational system? This thesis would show some connections with Gödel's statement concerning the non-finitistic character of the mental procedures that are involved in the choice of increasingly complex axioms in set theory, and with Turing's suggestions concerning non-organized machines.

4. Learning T-Machines

A *UTM* is capable of simulating the computational behavior of any *T*-Machine for any input. However, once the *UTM* has received as input the Gödel number (the program) of the specific *TM*, the Universal Machine is unable to modify it, since we can introduce a modification only as a different input (as a different Gödel number, a different program) that produces a different computation. There would be no reason for the *UTM* to modify its program, because it does not receive any additional information during the computation.

How can we describe a (semi-)classical computational device that, to cope with the fact that new information—neither present in, nor implied by the input—would be capable of generating modifications to the program non deterministically, hence through non-classical computational processes? How can we describe an interactive Universal Machine, a machine that interacts with intervening information?

In 1937, Turing described the not further reducible procedures that an abstract "computor" (human or artificial) performs as it computes a numerical function. Instead of developing an inter-linguistic characterization of the formal features of the effectively and finitely computable functions that would be similar to Gödel's and Church's (Gandy, 1980), he provided a mathematical (axiomatic) model of the procedure. In 1948, Turing tried to follow a similar path. He described the elements and the system of relations characterizing the abstract behavior of a "computor" that—faced with new intervening information—modifies the system of rules governing the computation.

We could describe this innovative procedure as an "education" of the computor that—through a process that is non-deterministic, possibly non-recursive, and non-T-computable—amounts to an increase of its "computational" resources.

Turing (1948) remarked that the "education" of the computor takes place through a complex interference that occurs among the information system, the intervening information, the environment information, and the modifications that the system implements in order to cope with the increased flow of information.

The Universal Machine computes everything that is finitely computable and that we can codify in discrete digital form. We could describe the device that modifies the program during computation in terms of a *system* that, through *learning,* could *evolve* its computational rules while being capable of performing all effective computational tasks. Such a procedure would not involve adding more hardware (see the "screwdriver interference" that Turing discussed in his 1948 report), but *programming steps* (Copeland and Proudfoot, 2004). The formal explication of "computation" in the classical form of the "computor" consists in the representation of a universal computing machine as a *set of rules*, or as a *universal paper machine.* Therefore, the issue comes down to the formal description of the *paper procedure* that, facing interference and intervening information, would lead to the *"evolution" of systems of rules that are modified during the computational process.*

Turing's experience at Bletchley Park possibly helped him realize the feasibility of computing machines that are capable of solving problems through a search of solutions by means of heuristic principles and through the use of random procedures. At the 1947 Royal Astronomical Society conference, Turing characterized the brain as a digital machine, and insisted that machines are capable of acting intelligently. They can learn from experience; the mechanism whereby machines learn from the experience that *intervening information* brings forth is their capacity to modify the instructions they receive. These machines would require that we broaden the spectrum of the properties that are typical of the Universal Machine, so as to let the *UTM* modify the rules of the program during the computational process.

These *"unorganized machines,"* networks of Boolean elements similar to casually connected neurons, were Turing's first device capable of partially satisfy ing some of the requirements I mentioned above, thus bypassing the deterministic machines that compute only numerical functions. In his 1948 report, Turing developed a few ideas about the methods whereby we could organize neural networks that initially are unorganized through educational training and interference. Training and learning would make some amongst the neural paths effective, while others not. This hypothesis is functionally equivalent to incoming information that acquires the form of new connections in the network, consequently inducing the creation of modified instructions. If we follow the analogy with the "computor" as an abstract model—*in 1936, of finite computing; in 1948, of the evolution that instructions undergo because of intervening information*—we might say that we could structure the unorganized machine (or lead it to "learn") through an interference-based training process. The interference comes from information that the computational process initially did not imply but that enters it as it runs.

Back in 1939, Turing had asserted that mathematical procedures call for a cooperation between intuition and ingenuity, because in mathematics, in order to find and/or list demonstrations, we need both procedures that execute rules and those that interpret them. But we should reduce the role of intuition so that precise procedures, possibly effective mechanical procedures, could replace it. A decade later, in 1948, Turing rephrased the 1939 statement about "cooperation of intuition and ingenuity" in terms of "discipline" and "initiative." "Discipline" is the ability to finitely follow a rule that converts a brain into a Universal Machine. But discipline by itself is not enough for a formal description of intelligence: we need "initiative" too, especially for the procedures that lead us to the choice of rules, axioms, and instructions. "Initiative" broadens the range of 1939's "ingenuity." Turing's 1948 investigation aimed at a formal description of this last kind of procedures (if we can qualify them as definite), and toward a "mechanical" characterization of these forms. The 1936 notion of "mechanical" that Turing applied to recursive procedures is likely to have only a partial overlap with the one that Turing applied to intelligence in 1948.

The idea is to let the machine operate evermore choices and decisions: the procedure would consist in designing unorganized machines without tape and with gaps in the program, and to intervene in their "education" through additional information that we provide as interference. A non-deterministic procedure would equip the machine with forms of organization that let it carry out some form of initiative. The procedure that leads to the partial organization of an unorganized machine through a cooperation of discipline and initiative will unlikely be a finite, local, discrete, and mechanical process in the sense that Turing intended in 1936.

In a 1940 letter to Maxwell Newman, Turing stressed that behind the cooperation between ingenuity and intuition lies a much more intriguing issue: namely, the relation among different machines, that is, among different systems of proofs (and of rules). Turing maintained that, "One imagines different machines allowing different sets of proofs, and by choosing a suitable machine one can approximate 'truth' by 'provability' better than with a less suitable machine, and can in a sense approximate it as well as you please" (Turing, 2004a, p. 215). These words echo Gödel's results, especially his First Theorem, and the effects that these results had on Hilbert's program and on the *Entscheidungsproblem*. They also echo Turing's 1936 achievements about the undecidability of the halting problem, as well as perhaps the suggestions that Wittgenstein may have given him on the occasion of their Cambridge encounters. If we view the process of choosing the fit machine as a definite but unlimited procedure, we would end up with a thesis that presents similarities with the principle stating the incompletability of mathematics that stems from the Second Theorem.

During the 1940s, the choice amongst different machines led to the question of how to describe a procedure that, by combining discipline and initiative, would represent the evolution of systems of rules as a non-deterministic but definite process. In a letter to W. Ross Ashby (1946, see Numerico, 2005), Turing agreed that it would be possible for the machine to try variations of behavior, and accept or reject them through forms of behavior that were close to the adaptive behaviors of the systems with regulation that Ashby studied (1956). These systems modified their behaviors to cope with changes of information that were coming from the environment.

First Turing suggested random machines and then unorganized machines with interference. Turing tried to put together incompatible notions: the notion of "machine"—implying a deterministic process—and the notions of "random" and "unorganized"—implying non-effective processes. The notion of "machine" refers in principle to effective procedures in a system of definite mechanical rules. "Randomness" and "lack of organization" refer to the *gaps* that must occur *in the program* in order to make possible that a procedure can lead the machine from a system of rules to a different one. A growth of information (environmental information, Ashby would say) produces an interference that leads to the regulation that in turn leads the partial organization of the unorganized machine toward stability.

We can extend such a process indefinitely, because incoming information (interference) might lead from TM_i to TM_j, and so on in an unlimited progression. We could say, with Gödel, that the procedure tends to infinity while at any given time remaining finite. In 1972, Gödel drew an explicit distinction between the brain conceived as a digital computer, and the mind capable of hosting infinitary processes that deal with abstract terms of increasing complexity. As both researchers were probably convinced of the incompletability of mathematics, Turing's 1948 "*answer*" to 1972 Gödel would then consist in a procedure where effective mechanical systems undertake the modifications of their internal setting through an unlimited procedure that allows the intervening of new information and of external interference. Instead, the procedure that leads from TM_0 to TM_j seems non-recursive, non-effective, and non-T-computable. Given that the process transforms information that comes from outside the computational process and that is oriented toward a state of stability consisting in a system of rules non-deterministically produced by the increase in information, even if we could design this process as a definite procedure, it would none the less be doing something different than finitely and effectively computing a numerical function. The growth of computational complexity that Gödel would view as increasingly complex abstract terms entering the sphere of our understanding, Turing would conceive as a modification and possibly a greater complexity of the rules (the program in the tape) of TM_j with respect to TM_0. At stake here is a *paper complexity* that consists in a cooperation between mechanical rules and information processes whose aim is to modify the rules.

WORKS CITED

Abelson, Robert P. (1968) "Simulation of Social Behavior," in *Handbook of Social Psychology*. Edited by Gardner Lindzey and Elliot Aronson. Vol. II. Reading, Mass.: Addison-Wesley, pp. 274–356.

Abraham, Tara H. (2002) "(Physio)Logical Circuits: The Intellectual Origins of the McCulloch-Pitts Neural Networks," in *Journal of the History of the Behavioral Sciences* 38:1, pp. 3–25.

——. (2003) "From Theory to Data: Representing Neurons in the 1940s," in *Biology and Philosophy* 18:3, pp. 415–426.

——. (2004) "Nicolas Rashevsky's Mathematical Biophysics," in *Journal of the History of Biology* 37:2, pp. 333–385.

——. (2006) "Cybernetics and Theoretical Approaches in 20th Century Brain and Behavior Sciences," in *Biological Theory* 1:4, pp. 418–422.

Adam, Alison. (2005) "Knowing Subjects: AI From Feminist Philosophy," in *Mechanical Bodies, Computational Minds: Artificial Intelligence from Automata to Cyborgs*. Edited by Stefano Franchi and Güven Güzeldere. Cambridge, Mass.: MIT Press, pp. 327–344.

Agre, Philip E. (1997) *Computation and Human Experience*. Cambridge: Cambridge University Press.

Aiken, Howard, John von Neumann, and Norbert Wiener. (1944a) *Letter to E. H. Vestine*. 4 Dec. 1944, box 2, folder 66, Norbert Wiener Papers, collection MC-22, Institute Archives and Special Collections, Massachusetts Institute of Technology Archives, Cambridge, Mass.

——. (1944b) *Letter to H. H. Goldstine*. 28 Dec. 1944, box 2, folder 66, Norbert Wiener Papers, collection MC-22, Institute Archives and Special Collections, Massachusetts Institute of Technology Archives, Cambridge, Mass.

Albert, David Z. (1987) "A Quantum-Mechanical Automaton," in *Philosophy of Science* 54:4, pp. 577–585.

Alexander, Franz. (1948) *Fundamental of Psychoanalysis*. New York: W.W. Norton.

Aloisi, Massimo. (1953) "La cibernetica e il cervello umano," in *Società* 9, pp. 531–545.

Antonello, Pierpaolo. (2005) *Il ménage a quattro. Scienza, filosofia, tecnica nella letteratura italiana del Novecento*. Firenze: Le Monnier.

Arbib, Michael A. (1961) "Turing Machines, Finite Automata and Neural Nets," in *Journal of the ACM* 8:4, pp. 467–475.

Aristotle. (1984) *The Complete Works of Aristotle*. Edited by Jonathan Barnes. Princeton, N.J.: Princeton University Press.

Asaro, Peter. (2006) "Working Models and the Synthetic Method: Electronic Brains as Mediators Between Neurons and Behavior," in *Science Studies* 19:1, pp. 12–34.

Asaro, Peter. (2008) "From Mechanisms of Adaptation to Intelligence Amplifiers: The Philosophy of W. Ross Ashby," in *The Mechanical Mind in History*. Edited by Michael Wheeler, Philip Husbands, and Owen Holland. Cambridge, Mass.: MIT Press, pp. 149–184.

——. (2009) "Information and Regulation in Robots, Perception and Consciousness: Ashby's Embodied Minds," in *International Journal of General Systems* 38:2.

Ashby, W. Ross. (1947) "The Nervous System as a Physical Machine; with Special Reference to the Origin of Adaptive Behavior," in *Mind* 56, pp. 44–59.

——. (1948) "Design for a Brain," in *Electronic Engineering* 20, pp. 379–383.

——. (1951a) *Letter to Wiener*. 6 Feb. 1951, box 3, folder 134, Norbert Wiener Papers, collection MC-22, Institute Archives and Special Collections, Massachusetts Institute of Technology Archives, Cambridge, Mass.

——. (1951b) "Statistical Machinery," in *Thalés* 7, pp. 1–8.

——. (1952a) "Can a Mechanical Chess-player Outplay its Designer?" In *The British Journal for the Philosophy of Science* III:9, pp. 44–57.

——. (1952b) *Design for a Brain*. 1st ed. New York: John Wiley and Sons.

——. (1953) "Mechanical Chess Player," in *Cybernetics. Circular Causal and Feedback Mechanisms in Biological and Social Systems*. Edited by Heinz von Foerster, Margaret Mead, and Hans Lukas Teuber. New York: Josiah Macy, Jr. Foundation, pp. 151–154.

——. (1956a) *An Introduction to Cybernetics*. London: Chapman and Hall.

——. (1956b) "Design for an Intelligence Amplifier," in *Automata studies*. Edited by Claude E. Shannon and John McCarthy. Princeton, N.J.: Princeton University Press, pp. 215–234.

——. (1960) *Design for a Brain*. 2nd revised ed. London: Chapman and Hall.

——. (1961) "Brain and Computer," in *Proceedings of the 3rd International Congress of Cybernetics*. Association Internationale de Cybernétique. Nemur, Belgium.

——. (1962) "Simulation of a Brain," in *Computer Applications in the Behavioral Sciences*. Edited by H. Borko. New York: Plenum Press, pp. 452–466.

——. (1967) "The Place of the Brain in the Natural World," in *Currents in Modern Biology* 1:2, pp. 95–104.

——. (1968) "Information Processing in Everyday Human Activity," in *BioScience* 18:1, pp. 190–192.

——. (1970) "Connectance of Large Dynamic (Cybernetic) Systems: Critical Values for Stability," in *Mechanisms of Intelligence: Ross Ashby's Writings on Cybernetics*. Edited by Roger Conant. Seaside, Calif.: Intersystems Publications, pp. 89–91.

——. (1972) "Analysis of the System to be Modeled," in *The Process of Model-Building in the Behavioral Sciences*. Edited by Ralph M. Stogdil. New York: W.W. Norton, pp. 94–114.

——. (1981a) *Mechanisms of Intelligence: Ross Ashby's Writings on Cybernetics*. Edited by Roger Conant. Seaside, Calif.: Intersystems Publications.

——. (1981b) "Setting Goals in Cybernetics Systems," in *Mechanisms of Intelligence: Ross Ashby's Writings on Cybernetics*. Edited by Roger Conant. Seaside, Calif.: Intersystems Publications, pp. 115–126.

Aspray, William. (1990) *John Von Neumann and The Origins of Modern Computing*. Cambridge, Mass.: MIT Press.

Atti Trieste, (1987) *Atti del Convegno "Ricordo di Bruno de Finetti professore nell'Ateneo triestino" Trieste 30–31 maggio 1986*. Trieste: Dipartimento di Matematica applicata "Bruno de Finetti".

Atti Roma (1968) *Atti del convegno sul tema: L'automazione elettronica e le sue implicazioni scientifiche, tecniche e sociali (Roma, 16–19 ottobre 1967)*. Accademia Nazionale dei Lincei. Roma: Accademia Nazionale dei Lincei.

Atti Perugia (1965) *L'unificazione del sapere. Atti del XX Congresso nazionale di filosofia, Perugia, 1965*. Firenze: Sansoni.

Baars, Bernard. (1986) "Interview with George Miller," in *The Cognitive Revolution in Psychology*. Edited by Bernard Baars. New York: The Guilford Press, pp. 200–223.

Balbo, Felice. (1966) *Opere 1945–1964*. Torino: Einaudi.

Baldocchi, M.A. and F. Lenci, eds. (1971) *Atti del Congresso di cibernetica. Casciana Terme 11–13 ottobre 1971*. Pisa: Lito Felici.

Banfi, Antonio. (1945a) "Il problema di una sistematica del sapere," in *Scientia* 39:77;78, pp. 1–6; 1–8.

——. (1945b) "Introduzione," in *La scienza e il mondo moderno*. Milano: Bompiani.

Bara, Bruno G. (1978) "La validazione dei modelli di simulazione," in *Intelligenza artificiale*. Edited by Bruno G. Bara. Milano: Franco Angeli, pp. 67–92.

Barcroft, Joseph. (1932) "La fixité du milieu intérieur est la condition de la vie libre (Claude Bernard) [book review]," in *Biological Reviews and Biological Proceedings of the Cambridge Philosophical Society* 7, pp. 24–87.

Barone, Francesco. (1953) *Il neopositivismo logico*. Torino: Edizioni di *Filosofia*.

——. (1968) "L'automazione e i problemi filosofici," in *Atti del convegno sul tema: L'automazione elettronica e le sue implicazioni scientifiche, tecniche e sociali (Roma, 16–19 ottobre 1967)*. Accademia Nazionale dei Lincei. Roma: Accademia Nazionale dei Lincei, 495–510.

Bates, Elizabeth A. and Jeffrey L. Elman. (1993) "Connectionism and the Study of Change," in *Brain Development and Cognition: a Reader*. Edited by Mark Johnson. Oxford: Blackwell, pp. 623–642.

Bateson, Gregory. (1946) *memo to [invitees]*. 19 June 1946, box 2, folder 71, Norbert Wiener Papers, collection MC-22, Institute Archives and Special Collections, Massachusetts Institute of Technology Archives, Cambridge, Mass.

Bechtel, William and Robert C. Richardson. (1993) *Discovering Complexity: Decomposition and Localization As Strategies In Scientific Research*. Princeton, N.J.: Princeton University Press.

Beeman, William O. (1986) "Freedom to Choose: Symbols and Values in American Advertising," in *Symbolizing America*. Edited by Hervé Varenne. Lincoln, Nebr.: University of Nebraska Press.

Beer, Randall D. (1990) *Intelligence As Adaptive Behavior: An Experiment in Computational Neuroethology*. Boston, Mass.: Academic Press.

——. (In Press) "Dynamical Systems and Embedded Cognition," in *The Cambridge Handbook of Artificial Intelligence*. Edited by K. Frankish and W. Ramsey. Cambridge: Cambridge University Press.

Beeson, Michael J. (1985) *Foundations of Constructive Mathematics: Metamathematical Studies*. Berlin: Springer.

Bell, John S. (1973) "Subject and Object," in *The Physicist's Conception of Nature*. Edited by Jagdish Mehra. Dordrecht: Reidel, pp. 687–690.

Bennett, Max R. and Peter Michael Stephen Hacker. (2003) *Philosophical Foundations of Neuroscience*. Oxford: Blackwell.

Berdyaev, Nicholas. (1965) "Freedom," in *Russian Philosophy*. Edited by James M. Edie, James P. Scanlan, Mary-Barbara Zeldin, and with the collaboration of George L. Kline. Vol. III. Chicago, Ill.: Quadrangle Books.

Berg, Aksel' I. (1964) "Problemy upravleniia i kibernetika," in *Izbrannye trudy*. Vol. 2. Moscow: Energiia.

Berkeley, Edmund C. (1949) *Giant Brains: or Machines That Think*. New York: Science Editions Inc.

Bernard, Claude. (1869) *Discours de M. Claude Bernard prononcé à sa réception à l'Académie Française le 27 Mai 1899*. Paris: Didier.

——. (1966 [1878-1879]) *Leçons sur les phénomènes de la vie communs aux animaux et aux vegetaux*. Paris: Vrin.

——. (1974) *Lectures on the Phenomena of Life Common to Animals and Plants*. Edited by Hebbel E. Hoff, Roger Guillemin, and Lucienne Guillemin (tr.) Springfield, Ill.: Charles C. Thomas.

Bernstein, Jeremy. (1982) *Science Observed: Essays Out of My Mind*. New York: Basic Books.

Bernstein, Nicholas. (1967) *The Co-Ordination and Regulation of Movements*. Oxford: Pergamon Press.

Bienenstock, Elie L., Leon N. Cooper, and Paul W. Munro. (1982) "Theory for the Development of Neuron Selectivity: Orientation Specificity And Binocular Interaction in Visual Cortex," in *Journal of Neuroscience* 2, pp. 32–48.

Black, Harold Stephen. (1934) "Stabilized feedback amplifiers," in *Bell Systems Technical Journal* 13, pp. 1–18.

Bloor, David. (1976) *Knowledge and Social Imagery*. London: Routledge & K. Paul.

Bobrow, Daniel G. and Allan Collins, eds. (1975) *Representation and Understanding: Studies in Cognitive Science*. New York: Academic Press.

Boden, Margaret A. (2006) *Mind As Machine: a History of Cognitive Science*. Oxford: Oxford University Press.

Böhm, Corrado and Dino Dainelli. (1958) "Le attività dell'Istituto nazionale per le applicazioni del calcolo consentite dall'impiego della calcolatrice elettronica Ferranti

(FINAC)," in *Convegno internazionale sui problemi dell'automatismo. Milano, 8–13 aprile 1956*. Vol. I, Roma: Consiglio Nazionale delle Ricerche, pp. 669–676.

Boiko, E. I. et al. (1967) "Kibernetika i problemy psikhologii," in *Kibernetiku—sluzhbu kommunizmu*. Edited by A.I. Berg. Vol. 5. Moscow: Energiia.

Bonfanti, Corrado. (2005) "L'informatica italiana compie cinquant'anni," in *PRISTEM/Storia. Note di Matematica, Storia, Cultura* 12-13, pp. 3–27.

Bontadini, Gustavo. (1955) "Il processo di banalizzazione della filosofia contemporanea," in *Rivista di filosofia neo-scolastica* 47, pp. 291–301.

Boring, Edwin G. (1944) *Letter to Wiener*. 13 Nov. 1944, box 2, folder 66, Norbert Wiener Papers, collection MC-22, Institute Archives and Special Collections, Massachusetts Institute of Technology Archives, Cambridge, Mass.

——. (1945) *Letter to Wiener*. 8 Feb. 1945, box 2, folder 67, Norbert Wiener Papers, collection MC-22, Institute Archives and Special Collections, Massachusetts Institute of Technology Archives, Cambridge, Mass.

Borsellino, Antonio. (1968) "Neuroni, percezioni, intelligenza," in *Atti del convegno sul tema: L'automazione elettronica e le sue implicazioni scientifiche, tecniche e sociali (Roma, 16–19 ottobre 1967)*. Roma: Accademia Nazionale dei Lincei, pp. 219–226.

Botlanski, Jean-Elie. (1995) *La linguistique diachronique*. Paris: Presses Universitaires de France.

Bouquet, Simon. (1997) *Introduction à la lecture de Saussure*. Paris: Payot et Rivages.

Boyce, J. C. (1942) *Letter to Warren Weaver*. 23 Mar. 1942, Record Group 227, OSRD, Division 7, General Project Files, 1940–46, General Mathematical Theory of Prediction and Application, Massachusetts Institute of Technology, Wiener, NDCrc-83, National Archives, Library of Congress, Washington, D.C.

Braitenberg, Valentino. (1968) "Contributi tecnici e concettuali dell'elettronica allo studio dei cervelli viventi," in *Atti del convegno sul tema: L'automazione elettronica e le sue implicazioni scientifiche, tecniche e sociali (Roma, 16–19 ottobre 1967)*. Roma: Accademia Nazionale dei Lincei, pp. 211–217.

Braitenberg, Valentino, Eduardo R. Caianiello, C. Crocchiolo, G. Gambardella, F. Lauria, M. Saffiotti, C. Taddei, G. Trautteur, and U.Vota, eds. (1964) *Ricerche teoriche e sperimentali di cibernetica svolte nell'anno accademico 1961 1962 nel Laboratorio di cibernetica presso l'Istituto di fisica teorica dell'Università di Napoli per conto del Consiglio Nazionale delle Ricerche*. Roma: Consiglio Nazionale delle Ricerche.

Brandom, Robert. (1994) *Making It Explicit: Reasoning, Representing, and Discursive Commitment*. Cambridge, Mass.: Harvard University Press.

Breuer, Thomas. (2001) "Von Neumann, Gödel, and Quantum Incompleteness," in *John von Neumann and the Foundations of Quantum Physics*. Edited by Miklós Rédei and Michael Stöltzner. Dordrecht: Kluwer, pp. 75–82.

Brooks, Rodney. (1999) *Cambrian Intelligence*. Cambridge, Mass.: MIT Press.

Brooks, Rodney and Luc Steels. (1995) *The Artificial Life Route to Artificial Intelligence*. Hillsdale, N.J.: Lawrence Erlbaum Associates.

Bruner, Jerome S. (1973) "The Process of Concept Attainment," in *Beyond the Information Given: Studies in the Psychology of Knowing*. New York: W.W. Norton.

Brushlinskii, Andrei. (1979) "Pochemu nevozmozhen 'iskusstvennyi intellekt'," in *Voprosy filosofii* 2.

Brzozowski, Janusz A. (1964) "Derivatives of Regular Expressions," in *Journal of the ACM* 11:4, pp. 481–494.

Bukovsky, Vladimir. (1999) "Iz knigi 'I vozvrashchaetsia veter'," in *Filosofiia. Logika. Poeziia. Zashchita prav cheloveka: Izbrannoe*. Moscow: RGGU.

Bundy, Alan, ed. (1983) *Proceedings of the 8th International Joint Conference on Artificial Intelligence. Karlsruhe, FRG, August 1983*. Los Altos, Calif.: Morgan Kaufmann.

Burke, Thomas. (1995) "Dance Floor Blues: The Case for a Social AI," in *Stanford Humanities Review* 4:2, pp. 221–248.

Cabanac, Michel. (1971) "Physiological Role of Pleasure," in *Science*. New Series 173:4002, pp. 1103–1107.

——. (1995) *La quête du plaisir: étude sur le conflit des motivations*. Montréal, Canada: Liber.

——. (1996) "On the Origin of Consciousness, a Postulate and its Corollary," in *Neuroscience and Biobehavioral Reviews* 20:1, pp. 33–40.

——. (2002) "What is Emotion?" In *Behaviorial Processes* 60:2, pp. 69–83.

Cadiot, Pierre and Yves-Marie Visetti. (2001) *Pour une théorie des formes sémantiques. Motifs, profils, thèmes*. Paris: Presses Universitaires de France.

——. (2006) *Motifs et proverbs*. Paris: Presses Universitaires de France.

Caffè, Federico. (1968) "Gli aspetti sociali dell'automazione," in *Atti del convegno sul tema: L'automazione elettronica e le sue implicazioni scientifiche, tecniche e sociali (Roma, 16–19 ottobre 1967)*. Roma: Accademia Nazionale dei Lincei, pp. 431–443.

Caianiello, Eduardo R. (1959) "Prolusione," in *Il Nuovo Cimento*. Series X XIII:2 (Suppl.) *Rendiconti del VII Corso che nella Villa Monastero a Varenna dal 7 al 19 luglio 1958 fu tenuto a cura della Scuola internazionale di Fisica della Società italiana di Fisica sulla Teoria dell'informazione*.

——. (1961) "Outline of a Theory of Thought-Processes and Thinking Machines," in *Journal of Theoretical Biology* 2, pp. 204–235.

——. ed. (1965) *Cybernetics of Neural Processes: Course Held at the International School of Physics, Sponsored by Nato at the Istituto di fisica teorica, Università di Napoli April 26–May 13, 1962*. Roma: Consiglio Nazionale delle Ricerche.

——. ed. (1966) *Automata Theory*. New York-London: Academic Press.

Caizzi, Bruno. (1962) *Camillo e Adriano Olivetti*. Torino: UTET.

Calabi, Clotilde. (2005) "Spiegazione e riduzione: Leibniz e i filosofi della mente," in *Linguaggio, mente, conoscenza. Intorno a Leibniz*. Edited by Stefano Gensini. Roma: Carocci, pp. 193–214.

Caldwell, Samuel H. (1940) *Proposal to Section D-2, National Defense Research Committee*. 22 Nov. 1940, introduction to Wiener, "Principles Governing the Construction of Prediction and Compensating Apparatus," Record Group 227, Office of Science and Research Development, Division 7, General Project Files, 1940–46, General Mathematical Theory of Prediction and Application, Massachusetts Institute of Technology.

Calosi, Carlo L. (1989) "L'elettronica e i controlli automatici," in *Civiltà delle macchine. Antologia di una rivista 1953–1957*. Edited by Vanni Scheiwiller. Milano: Scheiwiller.

Calvino, Italo. (1980) *Una pietra sopra. Discorsi di letteratura e società*. Torino: Einaudi.

Cambi, Franco. (1983) *Razionalismo e prassi a Milano (1945–1954)*. Milano: La Goliardica.

Cannon, Walter. (1929) "Organization for Physiological Homeostasis," in *Physiological Reviews* 9, pp. 399–431.

——. (1939[1932]) *The Wisdom of the Body*. 2nd ed. New York: W.W. Norton.

Capehart, Barney Lee. (1967) "Digital Simulation of Homeostat Modified to Show Memory and Learning." PhD thesis. University of Oklahoma.

Caracciolo di Forino, Alfonso, Giuseppe Cecchini, Elio Fabri, and Sergio Sibani. (1958) "Programma e attività del 'Centro studi calcolatrici elettroniche' dell'Università di Pisa," in *Convegno internazionale sui problemi dell'automatismo. Milano, 8–13 aprile 1956*. Vol. I. Roma: Consiglio Nazionale delle Ricerche, pp. 677–687.

Carnap, Rudolf. (1937) *The Logical Syntax of Language*. London: Kegan Paul, Trench, and Trubner.

——. (1966[1932]) "The Elimination of Metaphysics Through Logical Analysis," in *Logical Positivism*. Edited by Alfred J. Ayer. Glencoe, Ill.: Free Press, pp. 60–81.

Carpenter, Brian E., Robert W. Doran, Alan M. Turing, and Michael Woodger. (1986) *A.M. Turing's ACE Report of 1946 and Other Papers*. Cambridge, Mass.: MIT Press, p. 140.

Cassirer, Ernst. (1923) *Philosophie der symbolischen Formen*. Vol. I. Berlin: Bruno Cassirer.

——. (1945) "Structuralism in Modern Linguistics," in *Word* 1:2, pp. 99–120.

——. (1953) *The Philosophy of Symbolic Forms*. New Haven, Conn.: Yale University Press.

Castellana, Mario. (1988–1989) "Materiali per una storia dell'epistemologia in Italia. a. L'esperienza di 'Analisi'," in *Il Protagora* 28/29:13–16, pp. 151–168.

Cavell, Stanley. (1988) "The Uncanniness of the Ordinary," in *In Quest of the Ordinary: Lines of Skepticism and Romanticism*. Chicago, Ill.: The University of Chicago Press, pp. 153–78.

Ceccato, Silvio. (1952) "Contra Dingler, Pro Dingler," in *Methodos* 4:15–16, pp. 223–290.

——. (1953) "Ricordi di colloqui," in *Methodos* 5:18, pp. 250–252.

——. (1968) *Cibernetica per tutti*. Milano: Feltrinelli.

——. (1987) "Leonardo Sinisgalli, Civiltà delle macchine e Adamo II," in *Atti del simposio di studi su Leonardo Sinisgalli. Matera-Montemurro 14–16 maggio 1982*. Matera: Liantonio, pp. 495–503.

——. (1989) "La grammatica insegnata alle macchine," in *Civiltà delle macchine*. Edited by Vanni Scheiwiller. Milano: Scheiwiller, pp. 167–190.

Cellucci, Carlo. (1998) *Le ragioni della logica*. Bari: Laterza.

Chaitin, Gregory J. (2002) *Conversations with a Mathematician. Math, Art, Science and the Limits of Reason*. Berlin: Springer.

Chaitin, Gregory J. (2005) "Omega and Why Maths Has No TOEs," in *Plus Magazine* 37. URL: http://plus.maths.org/issue37/features/omega/.

Chalmers, David J. (1996) *The Conscious Mind*. Oxford: Oxford University Press.

Chiesa, Lorenzo. (2007) *Subjectivity and Otherness. APhilosophical Reading of Lacan*. Cambridge, Mass.: MIT Press.

Chomsky, Noam. (1969[1966]) *La linguistique cartésienne*. Paris: Editions du Seuil.

——. (1981) *Lectures on Government and Bindings. The Pisa Lectures*. Dordrecht: Foris.

Churchland, Patricia. (1986) *Neurophilosophy: Toward a Unified Science of the Mind-Brain*. Cambridge, Mass.: The MIT Press.

——. (2002) *Brain-Wise: Studies in Neurophilosophy*. Cambridge, Mass.: The MIT Press.

Churchland, Paul M. (1992a) *A Neurocomputational Perspective*. Cambridge, Mass.: MIT Press, p. 348.

——. (1992b) "Reductionism, Connectionism, and the Plasticity of Human Consciousness," in *A Neurocomputational Perspective*. Cambridge, Mass.: MIT Press, pp. 129–136.

——. (1995) *The Engine of Reason, the Seat of the Soul: a Philosophical Journey into the Brain*. Cambridge. Mass.: MIT Press.

Colonnetti, Gustavo. (1958) "L'automazione nei suoi aspetti fondamentali," in *Convegno internazionale sui problemi dell'automatismo. Milano, 8–13 aprile 1956*. Vol. I. Roma: Consiglio Nazionale delle Ricerche, pp. 3–8.

Conference on Feedback Mechanisms and Circular Causal Systems in Biology and the Social Sciences. n. d., box 2, folder 71, Norbert Wiener Papers, collection MC-22, Institute Archives and Special Collections, Massachusetts Institute of Technology Archives, Cambridge, Mass.

Convegno internazionale sui problemi dell'automatismo. Milan, 8–13 aprile 1956 (1958). Consiglio Nazionale delle Ricerche. Roma: Consiglio Nazionale delle Ricerche.

Cooper, Steven J. (2008) "From Claude Bernard to Walter Cannon. Emergence of the Concept of Homeostasis," in *Appetite* 51:3, pp. 419–427.

Copeland, Jack and Diane Proudfoot. (2004) "The Computer, Artificial Intelligence, and the Turing test," in *Alan Turing: Life and Legacy of a Great Thinker*. Edited by Christof Teuscher. Frankfurt am Main: Springer, pp. 317–351.

Copi, Irving M., Calvin C. Elgot, and Jesse B. Wright. (1958) "Realization of Events by Logical Nets," in *Journal of the ACM* 5:2, pp. 181–196.

Cordeschi, Roberto. (2002) *Discovery of the Artificial: Behavior, Mind, and Machines Before and Beyond Cybernetics*. Dordrecht: Kluwer Academic Publishers.

——. (2008) "Step Toward the Synthetic Method: Symbolic Information Processing and Self-Organizing Systems in Early Artificial Intelligence Modeling," in *The Mechanical Mind in History*. Edited by Philip Husbands, Owen Holland, and Michael Wheeler. Cambridge, Mass.: MIT Press, pp. 219–258.

Corn, Joseph J. (1983) *The Winged Gospel: America's Romance with Aviation, 1900–1950*. New York: Oxford University Press.

Cox, Russ. (2007) *Regular Expression Matching Can Be Simple And Fast.* URL: http://swtch.com/~rsc/regexp/regexp1.html.

Crane, Conrad C. (1993) *Bombs, Cities, and Civilians: American Airpower Strategy in World War II.* Lawrence, Kans.: University Press of Kansas.

Crevier, Daniel. (1993) *AI: The Tumultuous Search for Artificial Intelligence.* New York: Basic Books.

Crowther-Heyck, Hunter. (2005) *Herbert A. Simon: The Bounds of Reason in Modern America.* Baltimore, Md.: Johns Hopkins University Press.

Cuzzer, Anna. (1954) "La cibernetica come teoria dell'informazione," in *Convegno di elettronica e televisione indetto dal CNR in occasione della XXXII Fiera di Milano (Giornate della scienza, 12–17 aprile 1954).* Vol. II. Roma: Consiglio Nazionale delle Ricerche, pp. 1164–1169.

——. (1993) "La diffusione dell'informatica in Italia," in *La cultura informatica in Italia. Riflessioni e testimonianze sulle origini 1950–1970.* Torino: Bollati Boringhieri, pp. 5–36.

Dadda, Luigi. (1968) "Situazione e prospettiva del calcolo elettronico in Italia," in *Atti del convegno sul tema: L'automazione elettronica e le sue implicazioni scientifiche, tecniche e sociali (Roma, 16–19 ottobre 1967).* Roma: Accademia Nazionale dei Lincei, pp. 35–54.

——. (1993) "Ricordi di un informatico," in *La cultura informatica in Italia. Riflessioni e testimonianze sulle origini 1950–1970.* Torino: Bollati Boringhieri, pp. 67–106.

——. (2005) "Il primo calcolatore del Politecnico di Milano nel 1954," in *PRISTEM/Storia. Note di Matematica, Storia, Cultura* 12–13, pp. 29–41.

Davis, Martin. (1965) *The Undecidable: Basic Papers on Undecidable Propositions, Unsolvable Problems and Computable Functions.* Hewlett, N.Y.: Raven Press.

——. (1982) "Why Gödel Didn't Have Turing's Thesis," in *Information and Control* 54, pp. 3–24.

——. (2000) *The Universal Computer.* New York: W.W. Norton.

Davis, Philip J. and Reuben Hersh. (1981) *The Mathematical Experience.* Boston, Mass.: Birkhäuser.

Dawkins, Richard. (1976) *The Selfish Gene.* Oxford: Oxford University Press.

de Finetti, Bruno. (1949a) "Come funzionano le calcolatrici elettroniche," in *Sapere* 339–340, pp. 38–41.

——. (1949b) "La possibilità di una nuova macchina statistica elettronica," in *Rivista italiana di demografia e statistica* 3, pp. 154–167.

——. (1952a) "Macchine 'che pensano' (e che fanno pensare)," in *Tecnica ed organizzazione. Rivista bimestrale di economia e tecnica dell'industria meccanica* 2–3, pp. 14–31, 14–28.

——. (1952b) "Progressi e prospettive nel campo delle calcolatrici elettroniche," in *Tecnica italiana. Rivista di ingegneria e scienze* 8:1, pp. 25–33.

——. (1968) "Idee da correggere circa l'elettronica," in *Atti del convegno sul tema: L'automazione elettronica e le sue implicazioni scientifiche, tecniche e sociali (Roma, 16–19 ottobre 1967).* Roma: Accademia Nazionale dei Lincei, pp. 321–324.

Dennett, Daniel C. (1978) "Artificial Intelligence as Philosophy and as Psychology," in *Brainstorms*. Cambridge: Cambridge University Press.

——. (1980) "The Milk of Human Intentionality," in *Behavioral and Brain Sciences* 3, pp. 428–430.

——. (1991) *Consciousness Explained*. 1st ed. Boston, Mass.: Little, Brown, and Co.

——. (1998) "The Logical Geography of Computational Approaches: a View from the East Pole," in *Brainchildren. Essays in Designing Minds*. Cambridge, Mass.: MIT Press, pp. 215–234.

——. (2005) *Sweet Dreams. Philosophical Obstacles to a Science of Consciousness*. Cambridge, Mass.: MIT Press.

Denoth, Franco. (2005) "I primi calcolatori: la CEP pisana," in *PRISTEM/Storia. Note di Matematica, Storia, Cultura* 12-13, pp. 60–83.

Der, Ralf. (2001) "Self-Organized Acquisition of Situated Behaviors," in *Theory in Biosciences* 120:3, pp. 179–187.

Der, Ralf, Frank Hesse, René Liebscher, and J. Michael Herrmann. (forthcoming) "Contingent Robot Behavior Generated by Self-Referential Dynamical Systems," in *Biological Cybernetics*.

Desai, Niraj S. (2003) "Homeostatic Plasticity in the Cns: Synaptic and Intrinsic Forms," in *Journal of Physiology-Paris* 97:4-6, pp. 391–402.

Descartes, René. (1984) *The Philosophical Writings of Descartes*. Edited by John Cottingham, Robert Stoothoff, and Dugald Murdoch. Cambridge: Cambridge University Press.

——. (1988) *Philosophical Writings*. Edited by John Cottingham, Robert Stoothoof, and Dugald Murdoch. Cambridge: Cambridge UP.

Descombes, Vincent. (1996) *Les institutions du sens*. Paris: Minuit.

——. (2001) *The Mind's Provisions. A Critique of Cognitivism*. Princeton, N.J.: Princeton University Press.

Deutsch, David. (1985) "Quantum Theory, the Church-Turing Principle and the Universal Quantum Computer," in *Proceedings of the Royal Society of London A* 400, pp. 97–117.

Deutsch, L. Peter and Butler W. Lampson. (1967) "An Online Editor," in *Communications of the ACM* 10:12, pp. 793–799.

Dewey, John. (2004) *Essays in Experimental Logic*. New York: Dover.

Di Giandomenico, Mauro, ed. (2000) *L'uomo e la macchina trent'anni dopo. Filosofia e informatica ieri e oggi. Atti del Convegno Nazionale 1997 della Società Filosofica Italiana Bari 24–26 Ottobre 1997*. Bari: Laterza.

Dingler, Hugo. (1952) "Zu der Kritik von Silvio Ceccato," in *Methodos* 4:15–16, pp. 291–300.

Di Paolo, Ezequiel. (2000a) "Behavioral Coordination, Structural Congruence and Entrainment in A Simulation of Acoustically Coupled Agents," in *Adaptive Behavior* 8:1, pp. 25 –46.

——. (2000b) "Homeostatic Adaptation to Inversion of the Visual Field and Other Sensorimotor Disruptions," in *From animals to animats 6: Proceedings of the 6th International Conference on the Simulation of Adaptive Behavior*. Edited by Jean-Arcady Meyer,

Alain Berthoz, Dario Floreano, Herbert L. Roitblat, and Stewart W. Wilson. Cambridge, Mass.: MIT Press, pp. 440–449.

——. (2003) "Organismically-inspired Robotics: Homeostatic Adaptation and Teleology Beyond the Closed Sensorimotor Loop," in *Dynamical Systems Approach to Embodiment and Sociality*. Edited by K. Murase and T. Akasura. Adelaide, Australia: Advanced Knowledge International, pp. 19–42.

——. (2005) "Autopoiesis, Adaptivity, Teleology, Agency," in *Phenomenology and the Cognitive Sciences* 4:4, pp. 429–452.

Di Paolo, Ezequiel, Marieke Rohde, and Hanneke De Jaegher. (2011) "Horizons for the Enactive Mind: Values, Social Interaction, and Play," in *Enaction: Towards a New Paradigm for Cognitive Science*. Edited by John Stewart, Olivier Gapenne, and Ezequiel Di Paolo. Cambridge, Mass.: MIT Press.

Dobruskin, Elektron and Viacheslav Fomenko. (2008) "Istinno svobodnyi chelovek," in *Electron's scientific seminar*. URL: http://www.elektron2000.com/dobruskin_fomenko_0096.html.

Donskoi, Mikhail. (1999) "Istoriia 'Kaissy'," in *PC Week (Russian edition)* 20–21. URL: http://www.pcweek.ru/themes/detail.php?ID=51235.

Dower, John W. (1986) *War Without Mercy: Race and Power in the Pacific War*. 1st ed. New York: Pantheon Books.

Dretske, Fred. (1986) "Misrepresentation," in *Belief: Form, Content, and Function*. Oxford: Clarendon Press, pp. 17–36.

Dreyfus, Hubert L. (1992) *What Computers Still Can't Do: a Critique of Artificial Reason*. 1st ed. Cambridge, Mass.: MIT Press.

Dumit, Joseph. (2000) "Artificial Participation: an Interview with Warren Sack," in *Zeroing in on the Year 2000: The Final Edition (Late Editions, 8)*. Edited by George E. Marcus. Chicago, Ill.: The University of Chicago Press, pp. 59–87.

Dupuy, Jean-Pierre. (1994) *Aux origines des sciences cognitives*. Paris: La découverte.

——. (2000) *The Mechanization of the Mind: On the Origins of Cognitive Science*. Princeton, N.J.: Princeton University Press.

Durkheim, Émile. (1912) *Les formes élémentaires de la vie religieuse. Le système totémique en Australie*. Paris: Presses Universitaires de France.

Eddington, Arthur. (1928) *The Nature of the Physical World*. London: J. M. Dent & Sons LTD.

Eden, Murray. (1973) "Foreword," in Mikhail Tsetlin, *Automaton Theory and Modeling of Biological Systems*. Trans. by Scitran. New York: Academic Press.

Edwards, Paul N. (1996) *The Closed World: Computers and the Politics of Discourse in Cold War America*. Cambridge, Mass.: MIT Press.

Eiben, Agoston E. and Gunther Rudolph. (1999) "Theory of Evolutionary Algorithms: a Bird's Eye View," in *Theoretical Computer Science* 229:1–2, pp. 3–9.

Eldredge, Niles. (1995) *Reinventing Darwin: The Great Debate at the High Table of Evolutionary Theory*. New York: John Wiley.

Eldridge, Alice. (2002) "Ashby's Homeostat in Simulation." URL: http://www.ecila.org/ecila_files/content/papers/ACEhom.pdf.

Elman, Jeffrey L. (1990) "Finding Structure in Time," in *Cognitive Science* 14:2, pp. 179–211.

Elman, Jeffrey L., Elizabeth A. Bates, Mark H. Johnson, Annette Karmiloff-Smith, Domenico Parisi, and Kim Plunkett. (1996) *Rethinking Innateness, a Connectionist Perspective on Development*. Cambridge, Mass.: MIT Press.

Esenin-Vol'pin, Aleksandr S. (1961) *A Leaf of Spring*. Trans. by George Reavey. New York: Praeger.

Faedo, Alessandro. (1968) "Origine ed evoluzione dei calcolatori elettronici," in *Atti del convegno sul tema: L'automazione elettronica e le sue implicazioni scientifiche, tecniche e sociali (Roma, 16–19 ottobre 1967)*. Roma: Accademia Nazionale dei Lincei, pp. 17–33.

Feferman, Solomon. (1977) "Categorical Foundations and Foundations of Category Theory," in *Logic, Foundations of Mathematics and Computability Theory*. Edited by Robert E. Butts and Jaakko Hintikka. Vol. I. Dordrecht/Boston: Reidel, pp. 149–169.

——. (1996) "Gödel's Program for New Axioms: Why, Where, How and What?" In *Gödel '96: Logical Foundations of Mathematics, Computer Science and Physics—Kurt Gödel's Legacy*. Edited by Petr Hájek. Berlin: Springer, pp. 3–22.

——. (2006a) "Are There Absolutely Unsolvable Problems? Gödel's Dichotomy," in *Philosophia Mathematica* 14:2, pp. 134–152.

——. (2006b) "Enriched Stratified Systems for the Foundations of Category Theory?" In *What is Category Theory*. Edited by G. Sica. Monza: Polimetrica, pp. 185–203.

——. (2006c) "Open-Ended Schematic Axiom Systems (abstract)," in *Bulletin of Symbolic Logic* 12, p. 145.

——. (2008) "Lieber Herr Bernays!, Lieber Herr Gödel! Gödel on Finitism, Constructivity and Hilbert's Program," in *Dialectica* 62:2, pp. 179–203.

Feferman, Solomon, Robert M. Soloway, and Judson C. Webb. (1990) "Introductory Note to 1972a," in Kurt Gödel, *Collected Works, Volume II: Publications 1938–1974*. Edited by Solomon Feferman, John W. Dawson, Stephen C. Kleene, Gregory H. Moore, Robert M. Solovay, and Jean van Heijenoort. Oxford: Clarendon Press, pp. 281–303.

Feferman, Solomon and Thomas Strahm. (2000) "The Unfolding of Non-Finitist Arithmetic," in *Annals of Pure and Applied Logic* 104:1–3, pp. 75–96.

Fehr, Johannes. (1997) *Linguistik und Semiologie: Notizen aus dem Nachlass Texte, Briefe und Dokumente*. Frankfurt am Main: Suhrkamp.

——. (2000) *Saussure entre linguistique et sémiologie*. Paris: Presses Universitaires de France.

Feinberg, E.L. (1996) "Dlia budushchego istorika," in *On mezhdu nami zhil: Vospominaniia o Sakharove*. Edited by B.L. Al'tshuler et al. Moscow: Praktika.

Fischbein, Efraim. (1987) *Intuition in Science and Mathematics: An Educational Approach*. Dordrecht: Reidel.

Fisher Servi, Gisèle. (2001) *Quando l'eccezione è la regola. Le logiche non monotone*. Milano: McGraw-Hill.

Fodor, Jerry A. (1975) *The Language of Thought*. New York: Crowell Company.

——. (1983) *The Modularity of Mind*. Cambridge, Mass.: MIT Press.

——. (1987) *Psychosemantics: The Problem of Meaning in the Philosophy of Mind*. Cambridge, Mass.: MIT Press.

——. (2000) *The Mind Does not Work that Way: the Scope and Limits of Computational Psychology*. Cambridge, Mass.: MIT Press.

Fondazione Adriano Olivetti. (1993) *La cultura informatica in Italia. Riflessioni e testimonianze sulle origini 1950–1970*. Torino: Bollati Boringhieri.

Fortun, Mike and Silvan S. Schweber. (1993) "Scientists and the Legacy of World War II: The Case of Operations Research (OR)," in *Social Studies of Science* 23:4, pp. 595–642.

Foucault, Michel. (1966) *Les mots et les choses*. Paris: Gallimard.

Franchi, Stefano. (1997) "Endgames. Game and Play at the End of Philosophy." PhD thesis. Stanford University.

——. (2004) "L'Intelligenza artificiale e la teoria dei giochi. Un'indagine storico-critica," in *Percezione, linguaggio, coscienza. Percorsi tra cognizione e intelligenza artificiale*. Edited by Maurizio Matteuzzi and Francesco Bianchini. Macerata: Quodlibet, pp. 63–88.

——. (2006) "Herbert Simon, Anti-Philosopher," in *Computing and Philosophy*. Edited by Lorenzo Magnani. Pavia: Associated International Academic Publishers, pp. 27–40.

——. (2007) "Intelligenza artificiale come scienza, metafisica, o epistemologia. Note in margine ad uno scambio con Herbert Simon," in *Verso un'archeologia dell'intelligenza artificiale*. Edited by Stefano Franchi, Francesco Bianchini, and Maurizio Matteuzzi. Macerata: Quodlibet, pp. 19–39.

——. (2011) "Jammed Machines and Contingently Fit Animals: Psychoanalysis's Biological Paradox," in *French Literature Studies* 38, in press.

Franchi, Stefano and Güven Güzeldere. (2005a) "Machinations of the Mind," in *Mechanical Bodies, Computational Minds*. Edited by Stefano Franchi and Güven Güzeldere. Cambridge, Mass.: MIT Press, pp. 1–135.

——. eds. (2005b) *Mechanical Bodies, Computational Minds: Artificial Intelligence from Automata to Cyborgs*. Cambridge, Mass.: MIT Press.

Frege, Gottlob. (1980) *Philosophical and Mathematical Correspondence*. Oxford: Blackwell.

Freud, Sigmund. (1953) *The Standard Edition of the Complete Psychological Works of Sigmund Freud*. Vol. 1: *Project for a Scientific Psychology*. London: Hogarth Press, pp. 283–387.

Friedl, Jeffrey. (2006) *Mastering Regular Expressions*. 3rd ed. Sebastopol, Calif.: O'Reilly Media, Inc.

Fritzsche, Peter. (1992) *A Nation of Fliers: German Aviation and the Popular Imagination*. Cambridge, Mass.: Harvard University Press.

Fubini, Eugenio G. (1968) "Il calcolatore elettronico nella società del futuro," in *Atti del convegno sul tema: L'automazione elettronica e le sue implicazioni scientifiche, tecniche e sociali (Roma, 16–19 ottobre 1967)*. Roma: Accademia Nazionale dei Lincei, pp. 395–404.

Gaaze-Rapoport, Modest and Dmitrii Pospelov. (1987) *Ot ameby do robota: modeli povedeniia*. Moscow: Nauka.

Gadamer, Hans-Georg. (1996) *Truth and Method*. Trans. by Joel Weinsheimer and Donald G. Marshall. 2nd ed. New York: Continuum.

Galison, Peter. (1993) "The Cultural Meaning of *Aufbau*," in *Scientific Philosophy: Origins and Developments*. Edited by Friedrich Stadler. Dordrecht: Springer, pp. 75–93.

——. (1994) "The Ontology of the Enemy: Norbert Wiener and the Cybernetic Vision," in *Critical Inquiry* 21, pp. 228–266.

——. (1997) *Image and Logic: A Material Culture of Microphysics*. Chicago, Ill.: The University of Chicago Press.

Gallino, Luciano. (2003) *La scomparsa dell'Italia industriale*. Torino: Einaudi.

Gandy, Robin. (1980) "Church's Thesis and the Principle for Mechanism," in *The Kleene Symposium*. Edited by Stephen C. Kleene, Jon Barwise, H. Jerome Keisler, and Kenneth Kunen. Amsterdam: North-Holland, pp. 123–148.

——. (1988) "The Confluence of Ideas in 1936," in *A Half-Century Survey on the Universal Turing Machine*. Edited by Rolf Herken. 2nd ed. Oxford: Oxford University Press, pp. 55–111.

Gardner, Howard. (1985) *The Mind's New Science: a History of the Cognitive Revolution*. New York: Basic Books.

Geertz, Clifford. (1983) "Common Sense as a Cultural System," in *Local Knowledge. Further Essays in Interpretive Anthropology*. New York: Basic Books, pp. 73–93.

Gentile, Marino. (1968) "Implicazioni della automazione nel costume e nella mentalità," in *Atti del convegno sul tema: L'automazione elettronica e le sue implicazioni scientifiche, tecniche e sociali (Roma, 16–19 ottobre 1967)*. Roma: Accademia Nazionale dei Lincei, pp. 485–494.

Gerovitch, Slava. (2002) *From Newspeak to Cyberspeak: a History of Soviet Cybernetics*. Cambridge, Mass.: MIT Press.

Geymonat, Ludovico. (1945) *Studi per un nuovo razionalismo*. Torino: Chiantore.

Gilbert, Martin. (1989) *The Second World War: a Complete History*. 1st American ed. New York: H. Holt.

g. j. (1967) "I filosofi, l'uomo e la macchina," in *Civiltà delle macchine* 15:3, pp. 2–6.

Gödel, Kurt. (193?) "Undecidable Diophantine Propositions," in *Collected Works, Volume III: Unpublished Essays and Lectures*. Edited by Solomon Feferman, John W. Dawson, Stephen C. Kleene, Gregory H. Moore, Robert M. Solovay, and Jean van Heijenoort. Oxford: Clarendon Press, pp. 164–173.

——. (1934) "On Undecidable Propositions of Formal Mathematical Systems," in *Collected Works, Volume I: Publications 1929–1936*. Edited by Solomon Feferman, John W. Dawson jr., Stephen C. Kleene, Gregory H. Moore, Robert M. Solovay, and Jean van Heijenoort. Oxford: Clarendon Press, pp. 346–371.

——. (1946) "Remarks Before the Princeton Bicentennial Conference on Problems nn Mathematics," in *Collected Works, Volume II: Publications 1938–1974*. Edited by Solomon Feferman, John W. Dawson, Stephen C. Kleene, Gregory H. Moore, Robert M. Solovay, and Jean van Heijenoort. Oxford: Clarendon Press, pp. 150–153.

——. (1951) "Some Basic Theorems on the Foundations of Mathematics and Their Implications," in *Collected Works, Volume III: Unpublished Essays and Lectures*. Edited by Solomon Feferman, John W. Dawson, Stephen C. Kleene, Gregory H. Moore, Robert M. Solovay, and Jean van Heijenoort. Oxford: Clarendon Press, pp. 304–323.

——. (1958) "Über eine bisher noch nicht benüzte Erweiterung des finiten Standpunktes," in *Dialectica* 12:3–4, pp. 280–287.

——. (1964) "Post Scriptum to 'Undecidable Propositions of Formal Mathematical Systems'," in *Collected Works, Volume I: Publications 1929–1936*. Edited by Solomon Feferman, John W. Dawson jr., Stephen C. Kleene, Gregory H. Moore, Robert M. Solovay, and Jean van Heijenoort. Oxford: Clarendon Press, pp. 369–371.

——. (1972a) "On an Extension of Finitary Mathematics Which Has Not Yet Been Used," in *Collected Works, Volume II: Publications 1938–1974*. Edited by Solomon Feferman, John W. Dawson, Stephen C. Kleene, Gregory H. Moore, Robert M. Solovay, and Jean van Heijenoort. Oxford: Clarendon Press, pp. 271–280.

——. (1972b) "Some Remarks on the Undecidability Results," in *Collected Works, Volume II: Publications 1938–1974*. Edited by Solomon Feferman, John W. Dawson, Stephen C. Kleene, Gregory H. Moore, Robert M. Solovay, and Jean van Heijenoort. Oxford: Clarendon Press, pp. 305–306.

——. (1986) *Collected Works, Volume I: Publications 1929–1936*. Edited by Solomon Feferman, John W. Dawson jr., Stephen C. Kleene, Gregory H. Moore, Robert M. Solovay, and Jean van Heijenoort. Oxford: Clarendon Press.

——. (1990) *Collected Works, Volume II: Publications 1938–1974*. Edited by Solomon Feferman, John W. Dawson, Stephen C. Kleene, Gregory H. Moore, Robert M. Solovay, and Jean van Heijenoort. Oxford: Clarendon Press.

——. (1995) *Collected Works, Volume III: Unpublished Essays and Lectures*. Edited by Solomon Feferman, John W. Dawson, Stephen C. Kleene, Gregory H. Moore, Robert M. Solovay, and Jean van Heijenoort. Oxford: Clarendon Press.

——. (2003a) *Collected Works, Volume IV: Correspondence A-G*. Edited by Solomon Feferman, jr John W. Dawson, Warren Goldfarb, Charles Parsons, and Wilfried Sieg. Oxford: Clarendon Press.

——. (2003b) *Collected Works, Volume V: Correspondence H-Z*. Edited by Solomon Feferman, John W. Dawson jr., Warren Goldfarb, Charles Parsons, and Wilfried Sieg. Oxford: Clarendon Press.

Godel, Robert, ed. (1957) *Le sources manuscrites du* Cours de Linguistique générale *de F. de Saussure*. Paris-Genève: Droz.

Goldstine, Herman H. and John von Neumann. (1987[1946]) "On the Principles of Large Scale Computing Machines," in *Papers of John von Neumann on Computers and Computing Theory*. Edited by William Aspray and Arthur Burks. Cambridge, Mass.: MIT Press.

Gordin, Michael D., Karl Hall, and Alexei Kojevnikov, eds. (2008) *OSIRIS, vol. 23: Intelligentsia Science: The Russian Century, 1860–1960*. Chicago, Ill.: The University of Chicago Press.

Gould, Stephen Jay. (2002) *The Structure of Evolutionary Theory*. Cambridge, Mass.: Belknap Press of Harvard University Press.

Greene, Brian. (2000) *The Elegant Universe: Superstrings, Hidden Dimensions, and the Quest for the Ultimate Theory*. New York: Random House.

Grossberg, Stephen. (1968) "Some physiological and biochemical consequences of psychological postulates," in *Proceedings of the National Academy of Sciences of the United States of America* 60:3, pp. 758–765.

——. (1969) "Some networks that can learn, remember, and reproduce any number of complicated space–time patterns, I," in *Journal of Mathematics and Mechanics* 19:1, pp. 53–91.

Gunderson, Keith. (1964) "The Imitation Game," in *Minds and Machines*. Edited by Alan Ross Anderson. Englewood Cliffs, N.J.: Prentice-Hall, pp. 60–71.

Guzzo, Augusto. (1955) *La scienza*. Torino: Edizioni di *Filosofia*.

——. (1962) "Congresso filosofico in Puglia e Replica," in *La filosofia di fronte alle scienze*. Bari: Adriatica editrice, pp. 9–23, 789–808.

——. (1968) "Conclusioni filosofiche su l'automazione elettronica e le sue implicazioni," in *Atti del convegno sul tema: L'automazione elettronica e le sue implicazioni scientifiche, tecniche e sociali (Roma, 16–19 ottobre 1967)*. Roma: Accademia Nazionale dei Lincei, pp. 511–527.

——. (1967a) "Le macchine," in Società Filosofica Italiana, *L'uomo e la macchina. Atti del XXI Congresso Nazionale di Filosofia, Pisa 22–25 Aprile 1967*. Torino: Edizioni di *Filosofia*, pp. 1–25.

——. (1967b) "Le profezie degli scienziati e la loro legittimità," in *L'uomo e la macchina*. Torino: Edizioni di *Filosofia*.

Habermas, Jürgen. (1985) "Hans-Georg Gadamer: Urbanizing the Heideggerian Province," in *Philosophical-Political Profiles*. Cambridge, Mass: MIT Press, pp. 189–197.

Haldane, John Scott. (1913) *Mechanism, Life and Personality: An Examination of the Mechanistic Theory of Life and Mind*. London: John Murray.

——. (1917) *Organism and Environment As Illustrated by the Physiology of Breathing*. New Haven, Conn.: Yale University Press.

——. (1922) *Respiration*. New Haven, Conn.: Yale University Press.

Hallett, Michael. (1989) "Physicalism, reductionism & Hilbert," in *Physicalism in Mathematics*. Edited by A. D. Irvine. Dordrecht: Kluwer, pp. 183–255.

Haraway, Donna J. (1991a) "A Cyborg Manifesto: Science, Technology, and Socialist-Feminism in the Late Twentieth Century," in *Simians, Cyborgs, and Women: The Reinvention of Nature*. New York: Routledge, pp. 143–181.

——. (1991b) *Simians, Cyborgs, and Women: The Reinvention of Nature*. New York: Routledge.

——. (1991c) "The Biological Enterprise: Sex, Mind, and Profit from Human Engineering to Sociobiology," in *Simians, Cyborgs, and Women: The Reinvention of Nature*. New York: Routledge, pp. 43–69.

Haroules, George G. and P. F. Haire. (1960) *Jenny: an Improved Homeostat*. Tech. rep. AFCRC-TN-60-379. Air Force Cambridge Research Center.

Harvey, Inman, Philip Husbands, Dave Cliff, Adrian Thompson, and Nick Jakobi. (1997) "Evolutionary Robotics: the Sussex Approach," in *Robotics and Autonomous Systems* 20, pp. 205–224.

Harvey, Inman, Ezequiel Di Paolo, Rachel Wood, Matt Quinn, and Elio Tuci. (2005) "Evolutionary Robotics: a New Scientific Tool for Studying Cognition," in *Artificial Life* 11:1–2, pp. 79–98.

Hastings, Max. (1979) *Bomber Command*. New York: Dial Press/J. Wade.

Haugeland, John. (1980) "Programs, Causal Powers, and Intentionality," in *Behavioral and Brain Sciences* 3, pp. 432–433.

——. (1981[1978]) "The Nature and Plausibility of Cognitivism," in *Mind Design*. Edited by John Haugeland. Cambridge, Mass.: MIT Press, pp. 243–281.

——. (1985) *Artificial Intelligence. The Very Idea*. Cambridge, Mass.: MIT Press.

Hayles, N. Katherine. (1990) *Chaos Bound: Orderly Disorder in Contemporary Literature and Science*. Ithaca, N. Y.: Cornell University Press.

Hazen, Harold L. (1934) "Theory of servomechanisms," in *Journal of the Franklin Institute* 218, pp. 283–331.

Hebb, Donald O. (1949) *The Organization of Behavior*. New York: John Wiley and Sons.

Heidbreder, Edna. (1933) *Seven Psychologies*. Student's ed. New York: The Century Co.

Heims, Steve J. (1977) "Gregory Bateson and the Mathematicians: From Interdisciplinary Interaction to Societal Functions," in *Journal of the History of the Behavioral Sciences* 13, pp. 141–59.

——. (1980) *John von Neumann and Norbert Wiener: From Mathematics to the Technologies of Life and Death*. Cambridge, Mass.: MIT Press.

——. (1991) *The Cybernetics Group*. Cambridge, Mass.: MIT Press.

——. (1993) *Constructing a Social Science for Postwar America: The Cybernetics Group, 1946–1953*. Cambridge, Mass.: MIT Press.

Henderson, Lawrence Joseph. (1928) *Blood: a Study in General Physiology*. New Haven, Conn.: Yale University Press.

Herken, Rolf, ed. (1995) *The Universal Turing Machine: A Half-Century Survey*. 2nd ed. Oxford: Oxford University Press.

Herrmann, J. Michael, Michael Holicki, and Ralf Der. (2004) "On Ashby's Homeostat: A Formal Model of Adaptive Regulation," in *From Animals to Animats 8. Proceedings of the Eighth International Conference on Simulation of Adaptive Behavior*. Edited by Stefan Schaal. Cambridge, Mass.: MIT Press, pp. 324–333.

Hertz, Heinrich. (1894) *Die Prinzipien der Mechanik*. Leipzig: Johann Ambrosius Warth (Arthur Meiner).

Hertz, Heinrich. (1956) *The Principles of Mechanics*. New York: Dover.

Hilbert, David. (1900) "Mathematische Probleme," in *Nachrichten von der Königlichen Gesellschaft der Wissenschaften zu Göttingen, Math.-Phys. Klasse*, pp. 253–297.

——. (1976) "Mathematical Problems," in *Mathematical Developments Arising From Hilbert Problems. Proceedings Of the Symposium on Pure Mathematics of the American Mathematical Society, Northern Illinois University, 1974*. Edited by Felix E. Browder. Vol. XXVIII. Providence, R. I.: American Mathematical Society, pp. 1–34.

Hilbert, David, Lothar Nordheim, and John von Neumann. (1927) "Über die Grundlagen der Quantenmechanik," in John von Neumann, *Collected Works*. Edited by Abraham H. Taub. Vol. 1. New York: Pergamon Press, pp. 104–133.

Hodges, Andrew. (2004) "What Would Alan Turing Have Done after 1954?" In *Alan Turing: Life and Legacy of a Great Thinker*. Edited by Christof Teuscher. Frankfurt am Main: Springer, pp. 43–58.

Hofstadter, Douglas R. (1985) *Metamagical Themas. Questing for the Essence of Mind and Pattern*. Harmondsworth, UK: Penguin Books.

Hofstadter, Douglas R. and Daniel C. Dennett. (1981) *The Mind's I. Fantasies and Reflections on Self and Soul*. New York: Basic Books.

Hofstadter, Douglas R. and the Fluid Analogies Research Group. (1995) *Fluid Concepts & Creative Analogies: Computer Models of the Fundamental Mechanisms of Thought*. New York: BasicBooks.

Holloway, David. (1999) "Physics, the State, and Civil Society in the Soviet Union," in *Historical Studies in the Physical and Biological Sciences* 30:1.

Holmes, Frederic L. (1969) "Joseph Barcroft and the Fixity of the Internal Environment," in *Journal of the History of Biology* 2, pp. 88–122.

Hopcroft, John E. and Jeffrey D. Ullman. (1969) *Formal Languages and Their Relation to Automata*. Boston, Mass.: Addison-Wesley.

Hopfield, John J. (1982) "Neural Networks and Physical Systems with Emergent Collective Computational Abilities," in *Proceedings of the National Academy of Science* 79, pp. 2554–2558.

——. (1984) "Neurons with a Graded Response Have Collective Computational Properties Like Those of Two-State Neurons," in *Proceedings of the National Academy of Science USA* 81, pp. 3088–3092.

Hösle, Vittorio. (2001) "Ontology and Ethics in Hans Jonas," in *Graduate Faculty Philosophy Journal* 23:1, pp. 31–50.

Hounshell, David A. (1997) "The Cold War, RAND, and the Generation of Knowledge, 1946–1962," in *Historical Studies in the Physical and Biological Sciences* 27:2, pp. 237–267.

Hughes, R. I. G. (1997) "Models and Representations," in *Philosophy of Science* 64 Supplement. Proceedings of the 1996 Biennial Meetings of the Philosophy of Science Association. S325–S336.

Hull, Clark Leonard. (1943) *Principles of Behavior, an Introduction to Behavior Theory*. New York: D. Appleton-Century.

Husbands, Philip and Owen Holland. (2008) "The Ratio Club: a Hub of British Cybernetics," in *The Mechanical Mind in History*. Edited by Philip Husbands, Owen Holland, and Michael Wheeler. Cambridge, Mass.: MIT Press, pp. 91–148.

Husbands, Philip, Owen Holland, and Michael Wheeler, eds. (2008) *The Mechanical Mind in History*. Cambridge, Mass.: MIT Press.

Iizuka, Hiroyuki and Ezequiel Di Paolo. (2007) "Toward Spinozist Robotics: Exploring the Minimal Dynamics of Behavioral Preference," in *Adaptive Behavior* 15:4, pp. 359–376.

Ikegami, Takashi and Keisuke Suzuki. (2008) "From a Homeostatic to a Homeodynamic Self," in *Biosystems* 91:2, pp. 388–400.

Ivanov, Viacheslav Vs. (1998) "Iz istorii kibernetiki v SSSR. Ocherk zhizni i deiatel'nosti M.L. Tsetlina," in *Ocherki istorii informatiki v Rossii*. Edited by Dmitrii A. Pospelov and Iakov I. Fet. Novosibirsk: OIGG SO RAN.

Jackson, Dunham. (1940) *Letter to Harry Bateman, E. J. McShane, M. H. Stone, J. H. Van Vleck, Wiener, and S. S. Wilks*. 14 Feb. 1940, box 1, folder 56, Norbert Wiener Papers, collection MC-22, Institute Archives and Special Collections, Massachusetts Institute of Technology Archives, Cambridge, Mass.

Jakobson, Roman. (1929a) "Principes de phonologie historique," in *Selected Writings*. Vol. 1. Berlin: Mouton de Gruyter, pp. 202–220.

——. (1929b) "Über die heutigen Voraussetzungen der russiche Slavistik," in *Slavische Rundschau* 1, pp. 629–646.

——. (1962) *Selected Writings*. Berlin: Mouton de Gruyter.

——. (1978) *Six Lectures on Sound and Meaning*. Edited by John Mepham. Hassocks, UK: Harvester Press.

——. (1988) *Semiotik. Ausgewählte Texte 1919–1982*. Edited by Elmar Holenstein. Frankfurt am Main: Suhrkamp.

Jonas, Hans. (1966) *The Phenomenon of Life: Towards a Philosophical Biology*. New York: Harper and Row.

——. (1984) *The Imperative of Responsibility. Foundations of an Ethics for the Technological Age*. Chicago, Ill.: The University of Chicago Press.

Jones, Matthew L. (2006) *The Good Life in the Scientific Revolution: Descartes, Pascal, Leibniz, and the Cultivation of Virtue*. Chicago, Ill.: The University of Chicago Press.

Jordan, Michael I. (1997) "Serial Order: a Parallel Distributed Processing Approach," in *Neural-Network Models of Cognition: Biobehavioral Foundations*. Edited by John W. Donahoe and Vivian Packard Dorsel. Amsterdam: Elsevier.

Kant, Immanuel. (1998) *Critique of Pure Reason*. Edited by Paul Guyer and Allen W. Wood. Cambridge: Cambridge University Press.

——. (2000) *Critique of the Power of Judgment*. Edited by Paul Guyer. Cambridge: Cambridge University Press.

Karsenti, Bruno. (2006) "Le 'dilemme durkheimien' en sociologie morale," in *La société en personne. Etudes durkheimiennes*. Paris: Economica.

Kay, Lily E. (2000) *Who Wrote the Book of Life?: a History of the Genetic Code*. Stanford, Calif.: Stanford University Press.

——. (2002) "From Logical Neurons to Poetic Embodiments of Mind: Warren S. McCulloch's Project in Neuroscience," in *Science in Context* 14:04, pp. 591–614.

Keller, Evelyn. (1996) *The Body of a New Machine: Situating the Organism between Telegraphs and Computers*. New York: Columbia University Press.

Kelty, Christopher. (2008) *Two Bits: The Cultural Significance of Free Software*. Durham, N.C.: Duke University Press.

Kelty, Christopher and Hannah Landecker. (2004) "A Theory of Animation: Cells, L-systems, and Film," in *Grey Room* 17, pp. 30–63.

Kharkhordin, Oleg. (1999) *The Collective and the Individual in Russia: A Study of Practices*. Berkeley, Calif.: University of California Press.

Khronika tekushchikh sobytii (1968). URL: http://www.memo.ru/history/DISS/chr/chr2.htm.

Kitcher, Patricia. (1992) *Freud's Dream: A Complete Interdisciplinary Science of Mind*. Cambridge, Mass.: MIT Press.

Kleene, Stephen C. (1956) "Representation of Events in Nerve Nets and Finite Automata," in *Automata Studies*. Vol. 34. Princeton University Press, pp. 3–41.

——. (1979) "Origins of Recursive Function Theory," in *Proceedings of the 20th Annual Symposium on Foundations of Computer Science*. IEEE Computer Society Washington, DC, USA, pp. 371–382.

Kohler, Ivo. (1955) "Experiments with Prolonged Optical Distortions," in *Acta Psychologica* 11, pp. 176–178.

——. (1964) *Formation and Transformation of the Perceptual World*. New York: International Universities Press.

Kojevnikov, Alexei. (1998) "Rituals of Stalinist Culture at Work: Science and the Games of Intraparty Democracy circa 1948," in *The Russian Review* 57, pp. 25–52.

Kornblith, Hilary, ed. (1985) *Naturalizing Epistemology*. Cambridge, Mass: MIT Press.

——. (2002) *Knowledge and Its Place in Nature*. Oxford: Clarendon Press.

Koyré, Alexandre. (1957) *From the Closed World to the Infinite Universe*. Baltimore, Md.: Johns Hopkins University Press.

Kreisel, G. (1972) "Which Number Theoretic Problems Can Be Solved in Recursive Progressions on Π_1^1-paths through *O*?" In *Journal of Symbolic Logic* 37, pp. 311–334.

Krementsov, Nikolai. (1997) *Stalinist Science*. Princeton, N.J.: Princeton University Press.

Kroeber, A. L. (1917) "The Superorganic," in *American Anthropologist* 19:2, pp. 162–213.

Kronrod, Aleksandr. (2001) *Besedy o programmirovanii*. Moscow: URSS.

Küpfmüller, Karl. (1928) "Über die Dynamik der selbsttätigen Verstärkungsregler," in *Elektrische Nachrichtentechnik* 5:11, pp. 459–467.

Kusch, Martin. (1995) *Psychologism: A Case Study in the Sociology of Philosophical Knowledge*. London: Routledge.

Labov, William. (1968) *Sociolinguistics Patterns*. Philadelphia, Pa.: University of Pennsylvania Press.

Lacki, Jan. (2000) "The Early Axiomatizations of Quantum Mechanics: Jordan, Von Neumann, and the Continuation of Hilbert's Program," in *Archive for the History of the Exact Sciences* 54, pp. 279–318.

Lakoff, George. (1987) *Women, Fire, and Other Dangerous Things. What Categories Reveal about the Mind*. Chicago, Ill.: The University of Chicago Press.

Lakoff, George and Mark Johnson. (1980) *Metaphors We Live By*. Chicago, Ill.: The University of Chicago Press.

Lakoff, George and Rafael E. Núñez. (2000) *Where Mathematics Comes From: How the Embodied Mind Brings Mathematics into Being*. New York: Basic Books.

Laks, Bernard. (1996) *Langage et cognition, l'approche connexionniste*. Paris: Hermès.

Landis, Evgenii M. and Isaak M. Yaglom. (2001) "About Aleksandr Semenovich Kronrod," in *Russian Mathematical Surveys* 56:5.

Langacker, Ronald. (1987) *Foundations of Cognitive Grammar*. Stanford, Calif.: Stanford University Press.

Langewiesche, Wolfgang and Leighton Holden Collins. (1972) *Stick and Rudder: An Explanation of the Art of Flying*. New York: McGraw-Hill.

Langley, Leroy L., ed. (1973) *Homeostasis. Origins of the Concept*. Stroudsburg., Pa.: Dowden, Hutchinson & Ross.

Latour, Bruno. (1987) *Science In Action: How to Follow Scientists and Engineers Through Society*. Cambridge Mass.: Harvard University Press.

——. (1990) "Drawing Things Together," in *Representation in Scientific Practice*. Edited by M. Lynch and S. Woolgar. Cambridge, Mass.: MIT Press, pp. 19–68.

Latour, Bruno and Stephen Woolgar. (1986) *Laboratory Life: the Construction of Scientific Facts*. Princeton, N.J.: Princeton University Press.

Laugier, Sandra. (2006) "Où se trouvent les règles?" In *Lire les Recherches Philosophiques de Wittgenstein*. Edited by Sandra Laugier and Christian Chauviré. Paris: Vrin, pp. 131–156.

Ledeneva, Alena. (1998) *Russia's Economy of Favours: Blat, Networking and Informal Exchange*. Cambridge: Cambridge University Press.

Lee, John A. N., Robeti Rosin, Fernando J. Corbato, Robert M. Fano, Martin Greenberger, Joseph C. R. Licklider, Douglas T. Ross, and Allan L. Scherr. (1992) "The Project MAC interviews," in *IEEE Annals of the History of Computing* 14:2, pp. 14–35.

Leibniz, Gottfried W. (1875-90) *Theodicy. Essays on the Goodness of God, the Freedom of Man and the Origin of Evil*. Edited by E. M. Huggard and C. J. Gerhardt. La Salle, Ill.: Open Court.

——. (1903) *Opuscules et fragments inédits de Leibniz: extraits des manuscrits de la Bibliothèque royale de Hanovre*. Edited by Louis Couturat. Paris: Felix Alcan.

——. (1970a) "Meditations on Knowledge, Truth, and Ideas," in *Philosophical Papers and Letters*. Edited by Leroy E. Loemker. Dordrecht: Reidel.

Leibniz, Gottfried W. (1970b) *Philosophical Papers and Letters*. Edited by Leroy E. Loemker. Dordrecht: Reidel.

——. (1985[1714]) *The Monadology, and Other Philosophical Writings*. Edited by Robert Latta. New York-London: Garland.

——. (1996[1705]) *New Essays on Human Understanding*. Edited by Peter Remnant and Jonathan Francis Bennett. Cambridge: Cambridge University Press.

——. (1999) *General Investigations Concerning the Analysis of Concepts and Truths*. Edited by Walter H. O'Briant. Athens, Ga.: University of Georgia Press.

Leonard, Robert J. (1995) "From Parlor Games to Social Science: von Neumann, Morgenstern, and the Creation of Game Theory, 1928–1944," in *Journal of Economic Literature* 33:2, pp. 730–761.

——. (2004) "'Between Worlds,' or an Imagined Reminiscence by Oskar Morgenstern about Equilibrium and Mathematics in the 1920s," in *Journal of The History of Economic Thought* 26:3, pp. 285–310.

Lestel, Dominique. (2001) *Les origines animales de la culture*. Paris: Flammarion.

Lettvin, Jerome Y., Humberto R. Maturana, Warren S. McCulloch, and Walter H. Pitts. (1959) "What the Frog's Eye Tells the Frog's Brain," in *Proceedings of the Institute of Radio Engineers* 47:11, pp. 1940–1951.

Levinas, Emmanuel. (1989) *The Levinas reader*. Edited by Seán Hand. Oxford: Blackwell.

Lévi-Strauss, Claude. (1968) *Mythologiques***. L'origine des manières de table*. Paris: Plon.

——. (1973) *Anthropologie Structurale Deux*. Paris: Plon.

——. (1983) *Le regard éloigné*. Paris: Plon.

——. (1995) *L'identité. Séminaire interdisciplinaire 1974–1975*. Paris: Presses Universitaires de France.

Lindström, Per. (2001) "Penrose's New Argument," in *Journal of Philosophical Logic* 30:3, pp. 241–250.

——. (2006) "Remarks on Penrose's "New Argument"," in *Journal of Philosophical Logic* 35:3, pp. 231–237.

Linz, Peter. (2001) *An Introduction to Formal Languages and Automata*. 3rd ed. Boston, Mass.: Jones and Bartlett.

Lipset, David. (1980) *Gregory Bateson: The Legacy of a Scientist*. Englewood Cliffs, N.J.: Prentice-Hall.

Lolli, Gabriele. (1994) "Introduzione," in *Intelligenza meccanica*. Torino: Bollati Boringhieri, pp. 7–23.

Lucas, John R. (1961) "Minds, Machines and Gödel," in *Philosophy* 36, pp. 112–137.

——. (1996) "Minds, Machines and Gödel: A Retrospect," in *Machines and Thought: The Legacy of Alan Turing*. Edited by P. J. R. Millican and A. Clark. Oxford: Oxford University Press, pp. 103–12.

Lyotard, Jean-François. (1984) *The Postmodern Condition*. Minneapolis, Minn.: University of Minnesota Press.

MacColl, Le Roy Archibald. (1945) *Fundamental Theory of Servomechanisms*. New York: D. Van Nostrand Company.

Mac Lane, Saunders. (1971) *Categories for the Working Mathematician*. New York: Springer.

Macrae, Norman. (1992) *John von Neumann*. New York: Pantheon Books.

Maestrini, Piero. (2002) *La calcolatrice elettronica pisana. Una storia che sembra una leggenda*. Pisa: CNR.

Maffei, Arianna and Gina Turrigiano. (2008) "The Age of Plasticity: Developmental Regulation of Synaptic Plasticity in Neocortical Microcircuits," in *Progress in Brain Research* 169, pp. 211–223.

Mancosu, Paolo, Klaus Frovin Jørgensen, and Stig Andur Pedersen. (2005) *Visualization, Explanation and Reasoning Styles in Mathematics*. Dordrecht: Springer.

Mandò, Manlio. (1949) "Le grandi calcolatrici moderne a successioni automatiche. I," in *Periodico di matematiche* 27:3, pp. 165–173.

——. (1950a) "Le grandi calcolatrici moderne a successioni automatiche. II," in *Periodico di matematiche* 28:1, pp. 28–43.

——. (1950b) "Le grandi calcolatrici moderne a successioni automatiche. III," in *Periodico di matematiche* 28:2, pp. 85–97.

Maniglier, Patrice. (2006a) "La langue comme milieux: une hypothèse sur la cognition culturelle." Seminar given at the *Lundis de la philosophie*, Paris, École Normale Supérieure, 11/30/2006. URL: http://www.diffusion.ens.fr/index.php?res=personnes&idpers=1552.

——. (2006b) *La vie énigmatique des signes. Saussure et la naissance du Structuralisme*. Paris: Léo Scheer.

Maretti, Enrico. (1958) "Modello meccanico di operazioni mentali," in *Convegno internazionale sui problemi dell'automatismo. Milano, 8–13 aprile 1956*. Vol. I. Roma: Consiglio Nazionale delle Ricerche, pp. 854–862.

——. (1989) "Adamo II," in *Civiltà delle Macchine*. Edited by Vanni Scheiwiller. Milano: Scheiwiller, pp. 191–203.

Martinoli, Gino. (1958) "L'automatismo e la necessità di una preparazione culturale," in *Convegno internazionale sui problemi dell'automatismo. Milano, 8–13 aprile 1956*. Vol. III. Roma: Consiglio Nazionale delle Ricerche, pp. 2308–2319.

Masani, Pesi Rustom. (1990) *Norbert Wiener, 1894–1964*. Basel: Birkhäuser.

Masani, Pesi Rustom and Ralph S. Phillips. (1985) "Antiaircraft Fire-Control and the Emergence of Cybernetics," in Norbert Wiener, *Collected Works with Commentaries*. Edited by Pesi R. Masani. Vol. 4. Cambridge, Mass.: MIT Press, pp. 141–79.

Matteuzzi, Maurizio. (1981) *La forma della teoria. Studio sull'espressione dell'invarianza*. Faenza, Italy: Faenza Editrice.

——. (2005) "Why AI Is not a Science," in *Mechanical Bodies, Computational Minds*. Edited by Stefano Franchi and Güven Güzeldere. Cambridge, Mass.: MIT Press, pp. 409–421.

Mayr, Otto. (1970) *The Origins of Feedback Control*. Cambridge, Mass.: MIT Press.

Mayr, Otto. (1986) *Authority, Liberty, & Automatic Machinery in Early Modern Europe*. Baltimore, Md.: Johns Hopkins University Press.

McCarthy, John. (1968) "Programs with Common Sense," in *Semantic Information Processing*. Edited by Marvin Minsky. Cambridge, Mass.: MIT Press, pp. 403–409.

McCarthy, John, Marvin Minsky, Nathaniel Rochester, and Claude E. Shannon. (1956) *A Proposal for the Dartmouth Research Project in Artificial Intelligence*. URL: http://www.formal.stanford.edu/jmc/history/dartmouth.html.

McCulloch, Warren S. (1946) *Memo to the Members of the "Macy Conference on Feedback Mechanisms, 17–19 Oct. 1946*. n. d., box 2, folder 71, Norbert Wiener Papers, collection MC-22, Institute Archives and Special Collections, Massachusetts Institute of Technology Archives, Cambridge, Mass.

——. (1989[1948]) "Through the Den of the Metaphysician," in *Embodiments of Mind*. Cambridge, Mass.: MIT Press, pp. 142–156.

Meschieri, Luigi. (1968) "Riflessi psicologici dell'automazione sul lavoratore," in *Atti del convegno sul tema: L'automazione elettronica e le sue implicazioni scientifiche, tecniche e sociali (Roma, 16–19 ottobre 1967)*. Roma: Accademia Nazionale dei Lincei, pp. 465–474.

Meyrowitz, Norman and Andries van Dam. (1982) "Interactive Editing Systems: Part II," in *ACM Computing Surveys* 14:3, pp. 353–415.

Mialet, Hélène. (2009) "Seeing Through Calculations." Unpublished manuscript.

Miller, Harvey and George G. Haroules. (1962) *On Stability in the Homeostat Jenny*. Tech. rep. Cambridge, Mass.: Air Force Cambridge Research Center.

Minsky, Marvin and Seymour Papert. (1967) *Perceptrons*. Cambridge, Mass.: MIT Press.

Mirowski, Philip. (1991) "When Games Grow Deadly Serious: the Military Influence on the Development of Game Theory," in *History of Political Economy* 23, pp. 227–255.

——. (1992) "What Were von Neumann and Morgenstern Trying to Accomplish?" In *Toward a History of Game Theory*. Edited by Roy Weintraub. Durham, N.C.: Duke University Press, pp. 113–147.

Montalenti, Giuseppe. (1968) "III seduta. 17 ottobre 1967," in *Atti del convegno sul tema: L'automazione elettronica e le sue implicazioni scientifiche, tecniche e sociali (Roma, 16–19 ottobre 1967)*. Roma: Accademia Nazionale dei Lincei, pp. 163–164.

Morpurgo, Enzo. (1948) "Dialoghi probabili," in *Sigma. Conoscenza unitaria* 6–7, pp. 368–378.

Mrosovsky, Nicholas. (1990) *Rheostasis. The Physiology of Change*. Oxford University Press.

Mrozek, Slawomir. (1967) "Strip-Tease," in *The Tulane Drama Review* 11:3. Trans. by Edward Rother, pp. 52–59.

Myhill, John, Electronic Technology Laboratory, University of Pennsylvania, and Wright Air Development Division. (1960) *Linear Bounded Automata*. Tech. rep. Ohio: Wright Air Development Division, Air Research and Technology Command, United States Air Force, pp. 60–165.

Nagel, Ernest and James Roy Newman. (1956a) "Gödel's Proof," in *Scientific American* 194, pp. 71–90.

——. (1956b) "Gödel's Proof," in *The World of Mathematics*. Edited by James R. Newman. Vol. 4. New York: Simon and Schuster, pp. 1668–1695.

——. (1958) *Gödel's Proof*. 1st ed. New York: New York University Press.

Nagel, Ernest, James Roy Newman, and Douglas R. Hofstadter. (2001) *Gödel's Proof*. Rev. ed. New York: New York University Press.

Nastasi, Pietro. (2005) "Picone, il calcolo automatico e FINAC: una storia lunga 30 anni," in *PRISTEM/Storia. Note di Matematica, Storia, Cultura* 12-13, pp. 121–171.

Nathans, Benjamin. (2007) "The Dictatorship of Reason: Aleksandr Vol'pin and the Idea of Rights under 'Developed Socialism'," in *Slavic Review* 66:4, pp. 630–663.

Neisser, Ulric. (1967) *Cognitive Psychology*. Englewood Cliffs, N.J.: Prentice-Hall.

Nerode, Anil. (1958) "Linear Automaton Transformations," in *Proceedings of the American Mathematical Society*. Vol. 9, pp. 541–544.

Newell, Allen. (1970) "Remarks on the Relationship Between Artificial Intelligence and Cognitive Psychology," in *Theoretical Approaches to Non-numerical Problem Solving*. Edited by Ranan B. Banerji and Mihajlo D. Mesarović. Berlin: Springer.

——. (1990) *Unified Theories of Cognition*. Cambridge, Mass: Harvard University Press.

Newell, Allen and Herbert A. Simon. (1956) "The Logic Theory Machine—a Complex Information Processing System," in *IRE Transactions on Information Theory* 2:3, pp. 61–79.

——. (1958) "Heuristic Problem Solving: The Next Advance in Operations Research," in *Operations Research* 6:1, pp. 1–10.

——. (1963) "GPS, a Program that Simulates Human Thought," in *Computers and Thought*. Edited by Edward Feigenbaum and Julian Feldman. 2nd ed. Cambridge, Mass.: MIT Press, pp. 279–293.

——. (1972) *Human Problem Solving*. Englewood Cliffs, N.J.: Prentice-Hall.

——. (1981[1976]) "Computer Science as Empirical Inquiry: Symbols and Search," in *Mind Design: Philosophy, Psychology, Artificial Intelligence*. Edited by John Haugeland. Cambridge, Mass.: MIT Press, pp. 35–66.

Newman, James Roy. (1956) *The World of Mathematics: A Small Library of the Literature of Mathematics From a'H-Mosé the Scribe to Albert Einstein*. New York: Simon and Schuster.

Newman, Ted. (1994) "Memories of the Pilot ACE," in *Computer Resurrection. The Bulletin of the Computer Conservation Society* 9. URL: http://www.cs.man.ac.uk/CCS/res/res09.htm.

Nolfi, Stefano and Dario Floreano. (2000) *Evolutionary Robotics: The Biology, Intelligence, and Technology of Self-Organizing Machines*. Cambridge, Mass.: MIT Press.

Norman, Jeremy M., ed. (2005) *From Gutenberg to the Internet: A Sourcebook on the History of Information Technology*. Novato, Calif.: historyofscience.com.

Northrop, Filmer Stuart Cuckow. (1949) *Ideological Differences and World Order: Studies in the Philosophy and Science of the World's Cultures*. New Haven, Conn.: Publications for the Viking Fund of Yale University Press.

Numerico, Teresa. (2005) *Alan Turing e l'intelligenza delle macchine*. Milano: Franco Angeli.

Nyquist, Harry. (1932) "Regeneration theory," in *Bell System Technical Journal* 11, pp. 126–147.

Okrent, Mark. (1991) *Heidegger's Pragmatism: Understanding, Being, and the Critique of Metaphysics*. Ithaca, N. Y.: Cornell University Press.

——. (2007) *Rational Animals: The Teleological Roots of Intentionality*. 1st ed. Columbus, Ohio: Ohio University Press.

Oldenbourg, Rudolf C. and Hans Sartorius. (1944) *Dynamik selbsttätiger Regelungen*. Munich: Oldenbourg.

Oliva, Carlo, ed. (2001) *Il carteggio Ceccato-Dingler 1949–1952*. Roma: Società Stampa sportiva.

Paci, Enzo. (1989) "La tecnica e la libertà dell'uomo," in *Civiltà delle macchine. Antologia di una rivista 1953–1957*. Edited by Vanni Scheiwiller. Milano: Scheiwiller.

Paige, Jeffrey M. and Herbert A. Simon. (1966) "Cognitive Processes in Solving Algebra Word Problems," in *Problem Solving: Research, Method and Theory*. Edited by Benjamin Kleinmuntz. New York: Wiley, pp. 51–148.

Parmenides. (1965) *Parmenides. A Text with Translation, Commentary and Critical Essays*. Edited by Leonardo Taran. Princeton, N.J.: Princeton University Press.

Penrose, Roger. (1989) *The Emperor's New Mind: Concerning Computers, Minds, and the Laws of Physics*. Oxford: Oxford University Press.

——. (1994) *Shadows of the Mind: A Search for the Missing Science of Consciousness*. Oxford: Oxford University Press.

——. (1996) "Beyond the Doubting of a Shadow," in *Psyche* 2, pp. 89–129. URL: http://psyche.cs.monash.edu.au/v2/psyche-2-23-penrose.html.

——. (2005) *The Road to Reality: A Complete Guide to the Laws of the Universe*. New York: Knopf.

Petitot, Jean. (1985a) *Les catastrophes de la parole. De Roman Jakobson à René Thom*. Paris: Maloine.

——. (1985b) *Morphogenèse du sens*. Paris: Presses Universitaires de France.

Petrovskii, Iu. (1956) "Na grani fantastiki," in *Znanie—sila* 7.

Piaget, Jean. (1950) *Introduction à l'Épistémologie Génétique*. Paris: Presses Universitaires de France.

——. (1971) *Biology and Knowledge*. Chicago: The University of Chicago Press.

Piccinini, Gualtiero. (2004) "The First Computational Theory of Mind and Brain: A Close Look at McCulloch and Pitts's "Logical Calculus of Ideas Immanent in Nervous Activity"," in *Synthese* 141:2, pp. 175–215.

Pickering, Andrew. (1995) *The Mangle of Practice: Time, Agency, and Science*. Chicago, Ill: The University of Chicago Press.

——. (2002) "Cybernetics and the Mangle: Ashby, Beer, and Pask," in *Social Studies of Science* 32:3, pp. 413–437.

——. (2004) "The Science of the Unknowable: Stafford Beer's Cybernetic Informatics," in *Kybernetes* 33, pp. 499–521.

——. (2005) "A Gallery of Monsters: Cybernetics and Self-Organization, 1940-1970," in *Mechanical Bodies, Computational Minds: Artificial Intelligence from Automata to Cyborgs*. Edited by Stefano Franchi and Güven Güzeldere. Cambridge, Mass.: MIT Press, pp. 229–245.

——. (2007) "Ontological Theatre: Gordon Pask, Cybernetics, and the Arts," in *Cybernetics & Human Knowing* 14:4, pp. 43–57.

——. (2010) *The Cybernetic Brain: Sketches of Another Future*. Chicago, Ill.: The University of Chicago Press.

Pitts, Walter H. and Warren S. McCulloch. (1943) "A Logical Calculus of Ideas Immanent in Nervous Activity," in *Bulletin of Mathematical Biophysics* 5, pp. 115–133.

——. (1947) "How We Know Universals: The Perception of Auditory and Visual Forms," in *Bulletin of Mathematical Biophysics* 9:3, pp. 127–147.

Pobojewska, A. (2001) "New Biology - Jakob von Uexkull's *Umweltlehre*," in *Semiotica* 134:1–4, pp. 322–339.

Pogliano, Claudio. (1993) "Leonardo Sinisgalli," in *Belfagor* 48, pp. 667–686.

——. (2007) "1967: Il congresso della *Società filosofica italiana*. Cronaca delle quattro giornate pisane," in *Quaderni di Athenet* 20, pp. 4–8.

Polanyi, Livia. (1985) *Telling The American Story: A Structural and Cultural Analysis of Conversational Storytelling*. Norwood, N.J: Ablex.

Poli, S. (1945) "Finalmente qualcosa fuori dall'idealismo. Una rivista di critica scientifica," in *Il Politecnico* 10, p. 4.

Pollock, Ethan. (2006) *Stalin and the Soviet Science Wars*. Princeton, N.J.: Princeton University Press.

Pollock, Jean-Yves. (1997) *Langage et cognition. Introduction au programme minimaliste de la grammaire générative*. Paris: Presses Universitaires de France.

Pólya, George. (1944) *How to Solve It*. Princeton, N.J.: Princeton University Press.

Pospelov, Dmitrii. (1986) *Situatsionnoe upravlenie: teoriia i praktika*. Moscow: Nauka.

——. (2003) "Istoriia iskusstvennogo intellekta do serediny 80-kh godov," in *Istoriia informatiki v Rossii: Uchenye i ikh shkoly*. Edited by Valerii Zakharov, Rimma Podlovchenko, and Iakov Fet. Moscow: Nauka.

Pospelov, Dmitrii and Veniamin Pushkin. (1972) *Myshlenie i avtomaty*. Moscow: Sovetskoe radio.

Preti, Giulio. (1948) "Lettera a 'Sigma'," in *Sigma. Conoscenza unitaria* 433–435:6–7.

Pribram, Karl. (1962) "The Neuropsychology of Sigmund Freud," in *Experimental Foundation of Clinical Psychology*. Edited by Arthur J. Bachrach. New York: Basic Books, pp. 442–468.

Price, Matthew "'Man Must First Conceive'—A Critical Philology of the Cyborg." Unpublished manuscript.

Prince, Alan and Paul Smolensky. (2004) *Optimality Theory: Constraint Interaction in Generative Grammar*. New York: Wiley-Blackwell.

Proust, Joëlle. (1987) "L'intelligence artificielle comme philosophie," in *Le Débat* 47, pp. 88–102.

Pushkin, Veniamin. (1971) *Psikhologiia i kibernetika*. Moscow: Pedagogika.

Putnam, Hilary. (1960) "Review of Nagel and Newman (1958)," in *Philosophy of Science* 27:2, pp. 205–207.

——. (1975) *Mind, Language and Reality. Philosophical Papers*. Vol. 2. Cambridge: Cambridge University Press.

Quaranta, Mario, ed. (1978) *Analisi. Milano 1945–1947*. Bologna: Arnaldo Forni Editore.

Quazza, Giorgio. (1968) "L'automazione elettronica nell'industria," in *Atti del convegno sul tema: L'automazione elettronica e le sue implicazioni scientifiche, tecniche e sociali (Roma, 16–19 ottobre 1967)*. Roma: Accademia Nazionale dei Lincei, pp. 379–394.

Rabin, Michael O. and Dana Scott. (1959) "Finite Automata and Their Decision Problems," in *IBM Journal of Research and Development* 3:2, pp. 114–125.

Rabinbach, Anson. (1990) *The Human Motor: Energy, Fatigue, and the Origins of Modernity*. New York: Basic Books.

Rabinow, Paul. (2008) *Marking Time: On the Anthropology of the Contemporary*. Princeton, N.J.: Princeton University Press.

Rabinowitch, Ithai and Idan Segev. (2008) "Two Opposing Plasticity Mechanisms Pulling a Single Synapse," in *Trends in Neurosciences* 31:8, pp. 377–383.

Rao, Giuseppe. (2005) "Mario Tchou e l'Olivetti Elea 9003," in *PRISTEM/Storia. Note di Matematica, Storia, Cultura* 12-13, pp. 85–119.

Rasetti, Mario "Brain and Life." Unpublished manuscript.

Rassadin, Stanislav. (2004) *Kniga proshchanii*. Moscow: Tekst.

Rastier, François. (1987) *Sémantique interprétative*. Paris: Presses Universitaires de France.

Richter, Curt. (1976) *The Psychobiology of Curt Richter*. Edited by Elliot M. Blass. Baltimore, Md.: York Press.

Rickert, Heinrich. (1910) *Kulturwissenschaft und Naturwissenschaft*. Tübingen: J. C. B. Mohr.

Rider, Robin E. (1988) *Early Development of Operations Research: British and American Contexts*. Paper presented at the Joint Meeting of the British Society for History of Science and the History of Science Society, Manchester, 11–14 July 1988.

——. (1992) "Operations Research and Game Theory: Early Connections," in *Toward a History of Game Theory*. Edited by E. R. Weintraub. Durham, N.C.: Duke University Press, pp. 225–239.

——. (1994) "Operational Research," in *Companion encyclopedia of the history and philosophy of the mathematical sciences*. Vol. 1. London: Routledge, pp. 837–841.

Robbins, Philip and Murat Aydede, eds. (2008) *The Cambridge Handbook of Situated Cognition*. Cambridge: Cambridge University Press.

Rosenblueth, Arturo and Norbert Wiener. (1950) "Purposeful and Non-Purposeful Behavior," in *Philosophy of Science* 17:4, pp. 318–326.

Rosenblueth, Arturo, Norbert Wiener, and Julian Bigelow. (1943) "Behavior, Purpose and Teleology," in *Philosophy of Science* 10:1, pp. 18–24.

Rubinacci, Leopoldo. (1968) "Discorso pronunciato dal sen. Leopoldo Rubinacci ministro per la Ricerca scientifica e tecnologica," in *Atti del convegno sul tema: L'automazione elettronica e le sue implicazioni scientifiche, tecniche e sociali (Roma, 16–19 ottobre 1967)*. Roma: Accademia Nazionale dei Lincei, pp. 13–15.

Ruelle, David. (2007) *The Mathematician's Brain*. Princeton, N.J.: Princeton University Press.

Rumelhart, David E., James L. McClelland, and the PDP Research Group. (1986a) *Parallel Distributed Processing: Explorations in the Microstructure of Cognition*. Cambridge, Mass.: MIT Press.

Rumelhart, David E., Paul Smolensky, James L. McClelland, and Geoffrey E. Hinton. (1986b) "Schemata and sequential thought processes in PDP models," in *Parallel Distributed Processing: Explorations in the Microstructure of Cognition*. Cambridge, Mass.: MIT Press, pp. 7–57.

Russell, Stuart J. and Peter Norvig. (1995) *Artificial Intelligence: a Modern Approach*. Englewood Cliffs, N.J.: Prentice-Hall.

——. (2003) *Artificial Intelligence: a Modern Approach*. 2nd ed. Englewood Cliffs, N.J.: Prentice-Hall.

Sacerdote, Gino. (1953) "Prefazione," in Norbert Wiener, *Introduzione alla cibernetica*. Torino: Edizioni Scientifiche Einaudi.

Said, Edward W. (1978) *Orientalism*. New York: Vintage Books.

Salanskis, Jean-Michel. (2003) *Herméneutique et cognition*. Lille: Presses Universitaires du Septentrion.

Salus, Peter H. (1994) *A Quarter Century of UNIX*. Reading, Mass: Addison-Wesley.

Sarkar, Sahotra. (1992) "'The Boundless Ocean of Unlimited Possibilities': Logic in Carnap's *Logical Syntax of Language*," in *Synthese* 93:1/2, pp. 191–237.

Saussure, Ferdinand de. (1967) *Cours de linguistique générale*. Edited by Rudolf Engler. Wiesbaden: Otto Harrassowitz.

——. (1983) *Course of General Linguistics*. Chicago, Ill.: Open Court.

——. (2002) *Ecrits de linguistique générale*. Paris: Gallimard.

——. (2003) "La légende de Siegfried et l'histoire burgonde (présentation et edition par Béatrice Turpin)," in *Saussure*. Edited by Simon Bouquet. Paris: L'Herne, pp. 351–429.

——. "Unpublished manuscripts." Bibliotèque Nationale Française, Manuscripts français 3959/3.

Schank, Roger C. (1984) *The Cognitive Computer on Language Learning and Artificial Intelligence*. Reading, Mass.: Addison-Wesley.

Schank, Roger C. and Robert P. Abelson. (1977) *Scripts, Plans, Goals and Understanding*. Hillsdale, N.J.: Lawrence Erlbaum Associates.

Scheiwiller, Vanni, ed. (1989) *Civiltà delle macchine. Antologia di una rivista 1953–1957.* Milano: Scheiwiller.

Scheutz, Matthias, ed. (2002) *Computationalism: New Directions.* Cambridge, Mass.: MIT Press.

Schlatter, Mark and Ken Aizawa. (2008) "Walter Pitts and 'A Logical Calculus'," in *Synthese* 162:2, pp. 235–250.

Schulkin, Jay. (2003) *Rethinking Homeostasis.* Cambridge, Mass.: MIT Press.

Searle, John. (1980) "Minds, Brains, and Programs," in *Behavioral and Brain Sciences* 3, pp. 417–424.

——. (1983) *Intentionality: An Essay in the Philosophy of Mind.* Cambridge: Cambridge University Press.

——. (1995) *The Social Construction of Reality.* New York: Free Press.

Segre, Beniamino. (1968) "Saluto," in *Atti del convegno sul tema: L'automazione elettronica e le sue implicazioni scientifiche, tecniche e sociali (Roma, 16–19 ottobre 1967).* Roma: Accademia Nazionale dei Lincei, pp. 9–12.

Selfridge, Oliver G. (1955) "Pattern Recognition and Modern Computers," in *Proceedings of 1955 Western Joint Computer Conference.* Los Angeles, Calif.: ACM, pp. 91–93.

——. (1959) "Pandemonium: A Paradigm for Learning," in *Proceedings of the Symposium on Mechanisation of Thought Processes.* Edited by D. V. Blake and Albert M. Uttley. London, pp. 511–529.

——. (2008) "An Interview with Oliver Selfridge," in *The Mechanical Mind in History.* Edited by Philip Husbands, Owen Holland, and Micheal Wheeler. Cambridge,Mass.: MIT Press, pp. 397–407.

Servadio, Emilio. (1948) "Discussioni. Padre Gemelli e la psicoanalisi," in *Sigma. Conoscenza unitaria* 4–5, pp. 332–334.

Shannon, Claude E. (1948) *A Mathematical Theory of Communication.* Cambridge, Mass.: MIT Press.

Shapiro, Stewart. (1998) "Incompleteness, Mechanism, and Optimism," in *The Bulletin of Symbolic Logic* 4:3, pp. 273–302.

——. (2003) "Mechanism, Truth, and Penrose's New Argument," in *Journal of Philosophical Logic* 32:1, pp. 19–42.

Sherry, Michael S. (1987) *The Rise of American Air Power: The Creation of Armageddon.* New Haven, Conn.: Yale University Press.

Sieg, Wilfried. (1999) "Hilbert's Program: 1917–1922," in *The Bulletin of Symbolic Logic* 5:1, pp. 1–44.

——. (2004) "Computability." Mimeographed.

——. (2006) "Gödel on Computability," in *Philosophia Mathematica* 14:2, pp. 189–207.

Siegelmann, Hava T. (1999) *Neural Networks and Analog Computation: Beyond the Turing Limit.* Boston-Basel-Berlin: Birkhäuser.

——. (2003) "Neural and Super-Turing Computing," in *Minds and Machines* 13, pp. 103–114.

Simon, Herbert A. (1955) "A Behavioral Model of Rational Choice," in *The Quarterly Journal of Economics* 69:1, pp. 99–118.

——. (1956) "Rational Choice and the Structure of the Environment," in *Psychological Review* 63:2.

——. (1991) *Models of My Life*. New York: Basic Books.

——. (1997) *Administrative Behavior*. 4th ed. New York: Simon & Schuster.

Sinisgalli, Leonardo. (1978) "Civiltà delle macchine," in *Il Mattino* 87, p. 3.

Smalheiser, Neil R. (2000) "Walter Pitts," in *Perspectives in Biology and Medicine* 43:2, pp. 217–226.

Smirnov, M.S., ed. (1975) *Modelirovanie obucheniia i povedeniia*. Moscow: Nauka.

Smith, Brian Cantwell. (1996) *On the Origin of Objects*. Cambridge, Mass.: MIT Press, p. 420.

Smith, Gerard P. (2008) "Unacknowledged Contributions of Pavlov and Barcroft to Cannon's Theory of Homeostasis," in *Appetite* 51, pp. 428–432.

Sobolev, Sergei, Anatolii Kitov, and Aleksei Liapunov. (1955) "Osnovnye cherty kibernetiki," in *Voprosy filosofii* 4.

Società Filosofica Italiana. (1967) *L'uomo e la macchina. Atti del XXI Congresso Nazionale di Filosofia, Pisa 22–25 Aprile 1967*. Torino: Edizioni di *Filosofia*.

Somenzi, Vittorio. (1947a) "Arte e conoscenza," in *Sigma. Conoscenza unitaria* 2, pp. 147–149.

——. (1947b) "Arte e conoscenza in Jean-Paul Sartre," in *Sigma. Conoscenza unitaria* 3, pp. 173–190.

——. (1948) "Arte e conoscenza in Paul Valéry," in *Sigma. Conoscenza unitaria* 6–7, pp. 380–386.

——. (1951) "La cibernetica," in *Il Mondo* 3:11, p. 4.

——. (1952) "Archimede '52," in *Il Mondo* 4:6, p. 6.

——. (1954–55) "Informazione e teoria dell'informazione," in *Convegno di elettronica e televisione indetto dal CNR in occasione della XXXII Fiera di Milano (Giornate della scienza, 12–17 aprile 1954)*. Vol. II. Roma: Consiglio Nazionale delle Ricerche, pp. 1197–1201.

——. (1955) "'Homunculi' e astronavi sono di là da venire?" In *Civiltà delle macchine* 3:3, pp. 40–41.

——. ed. (1965) *La filosofia degli automi*. Torino: Boringhieri.

——. (1967) "Uomini e macchine," in *L'uomo e la macchina*.

——. (1968) "Aspetti filosofici del problema della 'intelligenza artificiale'," in *Atti del convegno sul tema: L'automazione elettronica e le sue implicazioni scientifiche, tecniche e sociali (Roma, 16–19 ottobre 1967)*. Roma: Accademia Nazionale dei Lincei, pp. 227–232.

——. (1987) "La scuola operativa italiana," in *Methodologia* 1, pp. 7–14.

Somenzi, Vittorio and Roberto Cordeschi, eds. (1986 [1994]) *La filosofia degli automi. Origini dell'intelligenza artificiale*. Torino: Bollati Boringhieri.

Speidel, Carl Caskey. (1942) "Studies of Living Nerves. VII. Growth Adjustments of Cutaneous Terminal Arborizations," in *The Journal of Comparative Neurology* 76:1, pp. 57–73.

Sperry, Roger W. (1947) "Effect of Nerves to Antagonistic Limb Muscles in the Monkey," in *Archives of Neurology and Psychiatry* 58, p. 452.

Sportelli, Silvano. (2007) *Egoismo metafisico ed egoismo morale. Storia di un termine nella Francia del settecento.* Pisa: Edizioni ETS.

Starling, Ernest Henry. (1923) "The Wisdom of the Body. The Harveian Oration, delivered before the Royal College of Physicians of London on St. Luke's Day, 1923," in *British Medical Journal* 2:3272, pp. 685–690.

Statera, Gianni. (1967) "Uomini, macchine e automi pensanti al Congresso Nazionale di Filosofia," in *La Cultura* 4, pp. 535–556.

Steels, Luc. (2000) "The Puzzle of Language Evolution," in *Kognitionswissenschaft* 8:4, pp. 143–150.

Steels, Luc and Pierre-Yves Oudeyer. (2000) "The Cultural Evolution of Syntactic Constraints in Phonology," in *Artificial Life VII.* Edited by Mark A. Bedau, John S. MacCaskill, Norman H. Packard, and Steen Rasmussen. *Proceedings of the Seventh International Conference on Artificial Life.* Cambridge, Mass.: MIT Press, pp. 382–394.

Sterling, Peter and Joseph Eyer. (1988) "Allostasis: A New Paradigm to Explain Arousal Pathology," in *Handbook of Life Stress, Cognition and Health.* Edited by S. Fisher and J. Reason. New York: John Wiley and Sons, pp. 629–649.

Stibitz, George R. (1941) *Report on Visit to Prof. Norbert Wiener.* 28 Oct. 1941, Record Group 227, OSRD, Division 7, Records of Chiefs and Members of Sections, 1940–46, George R. Stibitz, Section 7.1, Land-Based Fire Control Problems, "Predictions I," Massachusetts Institute of Technology, NDCrc-83, National Archives, Library of Congress, Washington, D.C.

——. (1942) *Section 2 of Division D, Diary of Chairman.* 1 July 1942, Boston, Project no. 6, Record Group 227, OSRD, Division 7, General Project Files, 1940–46, General Mathematical Theory of Prediction and Application, Massachusetts Institute of Technology, Wiener, NDCrc-83, National Archives, Library of Congress, Washington, D.C.

Stiles, Bert. (1947) *Serenade to the Big Bird.* Lindsay Drummond.

Suchman, Lucy A. (1987) *Plans and Situated Actions: The Problem of Human-Machine Communication.* 2nd ed. Cambridge: Cambridge University Press.

Sundman, John. (2003) *Artificial Stupidity.* URL: http://www.salon.com/tech/feature/2003/02/26/loebner_part_one/.

Rome Symposium (1957) *Symposium on Modern Computers. Proceedings of the Rome Symposium (17–18 October 1956) organized by the Preparatory Committee of the International Computation Center under the auspices of Unesco.* Roma: Libreria Eredi Virgilio Vecchi.

Taylor, James G. (1962) *The Behavioral Basis of Perception.* New Haven, Conn.: Yale University Press.

Taylor, Richard. (1950a) "Comments on a Mechanistic Conception of Purposefulness," in *Philosophy of Science* 17:4, pp. 310–317.

——. (1950b) "Purposeful and Non-Purposeful Behavior: A Rejoinder," in *Philosophy of Science* 17:4, pp. 327–332.

Teller, Paul. (2001) "Twilight of the Perfect Model Model," in *Erkenntnis* 55:3, pp. 393–415.

Termini, Settimo, ed. (2006) *Imagination and Rigor. Essays on Eduardo R. Caianiello's Scientific Heritage*. Milan-Berlin-Heidelberg-New York: Springer.

Teuscher, Christof. (2002) *Turing's Connectionism: An Investigation of Neural Network Architectures*. New York: Springer.

——. ed. (2004) *Alan Turing: Life and Legacy of a Great Thinker*. Frankfurt am Main: Springer.

Thagard, Paul. (2005) *Mind: Introduction to Cognitive Science*. 2nd ed. Cambridge, Mass.: MIT Press.

Thomas-Stahle, Charles. (1940) *Letter to Wiener*. 23 July 1940, box 1, folder 57, Norbert Wiener Papers, collection MC-22, Institute Archives and Special Collections, Massachusetts Institute of Technology Archives, Cambridge, Mass.

Thompson, Ken. (1968) "Programming Techniques: Regular Expression Search Algorithm," in *Communications of the ACM* 11:6, pp. 419–422.

Tolman, Edward C. (1932) *Purposive Behavior in Animals and Men*. New York: The Century Co.

Tooby, John and Leda Cosmides. (1992) "The Psychological Foundations of Culture," in *The Adapted Mind: Evolutionary Psychology and the Generation of Culture*. Edited by Jerome H. Barkow, Leda Cosmides, and John Tooby. New York: Oxford University Press, pp. 19–136.

Travaglini, Volrico. (1968) "Le premesse e le conseguenze economiche dell'automazione elettronica," in *Atti del convegno sul tema: L'automazione elettronica e le sue implicazioni scientifiche, tecniche e sociali (Roma, 16–19 ottobre 1967)*. Roma: Accademia Nazionale dei Lincei, pp. 413–430.

Trenholme, Russell. (1994) "Analog Simulation," in *Philosophy of Science* 61:1, pp. 115–131.

Tsetlin, Mikhail. (1973) *Automaton Theory and Modeling of Biological Systems*. Trans. by Scitran. New York: Academic Press.

Turing, Alan M. (1937) "On Computable Numbers with an Application to the *Entscheidungsproblem*," in *Proceedings of the London Mathematical Society, second series* 42, pp. 230–265.

——. (1939) "Systems of Logic Based on Ordinals," in *Proceedings of the London Mathematical Society* 45, pp. 161–228.

——. (Nov. 1946) "Letter to W. Ross Ashby." Turing Archives, 11/20/46.

——. (1947) "Lecture on the Automatic Computing Engine," in *The Essential Turing*. Edited by Jack Copeland. Oxford: Clarendon Press, pp. 362–393.

——. (1948) "Intelligent Machinery. Report to the National Physics Laboratory," in *The Essential Turing*. Edited by Jack Copeland. Oxford: Clarendon Press, pp. 395–432.

——. (1950) "Computing Machinery and Intelligence," in *Mind* LIX, pp. 433–460.

Turing, Alan M. (1951) "Can Digital Computers Think?" In *The Essential Turing*. Edited by Jack Copeland. Oxford: Clarendon Press, pp. 482–486.

——. (1953) "Solvable and Unsolvable Problems," in *Science News* 31, pp. 7–23.

——. (1954) "Solvable and Unsolvable Problems," in *The Essential Turing. The Essential Turing*. Edited by Jack Copeland. Oxford: Clarendon Press, pp. 576–595.

——. (1965 [1936]) "On Computable Numbers, with an Application to the *Entscheidungsproblem*," in *The Undecidable: Basic Papers on Undecidable Propositions, Unsolvable Problems, and Computable Functions*. Edited by Martin Davis. Hewlett, N.Y.: Raven Press.

——. (1992a) "Intelligent Machinery, Report to the Executive Committee of the National Physical Laboratory (1948)," in *Collected Works of A. M. Turing*. Vol. 1: *Mechanical Intelligence*. Edited by Darrel C. Ince. New York: North-Holland, pp. 107–127.

——. (1992b) "Lecture to the London Mathematical Society on 20 February, 1947," in *Collected Works of A. M. Turing*. Vol. 1: *Mechanical Intelligence*. Edited by Darrel C. Ince. New York: North-Holland, pp. 87–105.

——. (1992c) *Collected Works of A. M. Turing*. Vol. 1: *Mechanical Intelligence*. Edited by Darrel C. Ince. New York: North-Holland.

——. (2004a) "Letter on Logic to Max Newman (c. 1940)," in *The Essential Turing*. Edited by Jack Copeland. Oxford: Clarendon Press, pp. 205–215.

——. (2004b) *The Essential Turing*. Edited by Jack Copeland. Oxford: Clarendon Press.

Turrigiano, Gina. (1999) "Homeostatic Plasticity in Neuronal Networks: The More Things Change, the More They Stay the Same," in *Trends in Neuroscience* 22:5, pp. 221–227.

——. (2007) "Homeostatic Signaling: The Positive Side of Negative Feedback," in *Current Opinion in Neurobiology* 17:3, pp. 318–324.

Turrigiano, Gina and Sacha B. Nelson. (2000) "Hebb and Homeostasis in Neuronal Plasticity," in *Current Opinion in Neurobiology* 10:3, pp. 358–364.

Urbani, R. (1997) "'L'uomo e la macchina.' Un congresso di trent'anni fa," in *Nuncius. Annali di storia della scienza* 12:2, pp. 497–524.

Utaker, Arild. (2002) *La Philosophie du langage: Une archéologie saussurienne*. Paris: Presses Universitaires de France.

Vaccarino, Giuseppe. (1947–1948) "Elementi per una teoria della conoscenza I–V," in *Sigma. Conoscenza unitaria* 1, 2, 3, 4–5, 6–7, pp. 6–27, 69–86, 215–233, 303–331, 387–433.

——. (1989) "Costruzione di un homunculus," in *Civiltà delle macchine*. Edited by Vanni Scheiwiller. Milano: Scheiwiller.

van Dam, Andries and David E. Rice. (1971) "On-line Text Editing: a Survey," in *ACM Computing Surveys* 3:3, pp. 93–114.

van Gelder, Timothy. (1998) "The Dynamical Hypothesis in Cognitive Science," in *Behavioral and Brain Sciences* 21, pp. 615–628.

van Gelder, Timothy and Robert Port, eds. (1995) *Mind As Motion*. Cambridge, Mass.: MIT Press.

van Vleck, Tom. (2008) *Multicians.org*. URL: http://multicians.org/.

Varela, Francisco J. (1979) *Principles of Biological Autonomy*. New York: North-Holland.

——. (1989) *Connaître les sciences cognitives. Tendances et perspectives*. Paris: Editions du Seuil.

——. (1997) "Patterns of Life: Intertwining Identity and Cognition," in *Brain and Cognition* 34:1, pp. 72–87.

Varela, Francisco J., Evan Thompson, and Eleanor Rosch. (1991) *The Embodied Mind: Cognitive Science and Human Experience*. Cambridge, Mass.: MIT Press.

Varenne, Hervé. (1986a) "Introduction," in *Symbolizing America*. Edited by Hervé Varenne. Lincoln, Nebr.: University of Nebraska Press.

——. ed. (1986b) *Symbolizing America*. Lincoln, Nebr.: University of Nebraska Press.

Varshavskii, Viktor and Dmitrii Pospelov. (1988) *Puppets Without Strings*. Moscow: Mir.

Viano, Carlo A. (1985) *Va' pensiero. Il carattere della filosofia italiana contemporanea*. Torino: Einaudi.

Virtanen, Reino. (1960) *Claude Bernard and His Place in the History of Ideas*. Lincoln, Nebr.: University of Nebraska Press.

von Foerster, Heinz, Margaret Mead, and Hans Lukas Teuber, eds. (1953) *Cybernetics. Circular Causal and Feedback Mechanisms in Biological and Social Systems*. New York: Josiah Macy, Jr. Foundation.

von Neumann, John. (1927) "Zur Hilbertschen Beweistheorie," in *Collected Works*. Edited by Abraham H. Taub. Vol. I. New York: Pergamon Press, pp. 256–300.

von Neumann, Johann. (1928) "Zur Theorie des Gesellschaftsspiele," in *Mathematische Annalen* 100:1, pp. 295–320.

von Neumann, John. (1932) *Mathematische Grundlagen der Quantenmechanik*. Berlin: Springer.

——. (1945a) *Letter to Wiener*. 1 Feb. 1945, box 2, folder 67, Norbert Wiener Papers, collection MC-22, Institute Archives and Special Collections, Massachusetts Institute of Technology Archives, Cambridge, Mass.

——. (1945b) *Memo to Aiken, Leland E. Cunningham, W. E. Deming, Goldstine, R. Lorente de No, W. S. McCulloch, Walter H. Pitts, Vestine, Wiener, and Wilks*. 12 Jan. 1945, box 2, folder 67, Norbert Wiener Papers, collection MC-22, Institute Archives and Special Collections, Massachusetts Institute of Technology Archives, Cambridge, Mass.

——. (1948) "The General and Logical Theory of Automata," in *Collected Works*. Edited by Abraham H. Taub. Vol. 5. New York: Pergamon Press, pp. 288–328.

——. (1951) "The General and Logical Theory of Automata," in *Cerebral Mechanisms in Behavior. The Hixon Symposium 1948*. Edited by Lloyd A. Jeffress. New York: Wiley, pp. 1–41.

——. (1958) *The Computer and the Brain*. New Haven, Conn.: Yale University Press.

——. (1961) *Collected Works*. Edited by Abraham H. Taub. New York: Pergamon Press.

——. (1993) "First Draft of a Report on the EDVAC (1945)," in *IEEE Annals of the History of Computing* 15:4, pp. 27–75.

von Neumann, John. (1997[1946]) "Letter to Norbert Wiener," in *Proceedings of Symposia in Applied Mathematics* 52, p. 507.

von Neumann, John and Oskar Morgenstern. (1944) *Theory of Games and Economic Behavior.* Princeton, N.J.: Princeton University Press.

Walker, Nigel. (1956) "Freud and Homeostasis," in *British Journal for the Philosophy of Science* 7:16 (25), pp. 61–72.

Walter, William Grey. (1961) *The Living Brain.* Harmondsworth, UK: Penguin Books.

Wang, Hao. (1974) *From Mathematics to Philosophy.* London: Routledge and Kegan Paul.

——. (1996) *A Logical Journey: From Gödel to Philosophy.* Cambridge, Mass.: MIT Press.

Weaver, Warren. (1942) *Letter to Boyce.* 24 Mar. 1942, Record Group 227, OSRD, Division 7, General Project Files, 1940–46, General Mathematical Theory of Prediction and Application, Massachusetts Institute of Technology, Wiener, NDCrc-83, National Archives, Library of Congress, Washington, D.C.

Webb, Judson C. (1990) "Introductory Note to Remark 3 of Gödel 1972," in Kurt Gödel, *Collected Works, Volume II: Publications 1938–1974.* Edited by Solomon Feferman, John W. Dawson, Stephen C. Kleene, Gregory H. Moore, Robert M. Solovay, and Jean van Heijenoort. Oxford: Clarendon Press, pp. 292–304.

Weber, Andreas and Francisco J. Varela. (2002) "Life after Kant: Natural Purposes and the Autopoietic Foundations of Biological Individuality," in *Phenomenology and the Cognitive Sciences* 1:2, pp. 97–125.

Weinreich, Uriel, William Labov, and Marvin Herzog. (1968) "Empirical Foundations for a Theory of Language Change," in *Directions for Historical Linguistics. A Symposium.* Edited by W. P. Lehmann and Yakov Malkiel. Austin, Tex.: University of Texas Press, pp. 95–195.

Weizenbaum, Joseph. (1965) "ELIZA, a Computer Program for the Study of Natural Language Communication between Man and Machine," in *Communications of the ACM* 9, pp. 36–45.

——. (1978) *Computer Power and Human Reason: from Judgment to Calculation.* San Francisco, Calif.: Freeman.

Werner, Heinz and Seymour Wapner. (1955) "The Innsbruck Studies on Distorted Visual Fields in Relation to an Organismic Theory of Perception," in *Psychological Review* 62:2, pp. 130–138.

Weyl, Hermann. (1949) *Philosophy of Mathematics and Natural Science.* Princeton, N.J.: Princeton University Press.

Wiener, Norbert. (1914) "The Highest Good," in *The Journal of Philosophy, Psychology, and Social Methods* 11.

——. (1932) "Back to Leibniz! Physics Reoccupies an Abandoned Position," in *Technology Review* 34, pp. 201–3, 222, 224.

——. (1940a) *Letter to Vannevar Bush.* 20 September 1940, box 2, folder 58, Norbert Wiener Papers, collection MC-22, Institute Archives and Special Collections, Massachusetts Institute of Technology Archives, Cambridge, Mass., collection MC-22, Institute Archives and Special Collections, Massachusetts Institute of Technology Archives, Cambridge, Mass.

——. (1940b) *NDCrc-83*. National Archives, Library of Congress, Washington, D. C.

——. (1941) *Letter to Marston Morse*. 12 March 1941, box 2, folder 59, Norbert Wiener Papers, collection MC-22, Institute Archives and Special Collections, Massachusetts Institute of Technology Archives, Cambridge, Mass.

——. (1942a) *Letter to E. L. Bowles*. 22 Mar. 1942, box 2, folder 62, Norbert Wiener Papers, collection MC-22, Institute Archives and Special Collections, Massachusetts Institute of Technology Archives, Cambridge, Mass.

——. (1942b) *Letter to J. B. S. Haldane*. 22 June 1942, box 2, folder 62, Norbert Wiener Papers, collection MC-22, Institute Archives and Special Collections, Massachusetts Institute of Technology Archives, Cambridge, Mass.

——. (1942c) *Report to D. I. C. 5980 A. A. Directors, "Summary Report for Demonstration,"* 10 June 1942, Record Group 227, OSRD, National Defense Research Committee Contractors' Technical Reports, Division 7, Massachusetts Institute of Technology, NDCrc-83, National Archives, Library of Congress, Washington, D.C.

——. (1942d) *Statistical Method of Prediction in Fire Control*. Final Report on Section D-2, Project no. 6, submitted to Weaver, Section D-2 National Defense Research Committee, 1 Dec. 1942, Record Group 227, OSRD, Contractors' Reports, Division 7, NDCrc-83, OSRD Report No. 1863, Massachusetts Institute of Technology, National Archives, Library of Congress, Washington, D.C.

——. (1942e) *to J. R. Kline*. 20 Aug. 1942, box 2, folder 63, Norbert Wiener Papers, collection MC-22, Institute Archives and Special Collections, Massachusetts Institute of Technology Archives, Cambridge, Mass.

——. (1943a) *Letter to Weaver*. 15 Jan. 1943, Record Group 227, OSRD, Contractors' Reports, Division 7, NDCrc-83, enclosure with OSRD Report No. 1863, Massachusetts Institute of Technology, NA-LOC. See also Wiener, "Final Report," 1 Dec. 1942, National Archives, Library of Congress, Washington, D.C.

——. (1943b) *Letter to Weaver*. 28 Jan. 1943, box 2, folder 64, Norbert Wiener Papers, collection MC-22, Institute Archives and Special Collections, Massachusetts Institute of Technology Archives, Cambridge, Mass.

——. (1944) *Letter to von Neumann*. 17 Oct. 1944, box 2, folder 66, Norbert Wiener Papers, collection MC-22, Institute Archives and Special Collections, Massachusetts Institute of Technology Archives, Cambridge, Mass.

——. (1945a) *Letter to Giorgio de Santillana*. 16 Oct. 1945, box 2, folder 69, Norbert Wiener Papers, collection MC-22, Institute Archives and Special Collections, Massachusetts Institute of Technology Archives, Cambridge, Mass.

——. (1945b) *Letter to Karl T. Compton*. 18 Oct. 1945, box 2, folder 69, Norbert Wiener Papers, collection MC-22, Institute Archives and Special Collections, Massachusetts Institute of Technology Archives, Cambridge, Mass.

——. (1945c) *Letter to Rosenblueth*. 24 Jan. 1945, box 2, folder 67, Norbert Wiener Papers, collection MC-22, Institute Archives and Special Collections, Massachusetts Institute of Technology Archives, Cambridge, Mass.

Wiener, Norbert. (1945d) *Letter to von Neumann*. 24 Jan. 1945, box 2, folder 67, Norbert Wiener Papers, collection MC-22, Institute Archives and Special Collections, Massachusetts Institute of Technology Archives, Cambridge, Mass.

——. (1945e) *Operationalism—Old and New*. box 11, folder 570, Norbert Wiener Papers, collection MC-22, Institute Archives and Special Collections, Massachusetts Institute of Technology Archives, Cambridge, Mass.

——. (1948) *Cybernetics*. 1st ed. Cambridge, Mass.: MIT Press.

——. (1953a) *Delivered to the Operations Research Society*. 23 Nov. 1953, box 12, folder 738, Norbert Wiener Papers, collection MC-22, Institute Archives and Special Collections, Massachusetts Institute of Technology Archives, Cambridge, Mass.

——. (1953b) *La cibernetica*. Milano: Bompiani.

——. (1954) *The Human Use of Human Beings: Cybernetics and Society*. 2nd ed. Boston, Mass.: Houghton Mifflin.

——. (1956) *I Am a Mathematician: the Later Life of a Prodigy; an Autobiographical Account of the Mature Years and Career of Norbert Wiener and a Continuation of the Account of His Childhood in Ex-Prodigy*. Garden City, N.Y.: Doubleday.

——. (1961a) *Cybernetics: or, Control and Communication in the Animal and the Machine*. 2nd ed. Cambridge, Mass.: MIT Press.

——. (1961b) "Über Informationstheorie," in *Die Naturwissenschaften* 48, pp. 174–176.

——. (1964) *God and Golem, Inc.: a Comment on Certain Points Where Cybernetics Impinges on Religion*. Cambridge, Mass.: MIT Press.

——. (1985) *Collected Works with Commentaries*. Edited by Pesi R. Masani. Cambridge, Mass.: MIT Press.

——. *The Extrapolation, Interpolation and Smoothing of Stationary Time Series with Engineering Applications*. DIC Contract 6037, Massachusetts Institute of Technology, 1 Feb. 1942, declassified, box 11, folder 561 A, Norbert Wiener Papers, collection MC-22, Institute Archives and Special Collections, Massachusetts Institute of Technology Archives, Cambridge, Mass.

Wiener, Norbert and J. P. Schadé. (1963) "Introduction to Neurocybernetics," in *Cybernetics of the Nervous System*. Edited by Norbert Wiener and J. P. Schadé. Vol. 2. Amsterdam: Elsevier, pp. 1–7.

Wigner, Eugene P. (1967) *Symmetries and Reflections*. Bloomington, Ind.: Indiana University Press.

Wilkins, Michael G. (1968) *A New Homeostat*. Tech. rep. 8.3. BCL (Biological Computer Laboratory, University of Illinois at Urbana-Champaign).

Williams, Hywel. (2006) "Homeostatic Adaptive Networks." PhD thesis. University of Leeds.

Williams, Hywel and Jason Noble. (2007) "Homeostatic Plasticity Improves Signal Propagation in Continuous-Time Recurrent Neural Networks," in *Biosystems* 87:2-3, pp. 252–259.

Williams, Robert E. (1961) *Static and Dynamic Responses of the Homeostat Jenny*. Tech. rep. Air Force Cambridge Research Laboratory.

Winograd, Terry. (1972) *Understanding Natural Language*. New York: Academic Press.

——. (1973) "A Procedural Model of Language Understanding," in *Computer Models of Thought and Language*. Edited by Roger Schank and Kenneth Colby. San Francisco, Calif.: Freeman.

Winsberg, Eric. (2001) "Simulations, Models, and Theories: Complex Physical Systems and Their Representations," in *Philosophy of Science* 68, S442–S454.

——. (2003) "Simulated Experiments: Methodology for a Virtual World," in *Philosophy of Science* 70, pp. 105–125.

Winston, Patrick. (1984) *Artificial Intelligence*. 2nd ed. Reading, Mass.: Addison-Wesley.

——. (1987) "Artificial Intelligence: A Perspective," in *AI in the 1980s and Beyond: an MIT Survey*. Edited by W. Eric L. Grimson and Ramesh S. Patil. Cambridge, Mass.: MIT Press.

Wittgenstein, Ludwig. (1973) *Philosophical Investigations*. Englewood Cliffs, N.J.: Prentice-Hall.

Woodworth, Robert Sessions and Mary Rose Sheehan. (1964) *Contemporary Schools of Psychology*. 3rd ed. New York: Ronald Press Co.

Wright, Robert. (1988) *Three Scientists and Their Gods*. New York: Times Books.

Yandell, Ben. (2002) *The Honors Class: Hilbert's Problems and Their Solvers*. Natick, Mass.: A. K. Peters.

Zawidzki, Tadeusz and William Bechtel. (2004) "Gall's Legacy Revisited. Decomposition and Localization in Cognitive Neuroscience," in *Mind as a Scientific Object: Between Brain and Culture*. Edited by Christina E. Erneling and David M. Johnson. Oxford: Oxford University Press, pp. 293–317.

Zemanek, Heinz. (1956) "La tortue de Vienne et les autres travaux cybernètiques," in *1er Congrès international de Cybernétique, 1st International Congress on Cybernetics*. Paris: Gauthier-Villars, pp. 770–780.

INDEX

ABOUT THE CONTRIBUTORS

Peter Asaro is a member of the Department of Media Studies and Film, The New School. His research focuses on the history of cybernetics and brain modeling, on natural and synthetic brains and their relationships with and through computational, and on information and media technologies. He is the author of several essays on these topics and is the writer and director of *Love machine* (2001), a documentary film on the development of robots capable of entering human social relations of love and caring.

Francesco Bianchini is an associate researcher in the Department of Philosophy, University of Bologna. His research focuses on computational modeling and cognitive architectures; epistemological problems relative to artificial intelligence and cognitive science; and issues in the philosophy of mind such as mental causation and knowledge representation, and the neuro-physiological and cognitive bases of language. Among other volumes, he is the author of *Concetti analogici. L'approccio subcognitivo allo studio della mente* (Macerata: Quodlibet, 2008) and has edited, with Alfio Gliozzo and Maurizio Matteuzzi, *Instrumentum vocale: intelligenza artificiale e linguaggio* (Bologna: Bononia University Press, 2007).

Solomon Feferman is the Patrick Suppes Family Professor of Humanities and Sciences, Em. and Professor of Mathematics and Philosophy, Em. at Stanford University. He has published extensively on issues in proof theory and theory of computation, the foundations of mathematics (especially constructive and predicative foundations), the philosophy of mathematics, and the history of modern logic. He is the editor-in-chief of Kurt Gödel's *Collected Works* (5 volumes, Oxford, Oxford University Press, 1986–2003), the author of *In the Light of Logic* (Oxford: Oxford University Press, 1998), and, with Anita Burdman Feferman, *Alfred Tarski. Life and Logic* (Cambridge: Cambridge University Press, 2004).

Stefano Franchi is Research Associate Professor in the Department of Hispanic Studies, Texas A&M University. He has published extensively on twentieth century European philosophy, the history and philosophy of science in the twentieth century, and the history of Artificial Intelligence and cognitive science. He is the editor (with Güven Güzeldere) of *Mechanical Bodies, Computational Minds* (Cambridge, Mass.: MIT Press, 2005).

Peter Galison is Professor of History of Science, Harvard University. His work explores the complex interaction between the three principal subcultures of twentieth century physics—experimentation, instrumentation, and theory. He is the author of *How Experiments End* (Chicago: Chicago University Press, 1987), *Image and Logic: A Material Culture of Microphysics* (Chicago: Chicago University Press, 1997), *Einstein's Clocks, Poincaré's Maps* (New York: Norton, 2003), and the editor of several collections of studies on the history and philosophy of science.

Slava Gerovitch is a Lecturer in the Department of Mathematics, Massachusetts Institute of Technology. His research focuses on the contemporary history of mathematics, science, and technology. He has published several essays on Russian scientific and technological culture during the Soviet era and is the author of *From Newspeak to Cyberspeak: A History of Soviet Cybernetics* (Cambridge, Mass.: MIT Press, 2002).

Christopher Kelty is Associate Professor in the Department of Information Studies, University of California at Los Angeles. His research focuses on internet culture and history, intellectual property, the public sphere, free and open source software, public domains, commons, authorship and ownership, and the history and philosophy of science and technology, in the US, Europe, and India. He has published extensively on these topics and is the author of *Two Bits: The Cultural Significance of Free Software and the Internet* (Durham, N.C.: Duke University Press, 2008).

Rossella Lupacchini is Associate Professor of Methodology of Natural Science and Philosophy of Physics, University of Bologna. She has published extensively on issues pertaining the foundations of physics, on quantum computing, and on the philosophy of symbolic forms. She is the author of *Esperimenti mentali e realtà fisica* (Bologna: CLUEB, 1995), *Risorse fisiche nella matematica. Dal finitismo hilbertiano ai calcolatori quantistici* (Cesena: Il Ponte Vecchio, 1999), *Elementi di computazione quantistica* (Bologna: CLUEB, 2004), and co-editor, with Giovanna Corsi, of *Deduction, Computation, Experiment. Exploring the Effectiveness of Proof* (Milan: Springer, 2008).

Patrice Maniglier is a Lecturer in the Department of Philosophy, Essex University. He has published extensively on the history of structuralism and post-structuralism in the social sciences (especially linguistics, anthropology, and psychoanalysis) and in philosophy, and more particularly in twentieth century French philosophy (from Sartre to Deleuze). He is the author of several volumes including *La vie énigmatique des signes. Saussure et la naissance du structuralisme* (Paris: Léo Scheer, 2006), *La Culture* (Paris: Ellipses, 2003), and (with Alain Badiou, Thomas Bénatouil, Elie During, David Rabouin and Jean-Pierre Zarader) *Matrix, machine philosophique* (Paris: Ellipses, 2003).

Maurizio Matteuzzi is Professor of Philosophy of Language at the University of Bologna. Among his interest: the philosophy of Leibniz (*Kalkulierbarkeit und Nicht-Kalkulierbarkeit nach Leibnizschen Wunschtraum*, in *Leibniz Werk und Wirkung*, Hannover 1983; *Individuare per caratteri—Saggio su Leibniz*, Bologna: Cappelli Editore, 1983; "Leibniz e i sincategoremi," in Klaus D. Dutz and Stefano Gensini, eds., *Im Spiegel des Verstandes*, Münster: Nodus Publikationen, 1996); The mind/body problem ("La melodia e il reggimento: la mente come rilevamento di morfismi," in *Percezione linguaggio coscienza*, Macerata: Quodlibet, 1999; "L'occhio della mosca e il ponte di Brooklyn—Quali regole per gli oggetti del second'ordine?" in Marco Carapezza and Franco Lo Piparo, eds., *La regola linguistica*, Palermo: Novecento, 2000); Artificial Intelligence from an epistemological standpoint ("A note on the notion of 'theory'," *Quality and Quantity*, 11, 1977; "Why Artificial Intelligence is not a science," in Stefano Franchi and Güven Güzeldere, eds., *Mechanical Bodies, Computational Minds. Artificial Intelligence from Automata to Cyborgs*, Cambridge, Mass: MIT Press 2005).

Claudio Pogliano is Professor of History of Science, University of Pisa. He has been a visiting scholar at the Wellcome Institute for the History of Medicine (London), Harvard University, University of California (Berkeley), Massachusetts Institute of Technology, Max Planck Institut für Wissenschaftsgeschichte in Berlin. He is the author of several articles and volumes on the history of science and history of technological culture, including *L'ossessione della razza. Antropologia e genetica nel XX secolo* (Pisa: Edizioni della Normale, 2005), and is the editor of *Scienze e storia nell'Italia del Novecento* (Pisa: Edizioni Plus—Pisa University Press, 2007).

Giorgio Sandri is Professor of Philosophy of Science, University of Bologna. He has published extensively on logic and the philosophy of science, with particular regards toward natural language semantic, theory of computation, and knowledge representation. He is the author of *Logica e grammatica in Richard Montague* (Bologna: CLUEB, 1979) and *Grammatiche, forma logica e linguaggio naturale* (Bologna: CLUEB, 1983). He is the co-editor, with Rema Rossini Favretti and Roberto Scazzieri, of *Incommensurability and translation: Kuhnian perspectives on scientific communication and theory change* (Cheltenham, UK: Edward Elgar, 1999).

VIBS

The **Value Inquiry Book Series** is co-sponsored by:

Adler School of Professional Psychology
American Indian Philosophy Association
American Maritain Association
American Society for Value Inquiry
Association for Process Philosophy of Education
Canadian Society for Philosophical Practice
Center for Bioethics, University of Turku
Center for Professional and Applied Ethics, University of North Carolina at Charlotte
Central European Pragmatist Forum
Centre for Applied Ethics, Hong Kong Baptist University
Centre for Cultural Research, Aarhus University
Centre for Professional Ethics, University of Central Lancashire
Centre for the Study of Philosophy and Religion, University College of Cape Breton
Centro de Estudos em Filosofia Americana, Brazil
College of Education and Allied Professions, Bowling Green State University
College of Liberal Arts, Rochester Institute of Technology
Concerned Philosophers for Peace
Conference of Philosophical Societies
Department of Moral and Social Philosophy, University of Helsinki
Gannon University
Gilson Society
Haitian Studies Association
Ikeda University
Institute of Philosophy of the High Council of Scientific Research, Spain
International Academy of Philosophy of the Principality of Liechtenstein
International Association of Bioethics
International Center for the Arts, Humanities, and Value Inquiry
International Society for Universal Dialogue
Natural Law Society
Philosophical Society of Finland
Philosophy Born of Struggle Association
Philosophy Seminar, University of Mainz
Pragmatism Archive at The Oklahoma State University
R.S. Hartman Institute for Formal and Applied Axiology
Research Institute, Lakeridge Health Corporation
Russian Philosophical Society
Society for Existential Analysis
Society for Iberian and Latin-American Thought
Society for the Philosophic Study of Genocide and the Holocaust
Unit for Research in Cognitive Neuroscience, Autonomous University of Barcelona
Whitehead Research Project
Yves R. Simon Institute

Titles Published

Volumes 1 - 202 see www.rodopi.nl

203. Fernand Vial, *The Unconscious in Philosophy, and French and European Literature: Nineteenth and Early Twentieth Century.* A volume in **Philosophy and Psychology**

204. Adam C. Scarfe, Editor, *The Adventure of Education: Process Philosophers on Learning, Teaching, and Research.* A volume in **Philosophy of Education**

205. King-Tak Ip, Editor, *Environmental Ethics: Intercultural Perspectives.* A volume in **Studies in Applied Ethics**

206. Evgenia Cherkasova, *Dostoevsky and Kant: Dialogues on Ethics.* A volume in **Social Philosophy**

207. Alexander Kremer and John Ryder, Editors, *Self and Society: Central European Pragmatist Forum*, Volume Four. A volume in **Central European Value Studies**

208. Terence O'Connell, *Dialogue on Grief and Consolation.* A volume in **Lived Values, Valued Lives**

209. Craig Hanson, *Thinking about Addiction: Hyperbolic Discounting and Responsible Agency.* A volume in **Social Philosophy**

210. Gary G. Gallopin, *Beyond Perestroika: Axiology and the New Russian Entrepreneurs.* A volume in **Hartman Institute Axiology Studies**

211. Tuija Takala, Peter Herissone-Kelly, and Søren Holm, Editors, *Cutting Through the Surface: Philosophical Approaches to Bioethics.* A volume in **Values in Bioethics**

212. Neena Schwartz: *A Lab of My Own.* A volume in **Lived Values, Valued Lives**

213. Krzysztof Piotr Skowroński, *Values and Powers: Re-reading the Philosophical Tradition of American Pragmatism.* A volume in **Central European Value Studies**

214. Matti Häyry, Tuija Takala, Peter Herissone-Kelly and Gardar Árnason, Editors, *Arguments and Analysis in Bioethics.* A volume in **Values in Bioethics**

215. Anders Nordgren, *For Our Children: The Ethics of Animal Experimentation in the Age of Genetic Engineering*. A volume in **Values in Bioethics**

216. James R. Watson, Editor, *Metacide: In the Pursuit of Excellence*. A volume in **Holocaust and Genocide Studies**

217. Andrew Fitz-Gibbon, Editor, *Positive Peace: Reflections on Peace Education, Nonviolence, and Social Change*. A volume in **Philosophy of Peace**

218. Christopher Berry Gray, *The Methodology of Maurice Hauriou: Legal, Sociological, Philosophical*. A volume in **Studies in Jurisprudence**

219. Mary K. Bloodsworth-Lugo and Carmen R. Lugo-Lugo, *Containing (Un)American Bodies: Race, Sexuality, and Post-9/11 Constructions of Citizenship*. A volume in **Philosophy of Peace**

220. Roland Faber, Brian G. Henning, Clinton Combs, Editors, *Beyond Metaphysics? Explorations in Alfred North Whitehead's Late Thought*. A volume in **Contemporary Whitehead Studies**

221. John G. McGraw, *Intimacy and Isolation (Intimacy and Aloneness: A Multi-Volume Study in Philosophical Psychology, Volume One)*, A volume in **Philosophy and Psychology**

222. Janice L. Schultz-Aldrich, Introduction and Edition, "*Truth" is a Divine Name, Hitherto Unpublished Papers of Edward A. Synan, 1918-1997*. A volume in **Gilson Studies**

223. Larry A. Hickman, Matthew Caleb Flamm, Krzysztof Piotr Skowroński and Jennifer A. Rea, Editors, *The Continuing Relevance of John Dewey: Reflections on Aesthetics, Morality, Science, and Society*. A volume in **Central European Value Studies**

224. Hugh P. McDonald, *Creative Actualization: A Meliorist Theory of Values*. A volume in **Studies in Pragmatism and Values**

225. Rob Gildert and Dennis Rothermel, Editors, *Remembrance and Reconciliation*. A volume in **Philosophy of Peace**

226. Leonidas Donskis, Editor, *Niccolò Machiavelli: History, Power, and Virtue*. A volume in **Philosophy, Literature, and Politics**

227. Sanya Osha, *Postethnophilosophy*. A volume in **Social Philosophy**

228. Rosa M. Calcaterra, Editor, *New Perspectives on Pragmatism and Analytic Philosophy*. A volume in **Studies in Pragmatism and Values**

229. Danielle Poe, Editor, *Communities of Peace: Confronting Injustice and Creating Justice*. A volume in **Philosophy of Peace**

230. Thorsten Botz-Bornstein, Editor, *The Philosophy of Viagra: Bioethical Responses to the Viagrification of the Modern World*. A volume in **Philosophy of Sex and Love**

231. Carolyn Swanson, *Reburial of Nonexistents: Reconsidering the Meinong-Russell Debate*. A volume in **Central European Value Studies**

232. Adrianne Leigh McEvoy, Editor, *Sex, Love, and Friendship: Studies of the Society for the Philosophy of Sex and Love: 1993–2003*. A volume in **Histories and Addresses of Philosophical Societies**

233. Amihud Gilead, *The Privacy of the Psychical*. A volume in **Philosophy and Psychology**

234. Paul Kriese and Randall E. Osborne, Editors, *Social Justice, Poverty and Race: Normative and Empirical Points of View*. A volume in **Studies in Jurisprudence**

235. Hakam H. Al-Shawi, *Reconstructing Subjects: A Philosophical Critique of Psychotherapy*. A volume in **Philosophy and Psychology**

236. Maurice Hauriou, *Tradition in Social Science*. Translation from French with an Introduction by Christopher Berry Gray. A volume in **Studies in Jurisprudence**

237. Camila Loew, *The Memory of Pain: Women's Testimonies of the Holocaust.*. A volume in **Holocaust and Genocide Studies**

238. Stefano Franchi and Francesco Bianchini, Editors, *The Search for a Theory of Cognition: Early Mechanisms and New Ideas*. A volume in **Cognitive Science**

CPSIA information can be obtained at www.ICGtesting.com
Printed in the USA
BVOW070841300112

281587BV00002B/7/P

9 789042 034273